Managerial Techniques for Environmental Waste Management

Managerial Techniques for Environmental Waste Management

Edited by **Victor Bonn**

SYRAWOOD
PUBLISHING HOUSE

New York

Published by Syrawood Publishing House,
750 Third Avenue, 9th Floor,
New York, NY 10017, USA
www.syrawoodpublishinghouse.com

Managerial Techniques for Environmental Waste Management
Edited by Victor Bonn

© 2016 Syrawood Publishing House

International Standard Book Number: 978-1-68286-095-3 (Hardback)

Contents

Preface

The world is advancing at a fast pace like never before. Therefore, the need is to keep up with the latest developments. This book was an idea that came to fruition when the specialists in the area realized the need to coordinate together and document essential themes in the subject. That's when I was requested to be the editor. Editing this book has been an honour as it brings together diverse authors researching on different streams of the field. The book collates essential materials contributed by veterans in the area which can be utilized by students and researchers alike.

Effective environmental waste management strategies are crucial to ensure sustainable development. This book on waste management consists of contributions made by international experts. It provides significant information to help develop a good understanding of waste management techniques through lucid elaboration on topics such as optimization and control of pollution mitigation processes, modeling transport of contaminants in environmental systems, sustainable water management, etc. The book is appropriate for students seeking detailed information as well as for researchers seeking an apt reference material. In this book, using case studies and examples, constant effort has been made to make the understanding of the difficult concepts of environmental waste management techniques as easy and informative as possible.

Each chapter is a sole-standing publication that reflects each author´s interpretation. Thus, the book displays a multi-facetted picture of our current understanding of applications and diverse aspects of the field. I would like to thank the contributors of this book and my family for their endless support.

Editor

A hybrid MCDA-LCA approach for assessing carbon foot-prints and environmental impacts of China's paper producing industry and printing services

Wencong Yue[1], Yanpeng Cai[1,2*], Qiangqiang Rong[1], Lei Cao[3] and Xumei Wang[3]

Abstract

Background: Labeling of carbon foot-prints (CFPs) for products and services is regarded as a convenient and effective method for reducing greenhouse gas (GHG) emissions. Life cycle analysis (LCA) is a useful tool for examine CFP of relevant products and services. However, the corresponding standards for CFP of products and services can hardly be satisfactorily adopted. Also, most of the previous studies were based on an individual indicator, which can hardly reflect multiple dimensions of sustainable implications of products and services.

Results: Thus, in this research, a hybrid life cycle analysis (LCA) and multi-criteria decision analysis (MCDA) method was proposed for helping evaluate CFP of products and services under multiple environmental indicators. The results indicated: (a) Air pollution caused by coal consumption was the primary environmental impact in China's paper-production industry, and (b) in printing industry, air pollution caused by VOC was the primary environmental impact in China. At the same time, CFP of 1,000 kg copying paper was 1,415.39 kg CO_2e based on LCI data of a paper factory in China. CFP of printing services was varied from each printing activity.

Conclusions: When purchasing copying paper, consumers should pay attention on coal consumption of the product. In printing industry, VOC of printing services should be taken serious consideration in China.

Keywords: Multi-criteria decision analysis; Life cycle analysis; Carbon foot-print; Copying paper; Printing services

Background

In the past decades, global market is becoming sensitive and responsive to environment-friendly technologies and services (Pineda-Henson et al. 2002). Accompanying with global endeavor to meet international commitments to reduce greenhouse gas (GHG) emissions, many consumers are eager to adopt an environmentally responsible living style to make a contribution to GHG emission reductions. Generally, it is not convenient for consumers to identify and single out products and services with reduced environmental impacts and GHG emissions. Therefore, propose of effective tools for helping consumers evaluate environmental impacts and carbon credits of relevant products and services are desired.

Previously, many efforts were undertaken for estimating carbon foot-prints (CFPs) of products by the methods of LCA. For example, Zhao et al. (2011) established a CFPs analysis model based on energy consumption, and estimated the amount of carbon emissions due to the consumptions of fossil energy in many regions of China. Yuttitham et al. (2011) estimated CFPs for sugar production from sugarcane in eastern Thailand. Namy Espinoza-Orias et al. (2011) estimated CFPs of breads that were produced and consumed in UK. Also, a number of papers presented the quantification of the uncertainties in estimating CFPs of a food product (Röös et al. 2010), plastic trays (Dormer et al. 2013) and shopping bags (Muthu et al. 2011). Among them, according to the Publicly Available Specification 2050 (i.e., PAS 2050) advanced by British Standards Institution (BSI 2008) in 2008, it was recommended that the method of life cycle analysis (LCA) be effective in evaluating CFPs of

* Correspondence: yanpeng.cai@bnu.edu.cn
[1]State Key Laboratory of Water Environment Simulation, School of Environment, Beijing Normal University, No. 19, XinJieKouWai St., HaiDian District, Beijing 100875, China
[2]Institute for Energy, Environment and Sustainable Communities, University of Regina, 120, 2 Research Drive, Regina, Saskatchewan S4S 7H9, Canada
Full list of author information is available at the end of the article

Table 1 Activity emission factor

	Activity emission factor
China Southern Power Grid	0.9787 kg CO_2/kwh
Coal	0.974 kg CO_2/kwh
Diesel oil	0.8733 kg CO_2/kwh
Natural gas	0.9538 kg CO_2/kg

products and services. However, in the previous studies, CFP was merely adopted as an individual indicator which can hardly reflect multiple dimensions of sustainable implications of products and services (Schmoldt 2001; Schmidt 2009). There is a growing need for adopting effective tools that can be used to evaluate CFP of products and services considering multiple environmental indicators such as water pollution, air pollution, energy and resource consumption. To remedy this shortage, the Society for Environmental Toxicology and Chemistry (SETAC) favored the adoption of many Multi-Criteria Decision Making (MCDM) models (Pineda-Henson et al. 2002). It was recognized as an important tool in facilitating environmental decision making for formalizing and addressing the problem of competing decision objectives (Yatsalo et al. 2007; Regan et al. 2007; Linkov et al. 2006; Lahdelma et al. 2000). Among many MCDM methods, Analytic Hierarchy Process (AHP) was widely used by many researchers (Kaya and Kahraman 2011). However, there was a lack of studies that could hybrid AHP with LCA to help evaluate CFPs of products and services.

Therefore, in this paper, a hybrid LCA and AHP approach will be advanced for evaluating overall environmental impacts and CFPs of paper production and printing services in China. Firstly, AHP will be adopted to help rank environmental impacts of this industry and the relevant services, which will give an overall assessment on environmental impacts. The method of LCA will be used to estimate the detailed CFPs.

Overview of copying-paper production in China

Pulp and paper making, a wastewater discharge intensive sector, has long been among the major water polluters in China. In 2009, for example, this sector discharged 18.8%, 28.9%, and 11.2% of national industrial wastewater,

COD, and NH_3-N emission loadings, respectively (NBS and MEP 2010). These values are partially attributable to the fact that paper consumption has been soaring with China's rapid economic growth over the last decade. Gross consumption of paper and paper board, for example, increased from 35.75 million tons in 2000 to 110.11 million tons in 2011, with an average annual growth rate of 6.1% (Zhang et al. 2012; NBS 2012). More recently, a greater number of studies highlighted energy consumption and greenhouse gas (GHG) emissions from the pulp and paper sector at global or national levels because of increasing concern on climate change (Zhang et al. 2012; Szabóa et al. 2009; Möllersten et al. 2003; Kallio et al. 2004; Davidsdottir and Ruth 2004).

Results and discussion
CFP of copying paper and printing services

1) Activity emission factor
   According to relevant research result from IPCC (IPCC 2006), Ecoinvent databases and Chen S (Sha et al. 2012), the activity emission factors were listed in Table 1.
2) CFP of copying paper in China
   Using the model described in Section 2.2, the cradle-to-grave CFP of 1000 kg of copying paper was found to be 1415.39 kg CO_{2e}. The contribution of the various life cycle inventories were showed in Table 2.
3) CFP of Printing services in China
   Using the model described in Section 2.2, the cradle-to-grave CFP of printing services for one book was found to be 5.249 kg CO_{2e}. The contribution of the various life cycle inventories were showed in Table 3, and CFP of services for printing a book was showed in Table 4.

AHP of paper and printing industries

The hierarchy tree was presented in Figures 1 and 2. The goal, which was ranking environmental impacts of paper and printing industries, was given at the first level (level 1). There were four main criteria presented in level 2 of the hierarchy, which were water pollution, air

Table 2 Life cycle inventories of producing 1000 kg copying paper in a paper-making factory of China

		Technological process				Total
		Pulping	Beating	Forming & pressing	Treating wastewater	
	Coal- kg	—	7.58E + 01	6.20E + 02	1.27E + 01	7.76E + 02
	Crude oil- kg	—	—	—	—	4.38E + 01
Resource consumption	Diesel oil- kg	6.99	—	—	—	7.11E + 00
	Natural gas- m^3	1.43E + 01	—	—	—	1.43E + 01
	Timber- kg	4.20E + 02	1.32E + 03	—	—	1.74E + 03

Table 3 Life cycle inventory of electricity printing services in a printing factory of China

Technological process		Quantity		Unit	Functional unit
Pre-press	Designing	China Southern power grid	0.571	kwh	1 piece of printing plate
		Diesel generator of the factory itself	8.16E-04	kwh	
	Plate-making	China Southern power grid	0.715	kwh	
		Diesel generator of the factory itself	2.77E-03	kwh	
Printing		China Southern power grid	0.0302	kwh	1 sheet for printing paper
		Diesel generator of the factory itself	8.47E-05	kwh	
Post-press	Bookbinding	China Southern power grid	2.13E-03	kwh	
		Diesel generator of the factory itself	1.13E-05	kwh	1 sheet for printing paper
	Storage	China Southern power grid	1.01E-04	kg	

Table 4 CFP of printing services taking a book as an example

Stage	CE of 1 functional unit	Functional unit	Unit: kg CO_{2e} CE of a book
Pre-press	2.450	1 sheet of plate	26.95
Printing	0.339	1 sheet of paper	3.729
Post-press	0.0207	1 sheet of paper	0.2277
Total	5.249		

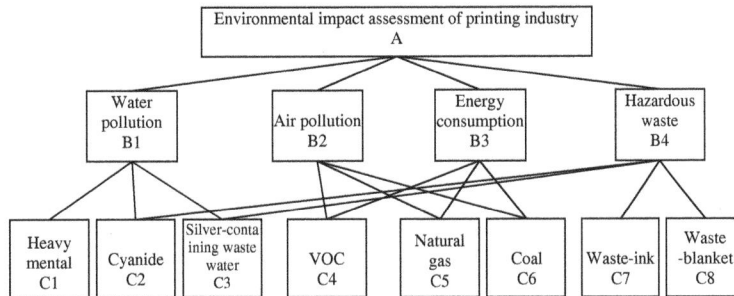

Figure 1 Hierarchy tree for environmental impact assessment of printing industry.

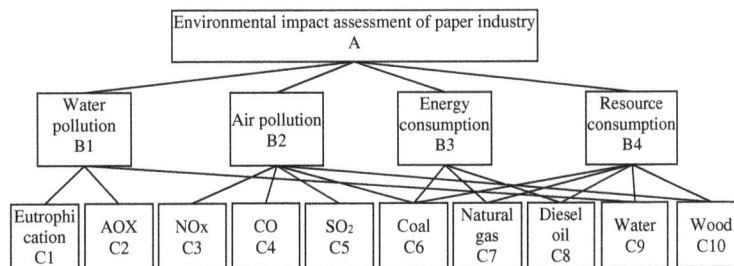

Figure 2 Hierarchy tree for environmental impact assessment of paper-making industry.

Table 5 Pair-wise evaluation of level 2 of paper-making industry

A	B1	B2	B3	B4	W'	Priorities
B1	1	3	0.5	0.33	0.84	0.155
B2	0.33	1.00	0.17	0.11	0.37	0.068
B3	2.00	6.00	1.00	0.67	1.68	0.311
B4	3.00	9.00	1.50	1.00	2.52	0.466

$\lambda_{max} = 4.25$, CI = 0.08, RI = 0.9, CR = 0.09.

Table 6 Pair-wise evaluation of level 3 of paper-making industry (1)

B1	C1	C2	C9	W'	Priorities
C1	1.00	0.60	3.00	1.22	0.33
C2	1.67	1	5	2.03	0.56
C9	0.33	0.20	1	0.41	0.11

$\lambda_{max} = 3$, CI = 0, RI = 0.58, CR = 0.

Table 7 Pair-wise evaluation of level 3 of paper-making industry (2)

B2	C3	C4	C5	C6	C10	W'	Priorities
C3	1.00	1.00	0.33	0.20	0.14	0.394	0.059
C4	1.00	1.00	0.33	0.20	0.14	0.394	0.059
C5	3.00	3.00	1.00	0.60	0.43	1.183	0.177
C6	5.00	5.00	1.67	1.00	0.71	1.971	0.294
C10	7.00	7.00	2.33	1.40	1.00	2.760	0.412

$\lambda_{max} = 5.001$, CI = 0.0003, RI = 1.12, CR = 0.00028.

Table 8 Pair-wise evaluation of level 3 of paper-making industry (3)

B3	C6	C7	C8	W'	Priorities
C6	1.00	1.67	5.00	2.027	0.555
C7	0.60	1.00	3.00	1.216	0.333
C8	0.20	0.33	1.00	0.405	0.111

$\lambda_{max} = 3$, CI = 0, RI = 0.58, CR = 0.

Table 9 Pair-wise evaluation of level 3 of paper-making industry (4)

B4	C6	C7	C8	C9	C10	W'	Priorities
C6	1.00	1.67	5.00	0.83	0.71	1.38	0.23
C7	0.60	1.00	3.00	0.50	0.43	0.83	0.14
C8	0.20	0.33	1.00	0.17	0.14	0.28	0.05
C9	1.20	2.00	6.00	1.00	0.86	1.65	0.27
C10	1.40	2.33	7.00	1.17	1.00	1.93	0.32

$\lambda_{max} = 5.001$, CI = 0, RI = 1.12, CR = 0.

Table 10 Pair-wise evaluation of level 2 of printing industry

a	B1	B2	B3	B4	W'	Priorities
B1	1	0.60	3.00	0.43	0.94	0.19
B2	1.67	1.00	5.00	0.71	1.56	0.31
B3	0.33	0.20	1.00	0.14	0.31	0.06
B4	2.33	1.40	7.00	1.00	2.19	0.44

$\lambda_{max} = 4$, CI = 0, RI = 0.9, CR = 0.

Table 11 Pair-wise evaluation of level 3 of printing industry (1)

B1	C1	C2	C3	W'	Priorities
C1	1.00	1.67	5.00	2.03	0.56
C2	0.60	1.00	3.00	1.22	0.33
C3	0.20	0.33	1.00	0.41	0.11

$\lambda_{max} = 3$, CI = 0, RI = 0.58, CR = 0.

Table 12 Pair-wise evaluation of level 3 of printing industry (2)

B2	C4	C5	C6	W'	Priorities
C4	1.00	2.33	7.00	2.537	0.636
C5	0.43	1.00	3.00	1.087	0.273
C6	0.14	0.33	1.00	0.362	0.091

$\lambda_{max} = 3$, CI = 0, RI = 0.58, CR = 0.

Table 13 Pair-wise evaluation of level 3 of printing industry (3)

B3	C4	C5	C6	W1	Priorities
C4	1.00	4.00	1.33	1.747161	0.518445
C5	0.25	1.00	0.33	0.43679	0.129611
C6	0.75	3.00	1.00	1.310371	0.388834

$\lambda_{max} = 3.003$, CI = 0.002, RI = 0.58, CR = 0.003.

Table 14 Pair-wise evaluation of level 3 of printing industry (4)

B4	C2	C3	C8	C10	W'	Priorities
C2	1.00	1.67	5.00	3.00	2.24	0.51
C3	0.60	1.00	3.00	1.00	1.16	0.26
C8	0.20	0.33	1.00	1.00	0.51	0.12
C10	0.20	0.33	1.00	1.00	0.51	0.12

$\lambda_{max} = 4.2$, CI = 0.07, RI = 0.9, CR = 0.07.

Table 15 Weighs of environmental factors of paper-making industry

Factor	Water pollution	Air pollution	Energy consumption	Resource consumption	Priorities
	0.155	0.068	0.311	0.466	
Eutrophication	0.33	-	-	-	0.051
AOX	0.56	-	-	-	0.087
NOx	-	0.059	-	-	0.004
CO	-	0.059	-	-	0.004
SO$_2$	-	0.177	-	-	0.012
Coal	-	0.294	0.556	0.23	0.300
natural gas	-	-	0.333	0.14	0.169
diesel oil	-	-	0.111	0.05	0.058
Wood	-	0.412	-	0.32	0.177
Water	0.11	-	-	0.27	0.143

pollution, energy consumption and resource consumption of paper-making industry and water pollution, air pollution, energy consumption and hazardous waste of printing industry. Level 2 of the hierarchy were further divided into several sub-criteria, which were showed in level 3. The results of comparison of level 2 in paper-making industry are shown in Table 5. Also, the results of level 3 in paper-making industry are shown in Tables 6 to 9. Moreover, the results of comparison of level 2 in printing industry are shown in Table 10. Correspondingly, the results of level 3 in printing industry are shown in Tables 11 to 14.

After pair-wise comparisons between elements at each level, the weights of environmental factors can be calculated (see Tables 15 and 16). Results of AHP of paper-making and printing industries in China were: (a) Air pollution caused by coal consumption was the primary environmental impact in China's paper-production industry, and (b) in printing industry, air pollution caused by VOC was the primary environmental impact in China.

Conclusions

This study demonstrated that the integrated LCA and MCDA approach provided a structured and comprehensive methodology for impact analysis and environmental decision making. In the background of growing concerns over global warming, carbon emission became an important factor. Carbon footprint, however, should not be merely one element in decision-making. The developed method could thus improve previous studies in comprehensive assessment on carbon footprints of products and service on multiple issues. The developed method was then applied to copying paper and printing services of China. The application indicated that the hybrid MCDA-LCA method can provide a structured and comprehensive methodology for accounting CFP as well as assessing environmental impacts of products and services. The results indicated that the most emergent environmental impacts caused by paper production and printing services were resource consumption and hazardous waste. At the same time, due to a lack of life-cycle inventory data of planting trees, carbon storage was not included in system boundary of copying paper. The next study would be furthered in detailed carbon emission of paper-making industry. Moreover, when purchasing copying paper, consumers should pay attention on coal consumption of the product. In printing

Table 16 Weighs of environmental factors of printing services

Factor	Water pollution	Air pollution	Energy consumption	Hazardous Waste	Priorities
	0.190	0.310	0.060	0.440	
heavy metal	0.560	-	-	-	0.106
Cyanide	0.330	-	-	0.250	0.173
silver-containing waste water	0.110	-	-	0.200	0.109
VOC	-	0.636	0.500	-	0.227
natural gas	-	0.273	0.125	-	0.092
Coal	-	0.091	0.375	-	0.051
waste-ink	-	-	-	0.500	0.220
waste-blanket	-	-	-	0.090	0.040

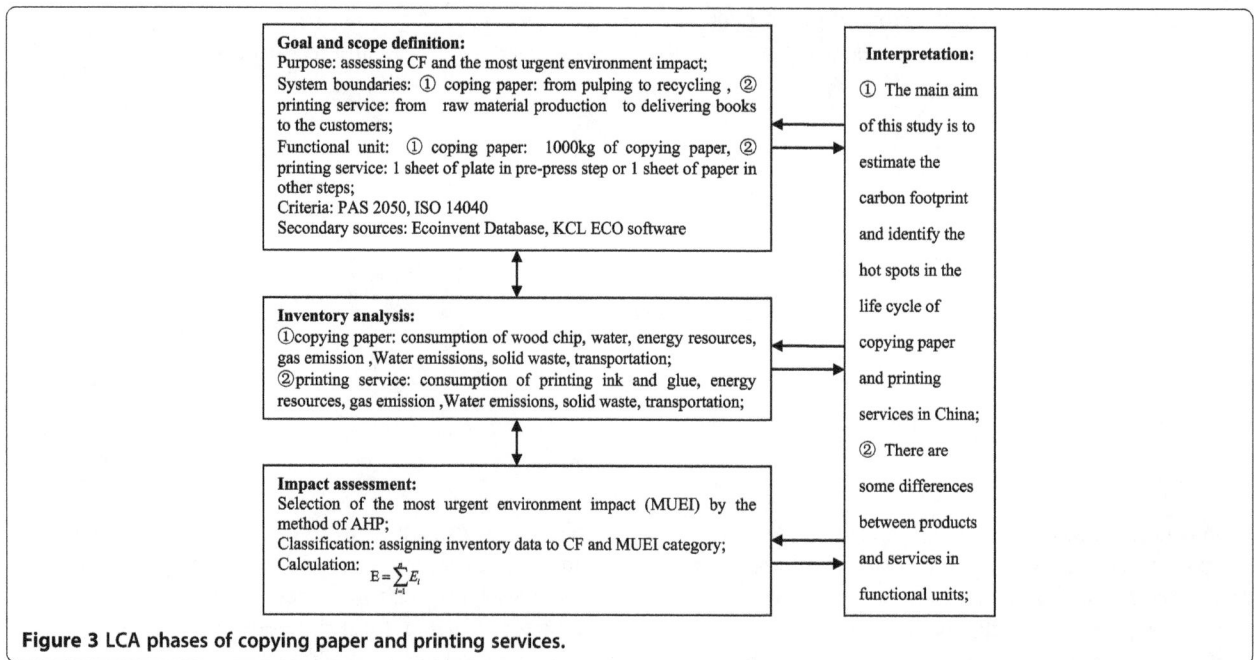

Goal and scope definition:
Purpose: assessing CF and the most urgent environment impact;
System boundaries: ① coping paper: from pulping to recycling , ② printing service: from raw material production to delivering books to the customers;
Functional unit: ① coping paper: 1000kg of copying paper, ② printing service: 1 sheet of plate in pre-press step or 1 sheet of paper in other steps;
Criteria: PAS 2050, ISO 14040
Secondary sources: Ecoinvent Database, KCL ECO software

Inventory analysis:
①copying paper: consumption of wood chip, water, energy resources, gas emission ,Water emissions, solid waste, transportation;
②printing service: consumption of printing ink and glue, energy resources, gas emission ,Water emissions, solid waste, transportation;

Impact assessment:
Selection of the most urgent environment impact (MUEI) by the method of AHP;
Classification: assigning inventory data to CF and MUEI category;
Calculation: $E = \sum_{i=1}^{t} E_i$

Interpretation:
① The main aim of this study is to estimate the carbon footprint and identify the hot spots in the life cycle of copying paper and printing services in China;
② There are some differences between products and services in functional units;

Figure 3 LCA phases of copying paper and printing services.

industry, VOC of printing services should be taken serious consideration in China.

Methods

This study was performed by a methodological framework based on hybrid LCA and AHP.

Life cycle assessment

Life-cycle assessment (LCA) is a method that builds on factual information and models of natural processes (Hertwich and Hammitt 2001). LCA is an increasingly important tool for environmental policy, and even for industry (Ayres 1995). The Society of Environmental Toxicology and Chemistry and the International Organization for Standardization developed the LCA methodology in the 1990s. The methodology is included under the international standards ISO 14040 series (Nanaki and Koroneos 2012), which are list in the following.

(1) ISO 14040:2006 Environmental management – Life cycle assessment – Principles and framework;
(2) ISO 14044:2006 Environmental management – Life cycle assessment – Requirements and guidelines;

Table 17 Functional units of printing services

Phases	Functional unit
Pre-press	1 piece of printing plate
Printing	1 sheet for printing paper
Post-press	1 sheet for printing paper

(3) ISO/TR 14047:2003 Environmental management – Life cycle impact assessment – Examples of application of ISO 14042;
(4) ISO/TS 14048:2002 Environmental management – Life cycle assessment – Data documentation format

There have also been developments on the standardization on the application of LCA-based methods for design purposes (Regan et al. 2007; ISO 2006). The Society of Environmental Toxicology and Chemistry's (SETAC) "Code of practice" originally distinguished four methodological components within LCA: goal and scope definition, life cycle inventory analysis, life cycle impact assessment, and life cycle improvement assessment. In ISO14040 (ISO 1997) life cycle improvement assessment is no longer regarded as a phase on its own, but rather as having an influence throughout the whole LCA methodology. In addition, life cycle interpretation has been introduced. This is a phase that interacts with all other phases in the LCA procedure, as illustrated in Figure 2.

Along with the increasing concerns over global warming, greenhouse gas (GHG) emissions arising from products (goods and services) are assessed by the help of life cycle assessment (LCA). Publicly Available Specification 2050:2008, Specification for the assessment of the life cycle greenhouse gas emissions of goods and services (BSI 2008) is regarded as a mechanism for simplifying and standardizing the methods for assessing the carbon footprint of products and services (Sinden 2009).

According to ISO 14044 (ISO 2006) and PAS2050, LCA phases of copying paper and printing service are divided into four phases: goal and scope definition,

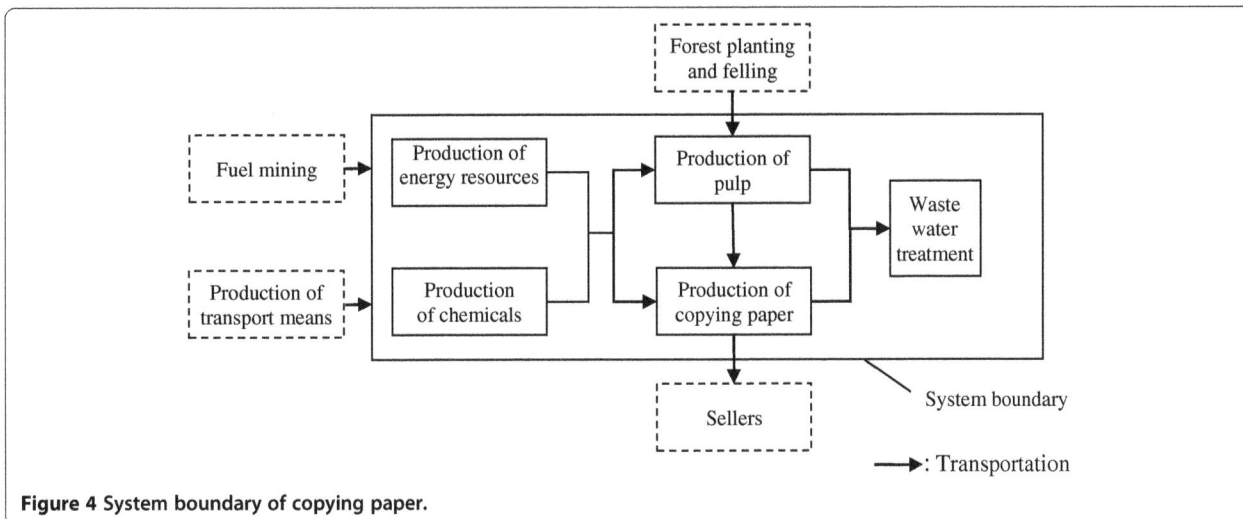

Figure 4 System boundary of copying paper.

inventory analysis, impact assessment and interpretation (see Figure 3).

Goal and scope of the study

The main aim of this study is to estimate the carbon foot-print and identify the hot spots in the life cycle of copying paper and printing services in China. Products of copying paper and printing services of study are presented. The functional unit of copying paper is 1000 kg. There are some differences between products and services in functional units. The functional units of services are not single and fixed. In another word, the functional units of services are changed along with the phases of services. In this paper, there are two functional units in printing services (see Table 17).

System boundaries and system definition

The system boundaries of copying paper, showed in Figure 4, include the following parts:

(1) Production of pulp
(2) Transport pulp to paper mills
(3) Paper-making process, including repulping, furnishing, forming and pressing, (1) (4) cutting and packing
(4) Waste water treatment
(5) Energy production
(6) Chemicals production
(7) Transport of copying paper to sellers

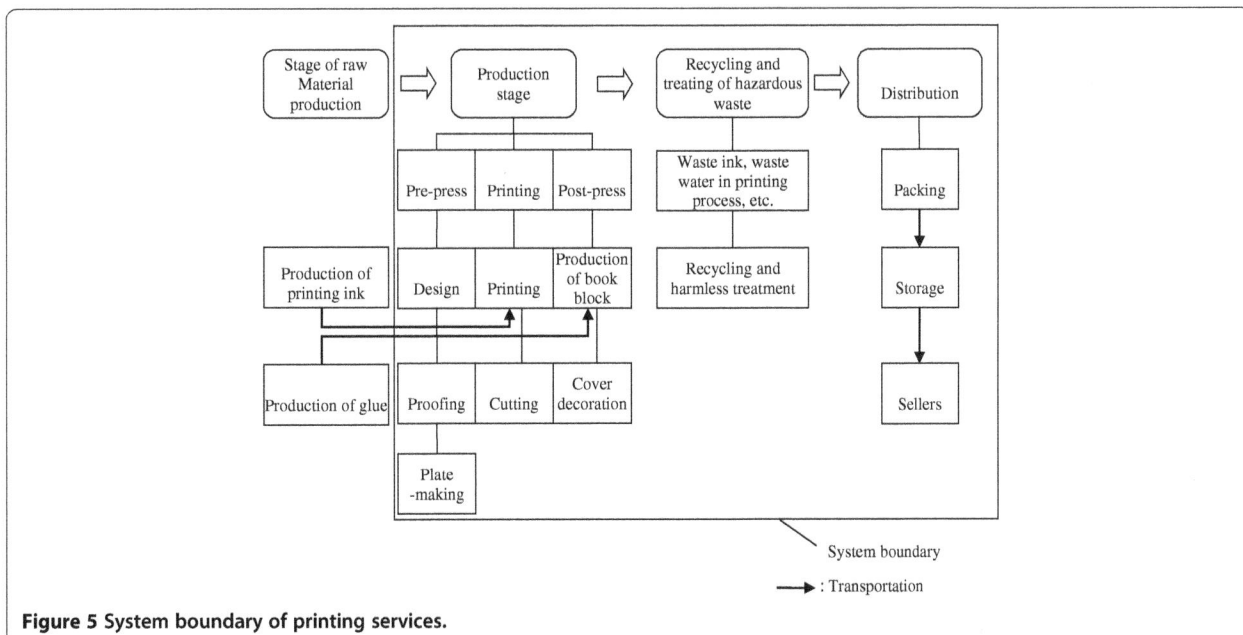

Figure 5 System boundary of printing services.

As shown in Figure 5, the following stages are included within the system boundary:

(1) Raw materials: producing printing ink and glue
(2) Processing: pre-press, printing and post-press stages in printing process
(3) Distribution stage: packing, storage, proofing and transporting to sellers
(4) Recycling and treating of hazardous waste

CFP calculation

The CFP of an activity is calculated by multiplying the activity data and the emission factor together (BSI 2008). The total CFP is calculated by then summing the individual CFPs for all activities within the specified life cycle as outlined in Eq. (1):

$$\text{Carbon footprint} = \sum \text{Activity data} \times \text{Activity emission factor} \qquad (1)$$

Analytic hierarchy process

The Analytic Hierarchy Process (AHP) is a general theory for measurement (Saaty 1987). The method of AHP is designed for multiple-criteria decisions (Schmoldt 2001). Three important components of AHP are: (1) the structuring of a problem into a hierarchy, which consisting of a goal and subordinate features (decomposition), (2) pair-wise comparisons between elements at each level (evaluation) and (3) propagation of level-specific, local priorities to global priorities (synthesis) (Schmoldt 2001). In methods of AHP, the elements in each level are compared pair wise with respect to their importance in making the decision that is under consideration (Dey 2002). The scale of integers in the range 1–9 is used for comparison (Schmoldt 2001; Saaty 1990). From the set of pair wise comparisons of the elements, a judgment matrix is generated with n rows and n columns, where n is the number of elements being considered (Pineda-Henson et al. 2002). In the matrix α_{ij} indicates how much more important row heading is than column heading j (Schmoldt 2001):

$$W_i{}' = \sqrt[n]{\prod_{i=1}^{n} \alpha_{ij}} \, (i = 1, 2, ..., n) \qquad (2)$$

$$W_i = \frac{W_i{}'}{\sum_{i=1}^{n} W_i{}'} \qquad (3)$$

The measure of consistency of an AHP judgment matrix is determined by considering the judgment matrix with n rows and n columns where $\alpha_{ij} = \frac{1}{\alpha_{ji}}$, all $\alpha_{ij} \geq 0$, and pi as the corresponding AHP priorities Ruby PH et al. (Pineda-

Henson et al. 2002) provides an approximate way of calculating the maximum eigenvalue λ_{max}:

$$\lambda_{max} = W_1 \sum_{i=1}^{n} \alpha_{i1} + W_2 \sum_{i=1}^{n} \alpha_{i2} + \cdots + W_n \sum_{i=1}^{n} \alpha_{in} \qquad (4)$$

The judgment matrix has an eigenvalue equal to n if the comparisons are perfectly consistent. The largest eigenvalue, λ_{max}, is greater than n if the comparisons are not perfectly consistent. The difference between λ_{max} and n is expressed by Saaty (2001) as the consistency index (CI), which is computed as:

$$CI = (\lambda_{max}\text{-}n)/(n\text{-}1) \qquad (5)$$

The CI is compared to the corresponding random consistency indices (RI) developed by Saaty (2001). The consistency ratio (CR) is computed from:

$$CR = CI/RI \qquad (6)$$

Saaty (Saaty 1990) recommends that the ratings from the experts may be accepted if the consistency ratio of the pair-wise comparison matrix is less than or equal to 0.10 (i.e., 90% consistent or 10% inconsistent). Otherwise, it is recommended that the pair-wise comparisons be revised to improve the consistency of these comparisons.

Competing interests
The authors declare that they have no competing interests.

Authors' contributions
The work presented here was carried out in collaboration between all authors. YUE WC and CAI YP explored the hybrid LCA-MCDA approach. YUE WC, RONG QQ, CAO L and WANG XM provided the case study for calculating CFP of the product and services. All authors have contributed to the paper preparation, and have seen and approved the manuscript.

Acknowledgements
This research was supported by the special fund of State Key Lab of Water Environment Simulation (11Z01ESPCN), the National Science Foundation for Innovative Research Group (No. 51121003) and the Twelfth Five-Year National Science and Technology Project (No. 2011BAC04B03).

Author details
[1]State Key Laboratory of Water Environment Simulation, School of Environment, Beijing Normal University, No. 19, XinJieKouWai St., HaiDian District, Beijing 100875, China. [2]Institute for Energy, Environment and Sustainable Communities, University of Regina, 120, 2 Research Drive, Regina, Saskatchewan S4S 7H9, Canada. [3]Environmental Development Centre of Ministry of Environmental Protection, No. 1 YuhuinanluChaoyang District, Beijing 100029, China.

References
Ayres RU (1995) Life-cycle analysis - a critique. Resour Conserv Recycling 14(3–4):199–223, Doi:10.1016/0921-3449(95)00017-d
BSI (2008) PAS 2050: 2008-Specification for the Assessment of the Life Cycle Greenhouse Gas Emissions of Goods and Services. BSI Group. London, London
Davidsdottir B, Ruth M (2004) Capital vintage and climate change policies: the case of US pulp and paper. Environ Sci Policy 7(3):221–233, doi:10.1016/j.envsci.2004.02.007

Dey PK (2002) Project risk management: a combined analytic hierarchy process and decision tree approach. Cost Eng 44(3):13–27

Dormer A, Finn DP, Ward P, Cullen J (2013) Carbon footprint analysis in plastics manufacturing. Journal of Cleaner Production 51:133–141, doi:10.1016/j.jclepro.2013.01.014

Espinoza-Orias N, Stichnothe H, Azapagic A (2011) The carbon footprint of bread. Int J Life Cycle Assess 16(4):351–365, doi:10.1007/s11367-011-0271-0

Hertwich EG, Hammitt JK (2001) A decision-analytic framework for impact assessment - part I: LCA and decision analysis. Int J Life Cycle Assess 6(1):5–12, doi:10.1065/lca2000.08.031

IPCC (2006) IPCC Guidelines for national greenhouse gas inventories. Institute for Global Environmental Strategies (IGES), Hayama, Japan

ISO (1997) 14040. Environmental management-Life cycle assessment-Principles and framework. International Organization for Standardization. ISO Central Secretariat, Switzerland

ISO (2006) 14044: environmental management—life cycle assessment—requirements and guidelines. International Organization for Standardization, ISO Central Secretariat, Switzerland

Kallio AMI, Moiseyev A, Solberg B (2004) The global forest sector model EFI-GTM: the model structure. European Forest Institute Joensuu, Joensuu

Kaya T, Kahraman C (2011) Fuzzy multiple criteria forestry decision making based on an integrated VIKOR and AHP approach. Expert Syst Appl 38(6):7326–7333, doi:10.1016/j.eswa.2010.12.003

Lahdelma R, Salminen P, Hokkanen J (2000) Using multicriteria methods in environmental planning and management. Environ Manage 26(6):595–605, doi:10.1007/s002670010118

Linkov I, Satterstrom FK, Kiker G, Batchelor C, Bridges T, Ferguson E (2006) From comparative risk assessment to multi-criteria decision analysis and adaptive management: recent developments and applications. Environ Int 32(8):1072–1093, doi:10.1016/j.envint.2006.06.013

Möllersten K, Yan J, Westermark M (2003) Potential and cost-effectiveness of CO_2 reductions through energy measures in Swedish pulp and paper mills. Energy 28(7):691–710, doi:10.1016/s0360-5442(03)00002-1

Muthu SS, Li Y, Hu JY, Mok PY (2011) Carbon footprint of shopping (grocery) bags in China, Hong Kong and India. Atmos Environ 45(2):469–475, doi:10.1016/j.atmosenv.2010.09.054

Nanaki EA, Koroneos CJ (2012) Comparative LCA of the use of biodiesel, diesel and gasoline for transportation. J Cleaner Prod 20(1):14–19, doi:10.1016/j.jclepro.2011.07.026

National Bureau of Statistics (NBS) of China and Ministry of Environmental Protection (MEP) of China (2010) China statistical yearbook on environment 2010. China Statistics Press, Beijing

National Bureau of Statistics (NBS) of China (2012) China statistical yearbook 2012, 2012-09-01st edn. China Statistics Press, Beijing

Pineda-Henson R, Culaba AB, Mendoza GA (2002) Evaluating environmental performance of pulp and paper manufacturing using the analytic hierarchy process and life-cycle assessment. Journal of Industrial Ecology 6(1):15–28, doi:10.1162/108819802320971614

Regan HM, Davis FW, Andelman SJ, Widyanata A, Freese M (2007) Comprehensive criteria for biodiversity evaluation in conservation planning. Biodivers Conserv 16(9):2715–2728, doi:10.1007/s10531-006-9100-3

Röös E, Sundberg C, Hansson P-A (2010) Uncertainties in the carbon footprint of food products: a case study on table potatoes. Int J Life Cycle Assess 15(5):478–488, doi:10.1007/s11367-010-0171-8

Saaty TL (1987) How to handle dependence with the analytic hierarchy process. Mathematical Modelling 9(3–5):369–376, doi:10.1016/0270-0255(87)90494-5

Saaty TL (1990) How to make a decision: The analytic hierarchy process. European Journal of Operational Research 48(1):9–26, doi:http://dx.doi.org/10.1016/0377-2217(90)90057-I

Saaty TL (2001) Decision Making for Leaders: The Analytic Hierarchy Process for Decisions in a Complex World: 1999/2000 Edition, vol 2. RWS Publications, Pittsburgh

Schmidt H-J (2009) Carbon footprinting, labelling and life cycle assessment. Int J Life Cycle Assess 14(1):6–9, doi:10.1007/s11367-009-0071-y

Schmoldt DL (2001) The analytic hierarchy process in natural resource and environmental decision making, vol 3. Springer, Dordrecht

Sha C, Li-juan R, Shui-yuan C, Zun-wen L, Cai-hua Z, Wen-cong Y (2012) Life cycle inventory study of thermal electric generation in china. Natural Gas 38931(15.3):56–51

Sinden G (2009) The contribution of PAS 2050 to the evolution of international greenhouse gas emission standards. Int J Life Cycle Assess 14(3):195–203, doi:10.1007/s11367-009-0079-3

Szabóa L, Soriaa A, Forsström J, Keränenb JT, Hytönenb E (2009) A world model of the pulp and paper industry: demand, energy consumption and emission scenarios to 2030. Environmental Science & Policy 12(3):257–269, doi:10.1016/j.envsci.2009.01.011

Yatsalo BI, Kiker GA, Kim SJ, Bridges TS, Seager TP, Gardner K, Satterstrom FK, Linkov I (2007) Application of multicriteria decision analysis tools to two contaminated sediment case studies. Integr Environ Assess Manag 3(2):223–233, doi:10.1897/ieam_2006-036.1

Yuttitham M, Gheewala SH, Chidthaisong A (2011) Carbon footprint of sugar produced from sugarcane in eastern Thailand. Journal of Cleaner Production 19(17–18):2119–2127, doi:10.1016/j.jclepro.2011.07.017

Zhang C, Chen J, Wen Z (2012) Alternative policy assessment for water pollution control in China's pulp and paper industry. Resources Conservation and Recycling 66:15–26, doi:10.1016/j.resconrec.2012.06.004

Zhao R, Huang X, Zhong T, Peng J (2011) Carbon footprint of different industrial spaces based on energy consumption in China. Journal of Geographical Sciences 21(2):285–300, doi:10.1007/s11442-011-0845-6

Water quality analysis using an inexpensive device and a mobile phone

Timo Toivanen[1*], Sampsa Koponen[2], Ville Kotovirta[1], Matthieu Molinier[1] and Peng Chengyuan[1]

Abstract

Background: Water transparency is one indicator of water quality. High water transparency is an indication of clean water. A common method for measuring water transparency is Secchi depth. In this paper, we present an approach to water quality (Secchi depth and turbidity) monitoring using mobile phones and a small device designed for water quality measurements.

Results: The water quality parameters were analysed automatically from the images taken using mobile phone cameras. During the summer of 2012, we conducted a field trial in which 100 test users gathered 1,146 observations using the system. The results of the automatic Secchi3000 depth analysis were compared against reference measurements, and they indicate that our approach can be used for quantitative water quality measurements.

Conclusions: Results show that overall the system performs well. Both Secchi depth and turbidity are estimated with excellent or good accuracy when the measurements are taken with care.

Keywords: Mobile application; Participatory sensing; Secchi depth; Water quality

Background

Water is a necessity for life, and clean water is usually a sign of a healthy environment. Human activities can have adverse effects on water quality. For example, the use of fertilizers in agriculture can lead to increased nutrient levels in nearby waters and cause increased eutrophication. Thus, monitoring water quality is an essential part of taking care of the environment.

Water transparency is one commonly used indicator of water quality. It is mentioned as a supporting factor for the biological elements in the Water Framework Directive (WFD) (Solimini et al. 2006) and as one of the physicochemical quality elements for the ecological classification of coastal waters in the Marine Strategy Framework Directive (MSD) (Piha and Zampoukas 2011). High transparency values indicate clear water, whereas low transparency values indicate turbid or highly absorbing water, which usually means poor water quality. Water transparency defines how deep sun light will penetrate and hence the depth at which plants can grow.

A common measure of water transparency is Secchi depth. It is measured using a circular white (or black and white) plate, known as a Secchi disk, which is lowered into the water until it is no longer visible. Secchi disk measurements have previously been used in volunteer monitoring, e.g. in the Secchi Dip-In –program (http://www.secchidipin.org/), in which over 41,000 records of more than 7,000 separate water bodies have so far been uploaded into a data base through a web form.

The measurement of transparency with a Secchi disk has some limitations. First, the measurement is subjective and depends on the eyesight and ability of the person making the measurements. Second, it is affected by environmental factors such as waves and weather. Finally, it cannot be used in shallow water where the disk hits the bottom before disappearing. Transparency can also be measured horizontally with a Secchi disk or a black disk. With this method the measurement requires a periscope and is thus more complicated.

Other devices for the assessment of water transparency have also been developed. One example is the Transparency Tube. To take a measurement it is filled with water and slowly emptied until the markings at the

* Correspondence: timo.toivanen@vtt.fi
[1]VTT Technical Research Centre of Finland, P.O. Box 1000, FI-02044 VTT Espoo, Finland
Full list of author information is available at the end of the article

bottom of the tube become visible when viewed from above. The measurement range of the tube is limited by the length of the tube (60 cm and 120 cm versions exist) and thus it is suitable only for low transparency waters such as rivers and streams.

Transparency can also be estimated by measuring turbidity. This requires a device that emits light and measures the amount of scattering from the sample. These turbidity meters are clearly more expensive than Secchi disks (in the range 800–1,000 USD).

The recent advances in mobile sensing and crowdsourcing provide potential means for collecting more data and observations from volunteers. Mobile phone sensing has become a hot topic in the areas of citizen science and crowdsourcing. Lane et al. (2010) presents four main reasons for the rapid increase in mobile sensing applications: 1) availability of cheap embedded sensors, 2) applications can be programmed to use these sensors, 3) app stores can be used to deliver new applications to large populations easily, and 4) back-end servers providing resources for computing.

Mobile sensing could extend the sensor networks of environmental institutes both spatially and temporally, and produce useful information for situation awareness, forecasting and scientific research. For example, D'Hondt et al. (2013) and Gallo and Waitt (2011) conclude that user-provided data can be used to gather useful datasets. Mobile sensing can be divided into two sensing paradigms (Lane et al. 2008). In opportunistic sensing the data collection is carried out automatically without actions by the users. Participatory sensing relies on the fact that users actively make and report their observations.

In this paper, we present a novel approach to water quality monitoring, called Secchi3000. This approach includes a participatory sensing application for mobile phones with which users take pictures of target areas inside a cheap and simple measurement device filled with water and send them with related metadata (e.g. location) to a central server. On the server, water turbidity and Secchi depth are automatically post-processed from the images using pattern recognition and computer vision approaches. The system was tested in the summer of 2012 and the results indicate that it provides a reliable method for water quality monitoring. Participatory water quality sensing can be a valuable tool either in providing measurement capabilities where none exist or in supplementing existing monitoring efforts with additional coverage. Also, Secchi3000 measurements can be used as an alert monitoring tool: if changes are detected in a water body, the authorities can send a professional water quality expert to assess the situation. In addition, the data collected using the system can be used e.g. in validation and completion of satellite image analysis.

Methods
Overall system architecture
The Secchi3000 water quality measurement system consists of the following parts: 1) mobile phone application for acquiring and sending observations, 2) Secchi3000 measurement device, 3) algorithms for analysing the water quality values from the images received, and 4) database, where all the data with analysed values are stored. The conceptual architecture of the Secchi3000 system is presented in Figure 1.

The mobile phone application the users can use to gather observations is called EnviObserver (Kotovirta et al. 2012). It is a tool for participatory sensing which utilizes people as sensors by enabling them to report environmental observations with a mobile phone. The current version of the Secchi3000 measurement device consists of a container and two measurement tags that are used in the analysis. The tags are located at different depths in order to derive water turbidity. After filling the container with water, the user takes a picture looking inside the container through a hole in the lid. GPS location and timestamp are automatically retrieved during the measurement. In addition to the picture, users can enter supplementary information, e.g. ID of the measurement site.

Finally, the user sends the data collected to a central server for automatic water quality analysis. The water quality analysis consists of two separate algorithms. After receiving a picture, the first algorithm automatically detects the locations of the tags in the picture and extracts pixel RGB values for the black, grey, and white areas of the two tags. The second algorithm carries out the actual water quality analysis based on the RGB values extracted by the tag recognition algorithm. After the automatic analysis, the result is sent back to the user and stored on a database.

Data collection
The system was tested in Finland during the summer of 2012. The test users were recruited by the Finnish Environment Institute and included people who collect water quality samples professionally, companies, water protection associations and private citizens. In the tests, 100 users sent 1,146 pictures for analysis. The samples were collected from lakes and coastal areas of Finland. The professional water quality experts took the Secchi3000 measurements during their routine measurement and sample collection trips. The other participants took measurements during their leisure activities mostly out of personal interest.

The results of the analysis are described in the next chapter. As the automatic analysis was not completely ready during the test period, the images received with related metadata were only stored in the database, and

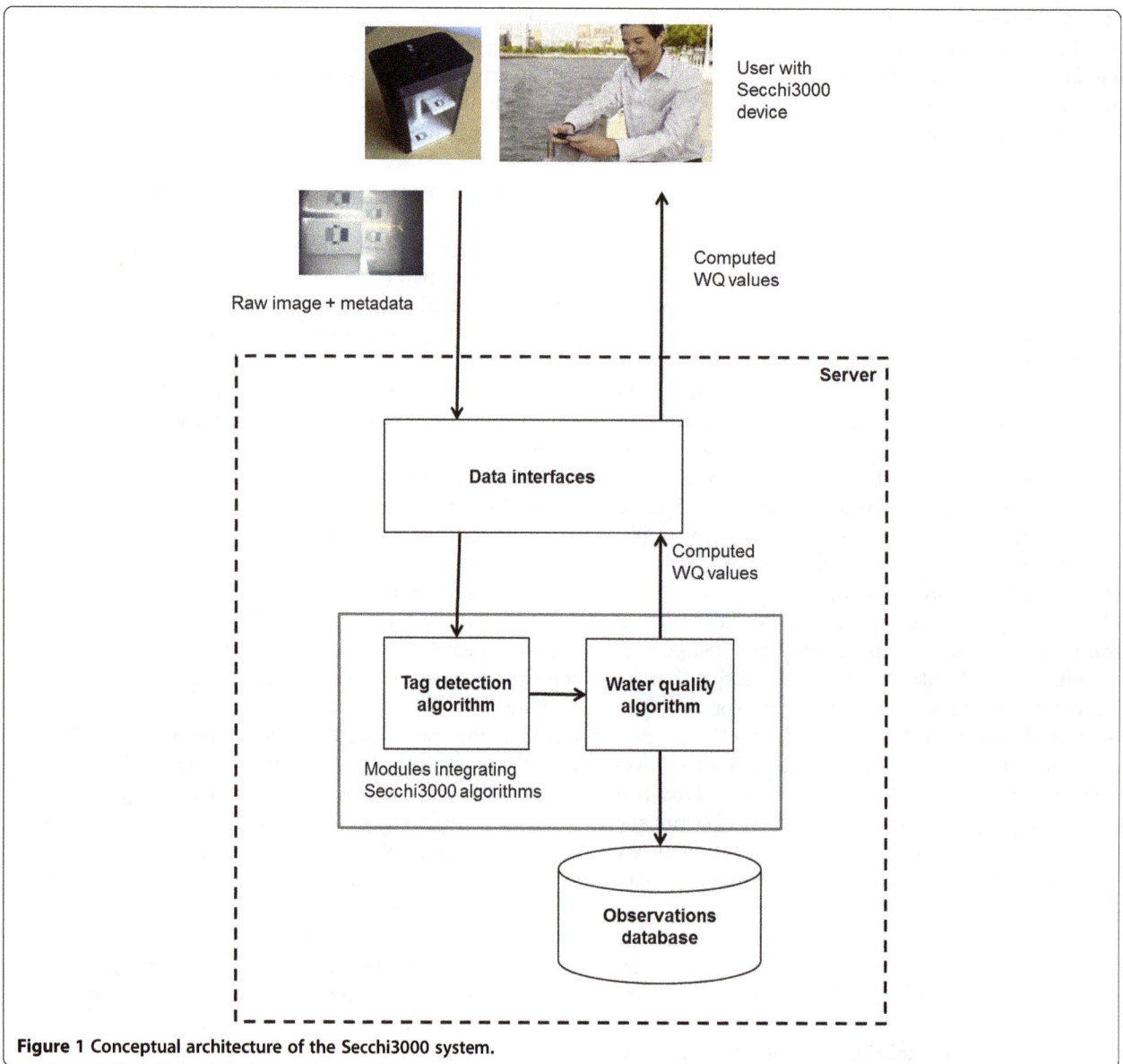

Figure 1 Conceptual architecture of the Secchi3000 system.

the analyses were carried out afterwards. The Secchi3000 devices were designed by the Finnish Environment Institute and VTT Technical Research Centre of Finland, and manufactured by a Finnish company. The system is currently being prepared for more large scale activities, including the commercial manufacture of the devices.

Target detection

Detecting accurate positions of white, grey and black targets in the images is challenging, because of many factors not related to the water quality:

- Varying illumination conditions create over-exposed areas, reflections and shadows in the inner structure of the measurement container. In the images, this translates into artefacts and undesired shapes that

can occlude, extend or overlap with the target rectangles. A purely shape-based approach will not always be robust to this kind of artefacts.
- Standard cameras from average mobile phones deliver mid- to low-end image quality, with noise, blur, varying resolutions (often rather poor). This affects the sharpness of target rectangles in the image and therefore their detectability.
- Variability during acquisitions made by different human operators using various devices introduces undesired discrepancies; in particular, the acquisition angles of the camera with regards to the scene vary all the time. Because the acquisition system is used in-situ and not in a controlled environment, one cannot assume a priori either the position of targets in the image or their orientation. The target

rectangles are not always oriented vertically or horizontally, they appear to be arbitrarily rotated in each image. Unlike in typical industrial computer vision applications, here the relative position and orientation of the camera and the scene cannot be known precisely.

On top of all these challenges, the lower the Secchi depth of the water is, the more difficult the target detection in the image. Because the water properties affect the reflectance and thus the pixel values, especially in cases of humic or very turbid waters, a target detection method based solely on colour or reflectance values in the image would most likely fail. Control points could be used to delineate the target areas; however, in turbid waters it is more reliable to detect larger homogeneous targets than control points.

For method development, a panel dataset of 20 images was manually chosen from a dataset of hundreds of water quality images acquired by users. The panel represents a wide variety of illumination conditions, water turbidity, and target detectability from the most favourable cases to the most difficult ones. Once fine-tuned on the panel, the method was applied to the whole image dataset for accuracy assessment.

The detection process was first to locate the surrounding areas of the lower and upper tags by the template matching method. The following processing is done in these two local tag areas (cf. Figure 2, the red rectangles are the areas detected using template matching).

The second step is to detect the white rectangle shape using a contour-based approach within the red rectangle area (the blue rectangle shape in Figure 2). The processing of the upper tag and lower tag is the same. The grey and black rectangle positions were obtained by similarity transformation. That is, by moving the rotated blue

rectangle leftward and rightward we get grey and black rectangles.

A mobile phone camera looking at the water tank through a hole approximates to a central projection in geometry computer vision. A central projection of a plane (tag) onto a parallel plane (image) is also called similarity transformation, which preserves parallelism, concurrence, ratio of division, etc. (Hartley and Zisserman 2004). That is angles between lines are not affected by rotation, translation or isotropic scaling. In particular parallel lines are mapped to parallel lines. The ratio of two lengths is an invariant. Similarly, a ratio of areas is an invariant.

When the image is extremely blurred and the camera is rotated, the white rectangle shape of the lower tag cannot be detected, but four corner image features can be extracted. Using the rectangle formed by these four corner feature points, the other two tags positions can be found, based on the similarity invariant.

Results

The results of the image analysis are shown in Figures 3 and 4. The Secchi3000 images were analysed both automatically and manually, and the results were compared against reference measurements. Our conclusions from the analysis are that the Secchi depth estimation works very well, while the turbidity estimates are a bit more scattered. Also, with larger turbidity values (> 6 FNU (Formazin Nephelometric Unit)) the difference from the reference value starts to increase (the order of magnitude is still correct).

The Secchi depth results shown in Figure 3 were derived from observations carefully made by one person using 3 mobile phones (Nokia E5, Nokia E7 and Nokia N500, depending on the location). The figure also shows that the difference between automatic and manual analysis is very small. Two to five observations were made with each phone from each location, and this explains the groups of data points visible in the image and gives an indication of the repeatability of the observations and differences between cameras. Overall, the variation of the measurements at each site was less than 10% of the reference value.

When the observations of the other test users are included in the results, there is significantly more scatter and the coefficient of determination is clearly lower. The likely explanations for this are: 1) the poor quality of the cameras in the mobile phones used by the test users (some were quite old models), 2) the deficient quality of the observations made by some of the users (the pictures sometimes included air bubbles, shadows and other error sources), 3) the difficulty of using the system in a rocking boat (as commented on by users), and 4) the lack of feedback from the system about the success of the measurement. The automated processing chain and the related feedback mechanism were implemented later based on

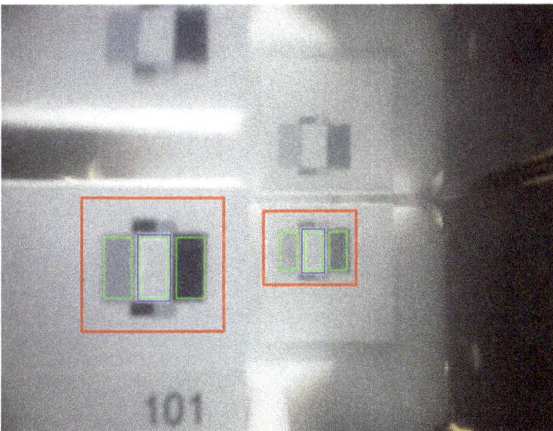

Figure 2 Locating the surrounding areas of lower and upper tags by the template matching method.

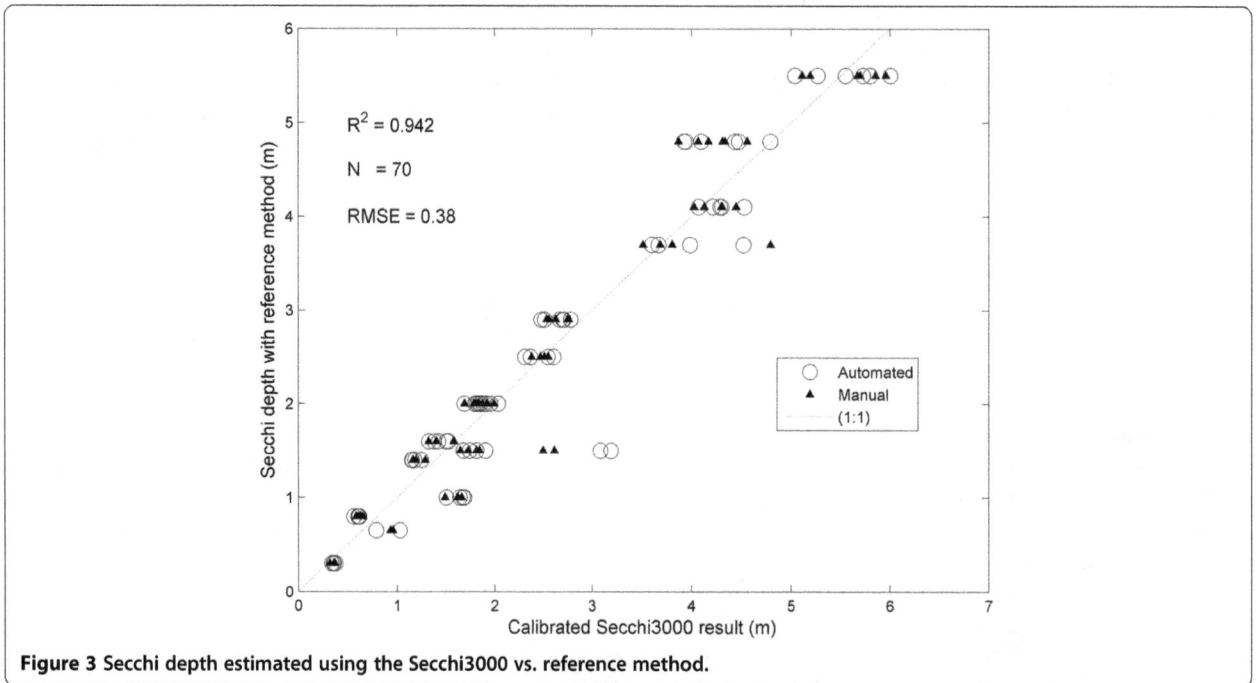

Figure 3 Secchi depth estimated using the Secchi3000 vs. reference method.

the user feedback. The average analysis time of the automatic processing is 2.09 seconds (N = 1146) per image using a Pentium 4 3.2 GHz processor with 4 GB memory.

Discussion

A measurement with the Secchi3000 has many advantages when compared to a measurement with a Secchi disk. For example, the measurement is not dependent on the eyesight of the user and the results are available quickly (with positioning) in a centralized system. Furthermore, the Secchi3000 measurement image can be re-analysed later if improved estimation algorithms become available. We are also planning to add more water quality parameters to be analysed automatically from the images such as total suspended matter (TSM) and absorption by humic substances (CDOM). This will give an indication of what

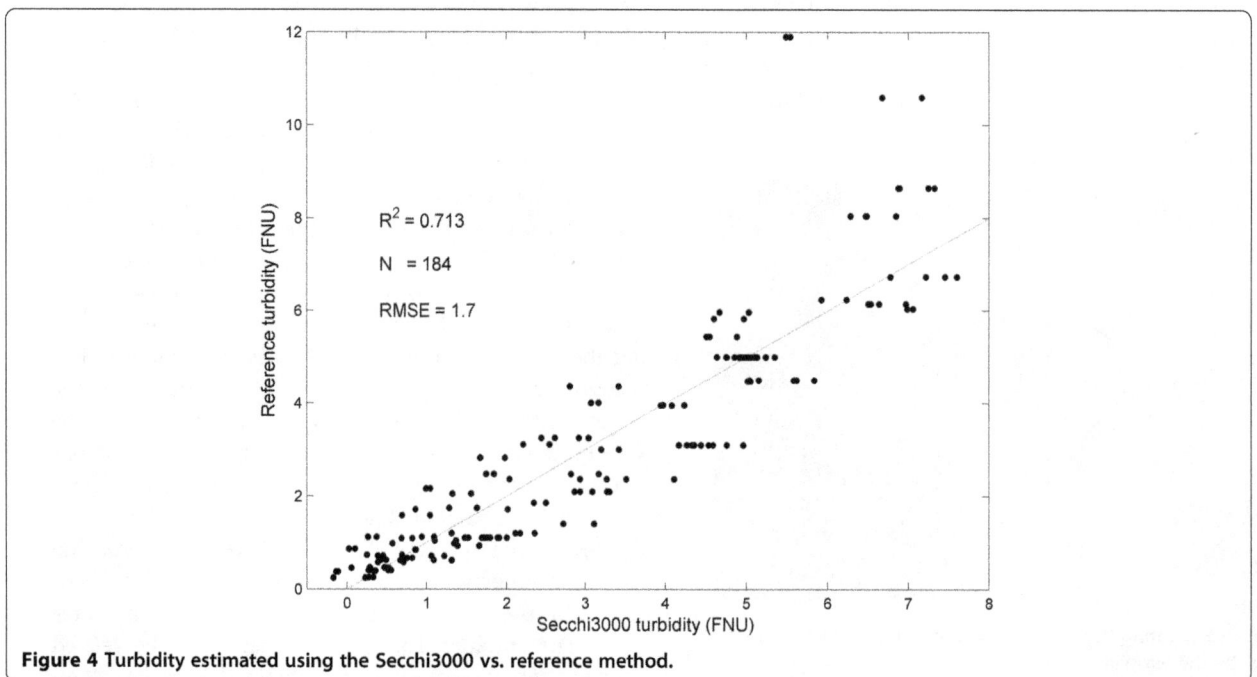

Figure 4 Turbidity estimated using the Secchi3000 vs. reference method.

causes the reduced transparency instead of only giving an estimate of the value of the transparency.

When compared to Transparency Tube, the Secchi3000 has the advantage of a greater estimation range. The tubes are limited to a maximum transparency of 1.2 m, while the Secchi3000 has successfully measured transparencies with a range from 0.3 m to over 5 m.

The turbidity meters mentioned above are more accurate but also much more expensive and thus not suitable for participatory monitoring.

Although the methods presented in this paper work quite well, the real benefits for water quality management would come from a wide use of the system. Participatory sensing tools have proved to be technically feasible, but motivating potential observers to collect data is still an open issue which has been discussed by e.g. Reddy et al (2010) and Juong-Sik and Hoh (2010). The popularization of the Secchi3000 method would require a business model which enables the manufacture and distribution of the devices, improvements and maintenance of the mobile software for multiple mobile devices, and a way of motivating people to make regular observations in their neighbourhood.

It is still necessary to develop the system further in order to make the measurements easier and less prone to errors. For example, the immediate delivery of the result of the measurement to the user, which has now been implemented, will help in reducing the number of poor quality observations. Also, the user instructions need to be improved and possibly moved into the app (the app should guide the user in making the measurement). In addition to the Secchi3000 Java ME mobile application that was used in data gathering during the summer of 2012, versions of the mobile phone application are now available for iOS, Android and Windows platforms. This greatly increases the potential user base of the system, which is essential for popularizing the method. The current price of a Secchi3000 measurement device is slightly higher than the cost of a Secchi disk. Presumably, this will change with larger scale production of the devices.

Conclusions

In this paper, we have presented a novel approach to automatic water quality analysis based on images from the Secchi3000 measurement device filled with water. The EnviObserver participatory sensing tool was used for collecting the data from users and for analysing the data received.

For tag detection, template matching can accurately locate the tags in an image with different size and orientation and separate them from clutter (such as tag reflection, text and the tag platform borders). Using image feature- and shape-based recognition increased the accuracy and robustness of the tag positions.

Results show that, overall, the system performs well. Both Secchi depth and turbidity are estimated with excellent or good accuracy when the measurements are taken with care. However, further development of the system is needed to ease the making of observations, and user instructions need to be improved so that users are able to provide better quality images for analysis.

Competing interests
The authors declare that they have no competing interests.

Authors' contributions
TT implemented the system architecture and drafted the manuscript, which all the authors helped to edit and polish. SK developed the algorithm for water quality estimation using target RGB-values extracted from mobile phone images and compared the measurements against reference measurements. VK provided domain-related guidance and helped design the system. MM and PC implemented the target detection algorithm. All the authors read and approved the final manuscript.

Acknowledgements
The work was supported by the Measurement, Monitoring and Environmental Assessment research programme funded by Tekes, the Finnish Funding Agency for Technology and Innovation. We are grateful to the Maa- ja vesitekniikan tuki ry for financial support.

Author details
[1]VTT Technical Research Centre of Finland, P.O. Box 1000, FI-02044 VTT Espoo, Finland. [2]Finnish Environment Institute, P.O. Box 140, Helsinki, Finland.

References

D'Hondt E, Stevens M, Jacobs A (2013) Participatory noise mapping works! An evaluation of participatory sensing as an alternative to standard techniques for environmental monitoring. Perv Mobile Comp 9(5):68–694. ISSN 1574–1192, doi:10.1016/j.pmcj.2012.09.002

Gallo T, Waitt D (2011) Creating a successful citizen science model to detect and report invasive species. Bio Sci 61(6):459–465. 10.1525/bio.2011.61.6.8

Hartley R, Zisserman A (2004) Multiple View Geometry in Computer Vision, 2nd edition. Cambridge University Press, Cambridge. ISBN 0521540518

Juong-Sik L, Hoh B (2010) Dynamic pricing incentive for participatory sensing. Perv Mobile Comp 6(6):693–708

Kotovirta V, Toivanen T, Tergujeff R, Huttunen M (2012) Participatory sensing in environmental monitoring – experiences. Proc Sixth Int Conf Innov Mobile Int Serv Ubiquit Comp (IMIS '12). doi:10.1109/IMIS.2012.70

Lane ND, Eisenman SB, Musolesi M, Miluzzo E, Campbell AT (2008) Urban sensing systems: opportunistic or participatory? In: Proceedings of the 9th workshop on Mobile computing systems and applications (HotMobile '08). ACM, New York, NY, USA. 10.1145/1411759.1411763, doi: 10.1145/1411759.1411763

Lane ND, Miluzzo E, Hong L, Peebles D, Choudhury T, Campbell AT (2010) A survey of mobile phone sensing. Comm Mag 48(9):140–150. doi:10.1109/MCOM.2010.5560598

Piha H, Zampoukas N (2011) Review of Methodological Standards Related to the Marine Strategy Framework Directive Criteria on Good Environmental Status. Publications Office of the European Union, Luxembourg. doi:10.2788/60512

Reddy S, Estrin D, Srivastava M (2010) Recruitment framework for participatory sensing data collections. Proceedings of the 8th International Conference on Pervasive Computing, Helsinki, Finland

Solimini A, Cadroso A, Heiskanen A (ed) (2006) Indicators and methods for the ecological status assessment under the Water Framework Directive. Office for Official Publications of the European Communities, Luxembourg. ISBN 92-79-02646-1

3

River water quality index for Morocco using a fuzzy inference system

Asmaa Mourhir[1][*], Tajjeeddine Rachidi[2] and Mohammed Karim[3]

Abstract

Background: The aim of this work is to propose a new river water quality index using fuzzy logic. The proposed fuzzy index combines quality indicators' prescribed thresholds extracted mainly from the Moroccan and the Quebec water legislations. The latter is reputed for its strict water quality assessment. The proposed index combines six indicators, and not only does it exhibit a tool that accounts for the discrepancy between the two base indices, but also provides a quantifiable score for the determined water quality. These classifications with a membership grade can be of a sound support for decision-making, and can help assign each section of a river a gradual quality sub-objective to be reached.

Results: To demonstrate the applicability of the proposed approach, the new index was used to classify water quality in a number of stations along the basins of Bouregreg-Chaouia and Zizi-Rhéris. The obtained classifications were then compared to the conventional physicochemical water quality index currently in use in Morocco. The results revealed that the fuzzy index provided stringent classifications compared to the conventional index in 41% and 33% of the cases for the two basins respectively. These noted exceptions are mainly due to the big disparities between the different quality thresholds in the two standards, especially for fecal coliform and total phosphorus.

Conclusions: These large disparities put forward an argument for the Moroccan water quality legislation to be upgraded to align water and environmental assessment methods with other countries in order to mitigate the risks of failing to achieve a good ecological status.

Keywords: Water quality index; Surface water; Fuzzy logic; Bouregreg-chaouia basin; Zizi-rhéris basin

Background

Water quality indices aim at turning several complex indicators into a single synthesized value that describes the water quality of a particular source, and which is understandable by a wide audience including non-experts like the public or decision and policy-makers. There exist a number of water quality indices based on different indicators and aggregation methods used today throughout the world, such as the U.S. National Sanitation Foundation Water Quality Index (NSFWQI); the Canadian Council of Ministers of the Environment Water Quality Index (CCMEWQI); the British Columbia Water Quality Index (BCWQI), and the Oregon Water Quality Index (OWQI).

A general flaw with conventional water quality indices is that they depend on human expertise with subjective and ambiguous information, which might raise a number of issues in water quality assessment. Another major problem is that quality indicator concentrations, no matter how far from or close to the limits have the same impact on the final score and might fall within the same classes (Icaga 2007). To deal with ambiguity and uncertainties, fuzzy logic, which was first introduced by (Zadeh 1965) is commonly used as a powerful formalism in environmental evaluation and assessment such as in water or air pollution issues. Ecological impact classification and environmental decision-making using fuzzy logic are discussed in detail by (Shepard 2005; Silvert 1997, 2000). According to the authors, using fuzzy logic is very convenient in the assessment of environmental issues because it can solve properly the ambiguities and subjectivity inherent in these problems. It also helps conciliating conflicting observations due to human expertise, and last but

* Correspondence: A.Mourhir@aui.ma
[1]School of Sciences and Engineering, Al Akhawayn University in Ifrane, PO.Box 2083, Ifrane 53000, Morocco
Full list of author information is available at the end of the article

not least, it can provide decision-makers with the ability to make well-informed decisions that are technically sound and legally defensible. Moreover, with fuzzy logic one can describe water quality in a location as being 10% excellent and 90% just good, which is not possible with classical approaches to water quality.

Fuzzy logic has shown a good promise in modeling new water quality indices. (Lu et al. 1999; Chang et al. 2001) studied the feasibility of applying Fuzzy Synthetic Evaluation (FSE) to water quality. The Fuzzy Comprehensive Assessment (FCA) method was used by (Shen et al. 2005) to investigate pollution and evaluate the soil environmental quality of the Taihu lake watershed. (Liou et al. 2003) applied a two-stage fuzzy set theory to river quality evaluation in Taiwan. (Ocampo-Duque et al. 2006) used fuzzy logic and a comprehensive multi-attribute decision-aiding method based on the Analytic Hierarchy Process (AHP) to estimate the relative importance of water quality indicators. A 6-step procedure to develop a fuzzy water quality index was described in (Icaga 2007), and was applied to lake water. In another study by (Sadiq and Tesfamariam 2008), a Weighted Averaging Operator (OWA) was used for aggregation in developing the water quality index. (Lermontov et al. 2009; Roveda et al. 2010) developed fuzzy water quality indices for Brazilian rivers, and compared their performance with the conventional WQIs. A different approach was carried out by (Nasiri et al. 2007); the authors proposed a fuzzy multi-attribute decision support system to compute the water quality index and to outline the prioritization of alternative plans based on the extent of improvements in water quality. (Mahapatra et al. 2011) used a Cascaded Fuzzy Inference System (CFS) to design a multi-input, multi-output water quality index. In a recent study, (Gharibi et al. 2012) developed a FWQI for which the water quality indicators were practical and easy to measure, including heavy metals, and used the index to assess water quality in the Mamloo dam for drinking purposes. A different approach based on hybrid fuzzy-probability models was adopted in (Ocampo-Duque et al. 2013; Nikoo et al. 2011). In a recent work, (Wang et al. 2014) used variable fuzzy set and the information entropy theory as an assessment model to evaluate water quality of the Meiliang Bay in Taihu Lake Basin in China.

The water quality in Moroccan rivers and streams is becoming more and more of a concern because of the significant amount of pollutants discharged into these ecosystems, in most cases without any treatment. Physico-chemical and microbiological measurements and analyses are performed regularly by a number of public administrations. However, these analyses remain insufficient given the diversity of chemicals and the variety of pollution sources. This said; considerable efforts have been deployed recently in Morocco to improve water quality. The adoption of the "10-95" Water Law in 1995, the establishment

of River Basin Management Agencies, the adoption of national plans for integrated water resource management, the adoption of a national program for sewerage and wastewater treatment in 2008, as well as the Water Strategy in 2009 are all actions in favor of better water quality outcomes. However, the application of the "10-95" texts is slow, and there is no improvement worthy to mention as yet in water quality, which remains generally degraded. In addition, the quality indicators used in analysis remain insufficient today, and pesticides intensified by the *Maroc Vert* plan are not taken into account. Only a simplified set of indicators that are realistic and easy to measure are actually used in the conventional index to estimate river water quality. For the other indicators (like heavy metals); the techniques used to evaluate these indicators are either complicated, time consuming, or costly, which results in discarding them from consideration in the final assessment of water quality.

Clearly, from the deficiencies cited above, there is a need to upgrade the Moroccan legislation on water quality to include a new index. Such an index must factor in the use of indicators with direct causes on human and animal health, as well as evaluation thresholds. Our approach is progressive, and consists of providing, through fuzzy logic, a new index that captures the essence of the existing one (the conventional physiochemical index designed by a Decree in 2002 and never revised since), and yet incorporates the benefits of another well standing index. This is an initial step to improve water quality assessment through enhancement of threshold values in order to account for real water pollution, the index still needs to be improved to take into consideration contamination with heavy metals. For this purpose, the overall water quality index of Quebec (IQBP) (Hébert 1997) was chosen. The IQBP is very comparable to the Moroccan conventional water quality index, which will be referred to in the remainder of the text as (IMBP). Both IMBP and IQBP estimate water quality based on a number of physicochemical and bacteriological indicators, and provide water quality classes for multiple usages. They are also similar in terms of the evaluation method used which is based on the "minimum operator" that assigns the lowest indicator quality to the overall water quality in a given location. Moreover, the approach used to design the IQBP index relied on a group of thirty water quality experts and professionals from different horizons (chemists, biologists, managers, etc..) who have been consulted according to the Delphi method (Linstone and Turoff 1975). However, a quick comparison between the Moroccan water quality indicator threshold values and the ones used in Quebec reveals a big disparity between the two standards.

In this work, fuzzy methodology is used to come up with a water quality index that we will be referring to as

MFWQI, by combining the Moroccan and Quebec standards. Membership functions for the different water quality indicators were developed considering boundaries from both water regulatory bodies. A fuzzy inference system was used to assess water quality with a membership grade in a number of stations along the basin under study. In the aforementioned studies about fuzzy indices, the generation of membership functions is usually done based on expert opinions, and hence remains difficult to maintain or update because they require expert consultation every time. In this paper, an easy-to-use method is introduced; it facilitates the generation of suitable membership functions based on a combination of already established standards by merging water quality thresholds in an automatic way. For any given indicator, the quality threshold values are determined in both standards; trapezoidal and triangular membership functions are then derived automatically based on interval intersection and linear interpolation. Furthermore, rules are derived automatically based on the number of variables as well as the number of membership functions, and the aggregation method used. If n is the number of required input variables, each represented with m membership functions, then the total number of rules required to model the system is $m^{^n}$. Our approach in this work is to automatically generate the fuzzy index with minimum need for expert involvement.

The concepts needed to model an index using fuzzy logic, including membership functions and fuzzy inference steps are described concisely in the Methods section.

Results and discussion

The results of comparison between the MFWQI and the conventional IMBP for each sampling site of the Bouregre-Chaouia basin over the period under study are illustrated in Figure 1.

Based on the IMBP simplified quality grid shown in Table 1, the results of the monitoring performed during four campaigns showed that 76% of the sampled surface water points had a poor to very poor quality in January-09, as compared to 23% of good to average quality. However, in July-09 the quality improved to 44% points of good to average quality as compared to 56% points of poor to very poor quality.

Based on the MFWQI, the results showed that 88% of the sampled surface water points had a poor to very poor quality in January-09, as compared to only 12% of good to average quality. In July-09 the quality improved to 32% points of good to average quality as compared to 68% points of poor to very poor quality.

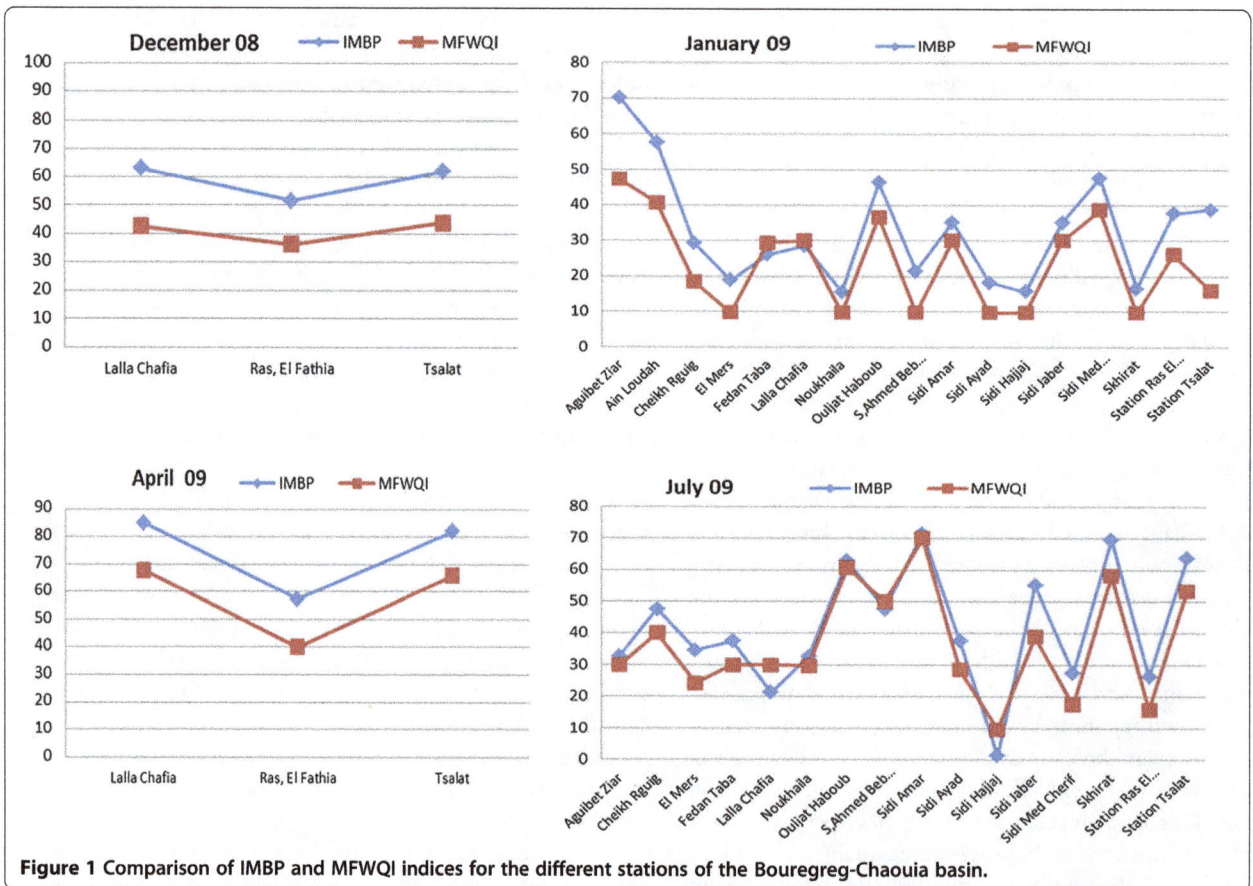

Figure 1 Comparison of IMBP and MFWQI indices for the different stations of the Bouregreg-Chaouia basin.

Table 1 Simplified rating table for the IMBP water quality indicators (Official Bulletin N°. 5062 of 5 December 2002)

Sub-index	DO (mg/L)	BOD5 (mg/L)	COD (mg/L)	NH4+ (mg/L)	TP (mg/L)	FC (cfu/100 ml)[1]	Index
Excellent	>7	<3	<30	<=0.1	<=0.1	<=20	80 – 100
Good	7 – 5	3 – 5	30 – 35	0.1 – 0.5	0.1 – 0.3	20 – 2000	60 – 80
Average	5 – 3	5 – 10	35 – 40	0.5 – 2	0.3 – 0.5	2000 – 20000	40 – 60
Poor	3 – 1	10 – 25	40 – 80	2 – 8	0.5 – 3	>20000	20 – 40
Very poor	<1	>25	>80	>8	>3	-	0 – 20

[1]*cfu* colony forming units.

As can be seen in Figure 1, both indices showed more or less correlated results with some sensitivity to water pollution. The IMBP and MFWQI had a correlation coefficient of 0.93, demonstrating the sound applicability of the index. In spite of this, the MFWQI resulted in a more severe assessment compared to the conventional physicochemical index. Indeed, the quality class obtained with the fuzzy index was lower for 41% of the sampled sites. In other cases, the evaluations of water quality carried out by the two indices were comparable.

The results of comparison between the MFWQI and the conventional IMBP for some sampling sites of the Ziz-Rhéris basin between 2007 and 2008 are illustrated in Figure 2. The water of Oued Ziz has a good quality upstream and becomes moderately polluted downstream near urban areas, due to urban discharges. The water of Oued Rhéris has a good quality most of the time. Here again it can be seen that the quality class obtained with the fuzzy index was lower for 33% of the sampled sites. This percentage is lower in the basin of Ziz-Rhéris since the unit is less polluted compared to the basin of Bouregreg-Chaouia.

By examining the relative distribution of these results by quality classes, it is obvious that disparities between the two indices appear throughout all of the quality classes, especially in the lower ones. It can also be noticed in Figures 1 and 2 that, when conventional and fuzzy indices do not produce a similar assessment, the gap between the quality classes is usually of a single class. These differences are observed at sites where a single indicator is always problematic (fecal coliforms or total phosphorus). Given the nature of the two indices and quality thresholds used for their calculation, this is quite normal. Fecal coliforms and total phosphorus are evaluated more harshly with the IQBP. For example, if we consider fecal coliforms, the value used in the IQBP to distinguish "poor" water quality from "very poor" quality is 3500 (cfu/100 mL), while the quality criterion used in the Moroccan index is "poor" when the FC is above 20000 (cfu/100 mL). Furthermore, there is no category for "very poor" quality for this indicator in the Moroccan index. It is important to note that individuals exposed to water contaminated with fecal coliforms have a potential health risk.

Unsurprisingly, the proposed fuzzy index revealed some discrepancies between the Moroccan water quality

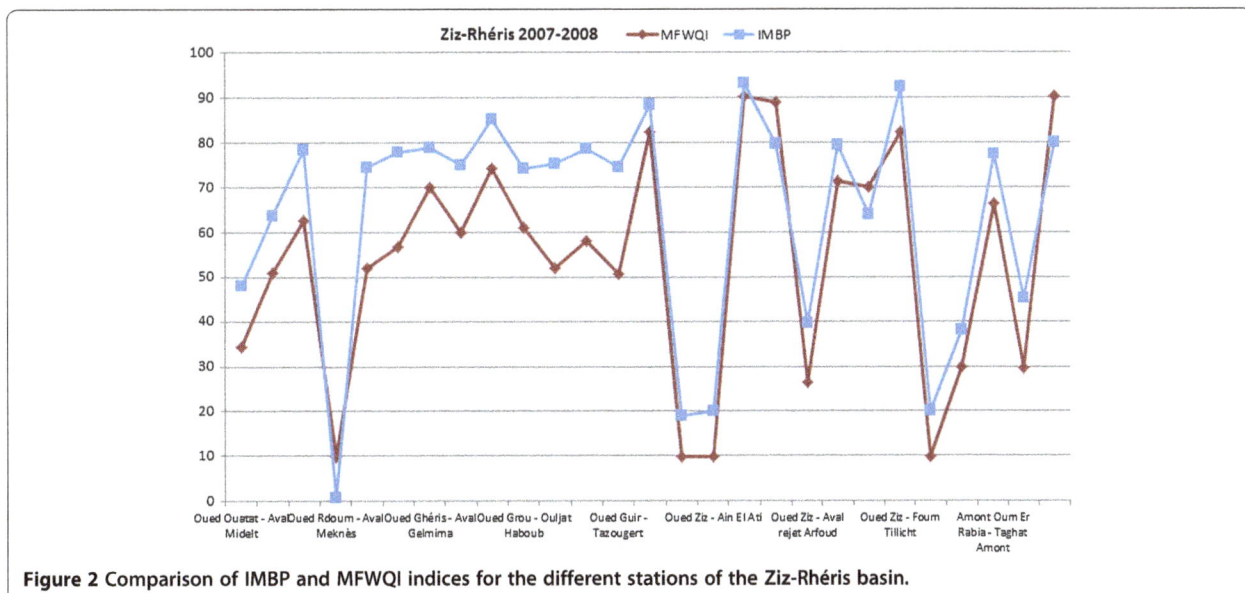

Figure 2 Comparison of IMBP and MFWQI indices for the different stations of the Ziz-Rhéris basin.

index and the other surface water quality standards. The fuzzy index is more effective in the sense that it is more accurate in detecting water pollution because it conciliates between water quality ranges as prescribed by the two water regulatory bodies, namely the Moroccan legislation and that of Quebec, which is considered to be stricter.

Moreover, the conventional assessment of water quality based on quality thresholds prescribed by the Moroccan legislation (IMBP) gives the results in the form of qualitative classes, such as "good", "average", or "poor", and therefore the information provided by the index is very limited. A weighted method was used to quantify the index and produced both a qualitative class and a score. With the fuzzy index, not only does water quality move from a linguistic description to a quantifiable representation without further computational overhead, but it produces a membership grade that shows to what strength a stream's water quality belongs to a class (examples are shown in Table 2). This approach gives an excellent quantitative insight, which can serve as a sound basis for further decision-making. Decision makers can assign different objectives to different parts of a river depending on the membership grade. A portion of a river qualified as being roughly halfway between "good", with a membership grade of 0.33, and "average" quality, with membership grade 0.30, can help professionals treat it differently from a portion qualified as having "average" quality with a membership grade of 1.

Conclusions

In the context of the ongoing efforts aimed at improving the environment in Morocco, a new quality index

Table 2 Fuzzy index scores with membership grades (GoM) for different classes

Period	Station	IMBP class	MFWQI class	GoM
Jan.-09	Cheikh Rguig	Poor	Very poor	0.85
			Poor	0.15
Jan.-09	Sidi Med Cherif	Average	Poor	0.86
			Average	0.14
Apr.-09	Ras, El Fathia	Average	Poor	0.5
			Average	0.5
July-09	Cheikh Rguig	Average	Poor	0.5
			Average	0.5
July-09	Ouljat Haboub	Good	Good	0.68
			Average	0.32
July-09	Sidi Jaber	Average	Poor	0.75
			Average	0.25
July-09	Skhirat	Good	Average	0.95
			Good	0.05

for surface waters using fuzzy logic has been developed (MFWQI). An application of the index was demonstrated for surface waters in the basins of Bouregreg-Chaouia and Ziz-Rhéris. Water quality was evaluated by means of six indicators (DO, BOD5, COD, FC, TP and NH4+). The proposed index can resolve problems of uncertainty and linguistic ambiguity inherent to this particular environmental problem. Moreover, unlike the conventional index, the new fuzzy index allows the results to be interpreted quantitatively or qualitatively along with membership grades. It also allows a better analysis since experts can describe a sampling station quality status as closer to its upper or lower limit.

A comparison between the conventional index and the new fuzzy index was carried out with the objective of being able to point out the weaknesses of the conventional approach, and to propose an upgrade to Moroccan water legislation in a simple and meaningful way. The proposed index has been shown to be more rigorous because it uses quality thresholds from the Quebec IQBP index (reputed to be very stringent), but yet it conserves the expert knowledge embodied in the Moroccan IMBP index. The conventional index does not fully comply with health expert knowledge about industrial and agricultural pollution known in the area. There is a clear need to review legislation about water quality that has been revealed by the proposed fuzzy index. The conventional index does not reflect the alarming situation of water quality; which minimizes the chances of triggering enough responses or the application of existing laws to handle the situation, and hence the use of water for drinking or for agriculture from the rivers without treatment may expose the population to health risks. While it is not expected that the proposed approach will be used for water quality assessment by local authorities, the intention is to draw attention to the need for Morocco to run an adjustment exercise to align its water and environmental assessment methods with other countries in order to mitigate the risks of failing to achieve a good ecological status.

Furthermore, the conventional IMBP is a state-oriented index, which somehow fails to reflect the socio-economic pressures that result in water quality degradation depending on different geographical zones. In Europe for example, a number of efforts have been deployed to account for the different pressures exerted by the socio-economic driving forces in addressing water problems (Borja et al. 2006). Thus, further efforts need to be taken to perform an integrated evaluation that takes into consideration socio-economic indicators besides ecological indicators.

Methods

Fuzzy logic formalism

Fuzzy logic is an extension of Boolean logic created by (Zadeh 1965) based on the theory of fuzzy sets, which is a generalization of the classical set theory. By allowing a condition to be partly true and partly false at the same time, fuzzy logic makes it suitable to take into account any ambiguities or uncertainties. A key concept in fuzzy logic is membership functions.

Membership functions

Let X be the universe of discourse and its elements denoted by x. A fuzzy set A in the universe X is characterized by a membership function $\mu_A: X \rightarrow [0, 1]$. The fuzzy set A can be represented by the set of pairs of an element $x \in X$ and its degree of membership defined by a membership function $\mu_A(x)$:

$$A = \{ (x, _A(x)) \ / x \in X \} \tag{1}$$

A degree of zero means that the value is not in the set, a degree of one means that the value is totally representative of the set, and a degree confined between zero and one means the value is partially in the set. The shape of the membership function is often chosen based on the advice of an expert or by statistical studies. A Sigmoid shape, Triangular, Trapezoidal, Gaussian or any other type can be used. The concept of membership functions discussed above allows the definition of fuzzy natural language systems that make use of linguistic variables, where the universe of discourse of a variable is divided into a number of fuzzy sets with a linguistic description attributed to each one.

Fuzzy operators

In order to easily manipulate fuzzy sets, the operators of the classical set theory are adapted to the membership functions specific to fuzzy logic, strictly allowing values between 0 and 1. Typically, the extension of the union operator (OR) to fuzzy sets A and B defined over the same set X is defined as:

$$\mu_{A \cup B}(x) = max \left[\mu_A(x), \mu_B(x) \right] \tag{2}$$

Where μ_A and μ_B are the membership functions for A and B respectively. Similarly, the fuzzy intersection (AND) is defined by:

$$\mu_{A \cap B}(x) = min \left[\mu_A(x), \mu_B(x) \right] \tag{3}$$

Fuzzy inference rules

In fuzzy logic, *if-then* rules and fuzzy set operators are used to describe the relationships between input variables and output variables of a system. Fuzzy rules are a collection of linguistic statements that describe how a fuzzy inference system should make a decision regarding classifying an input or controlling an output. A fuzzy rule has one or more antecedents, usually connected by linguistic operators such as "and" or "or". Rules are always written in the following form:

Ri : IF x is Ai and/or y is Bi THEN z is Ci

Where x and y are the input variables and z is the output variable. *Ai*, *Bi* and *Ci* are linguistic values for the variables x, y and z respectively.

Basic structure of a fuzzy inference system (FIS)

A fuzzy inference system (FIS) is an inference system based on fuzzy set theory, which maps input values to outputs. The fuzzy inference process involves four main steps (Ross 1995):

a) Fuzzification: in this step, crisp input values are mapped into linguistic variables using membership functions. This is required in order to activate rules that are in terms of linguistic variables. The *fuzzifier* takes input values and determines the degree to which they belong to each of the fuzzy sets via membership functions.

b) Rule evaluation: in this step, the consequence of a fuzzy *if-then* rule is computed. First, the rule strength is computed by combining the fuzzified inputs. Combination of multiple conjunctive antecedents is performed using the fuzzy intersection operation. Multiple disjunctive antecedents are combined using the fuzzy union operation. Then, the rule consequent is correlated with the strength value of the rule antecedent; the most common method for rule implication is to cut the consequent membership function at the level of the antecedent truth. This method is called clipping (alpha-cut).

c) Aggregation of rule outputs: outputs for all rules are then aggregated into a single fuzzy distribution. This is usually done by using the fuzzy union of all individual rule contributions.

d) Defuzzification: in this step, the aggregated output fuzzy set is mapped into a crisp number. Several methods are used in practice for *defuzzification*, including the "centroid", "maximum", "mean of maxima", "height", and "modified height". The most popular defuzzification method is the centroid method, which calculates the center of gravity of the aggregated fuzzy set:

$$COG = \int_a^b \mu_A(x) \, x \, dx \div \int_a^b \mu_A(x) \, dx \tag{4}$$

Development of the Moroccan fuzzy water quality index
The Moroccan conventional index (IMBP)

The assessment of surface water quality in Morocco is performed using the water quality index IMBP defined by water legislation (Official Bulletin N°. 5062 of 5 December 2002). The IMBP suggests recommended water quality using a number of physicochemical and bacteriological indicators, and aggregates them to produce a single quality class depending on a given usage such as fish life, irrigation, industrial uses, cooling, or raw water supply intended for drinking.

The IMBP index consists of six indicators: dissolved oxygen (DO), biological oxygen demand during five days (BOD5), chemical oxygen demand (COD), ammonium (NH4+), fecal coliforms (FC) and total phosphorus (TP). The simplified rating grid of surface waters shown in Table 1 establishes five dominant classes according to the utilization goals for which the water is intended. Each class is defined by a set of threshold values that the different physicochemical or bacteriological indicators, which are particularly important, must not exceed.

The IMBP index applies the concept of the lowest score, i.e., the "minimum operator" is used to produce the final index score. The approach dictates that the water quality in a sample corresponds to that of the indicator producing the lowest sub-index as computed for every indicator using threshold values determined by Table 1.

$$IMBP = minimum \begin{pmatrix} DO\ sub\text{-}index,\ BOD5\ sub\text{-}index,\ COD \\ sub\text{-}index,\ NH4+ \\ sub\text{-}index,\ FC\ sub\text{-}index,\ TP\ sub\text{-}index \end{pmatrix}$$

For example, if all the indicators have values corresponding to the "excellent" class, except one, which falls into the "average" class, the IMBP will assign the water body to the "average" class.

Preliminary indications show that the "minimum operator" approach is a more useful aggregation method than additive and multiplicative techniques. (Smith 1990) showed that most indices based on additive or multiplicative approaches were insensitive; i.e., they were little influenced by the poor quality associated with one or two descriptors because of the aggregation method used. Another advantage of this approach is that the index ensures that a certain number of basic indicators are assessed clearly before an overall classification is assigned to a given water body. The rationale behind the method is that it is very important to understand and specify the type of water pollution and the element that resulted in water quality alteration, in order to establish a clear diagnosis and identify the water quality problem.

Since the IMBP produces only qualitative classes such as "excellent" or "poor", a quantification method is adopted in order to obtain a quantitative value that can be easily compared to the crisp scores produced by the fuzzy index. To quantify the different classes, the values of the ranges set by the quality thresholds to evaluate water quality are transformed into dimensionless numbers ranging from 0 (for extremely poor water quality) to 100 (for absolutely excellent water quality).

The sub-index of a given indicator is elaborated by weighting, which means the index is obtained by producing a value that is proportional to its real position in a class range. The formula for calculating the weighted index (IP) is shown by Eq. 5.

$$IPpa = li + [(ls - li)/(bs - bi)] * (bs - pa) \tag{5}$$

Where:

- IP_{pa}: the weighted index for indicator pa
- li: the lower index
- ls: the upper index
- bi: the lower bound
- bs: the upper bound
- pa: the analyzed indicator value

The following example shows how to calculate the weighted index for BOD5 with value 3.5 mg/l. As shown in Table 1, the value 3.5 is between 3 and 5; and therefore it belongs to the "good" class. Thus, the lower bound bi is 3, the upper bound bs is 5, the lower index li is 60 and the upper index ls is 80. Hence IP $_{BOD5}$ is 75.

The Quebec water quality index (IQBP)

In the late 1990s, the Quebec Department of Environment developed a bacteriological and physicochemical water quality index (IQBP) for representing water quality throughout its river network. Based on this index, water bodies are grouped into five classes according to all the potential uses.

To classify a water body, water quality is examined using ten indicators, namely: total phosphorus (TP), fecal coliforms (FC), ammonium (NH4+), nitrates (NO_3^-)/nitrites (HNO_2), total chlorophyll a (Cha), dissolved oxygen percentage (%O2), biological oxygen demand in five days (BOD5), suspended solids (SS), turbidity (TU) and pH. Table 3 presents the criteria used to assign one of the five classes to a water body.

The IQBP requires, for each indicator analyzed, the transformation of measured concentrations into a sub-index, with a rating curve for assessing the water quality. Like the Moroccan index, the IQBP is a downgrading type index; that is to say, for a given sample, the index value corresponds to the lowest sub-index associated with the most problematic substance.

It can be noticed from the rating grids (Table 1 and Table 3) that there are differences in quality ranges

Table 3 Class boundaries of water quality for some indicators used in the IQBP (Hébert 1997)

Class	FC (cfu/100 ml)	BOD5 (mg/L)	TP (mg/L)	O2 (%)	SS (mg/L)	NOx- (mg/L)	NH4+ (mg/L)
A	≤ 200	≤ 1,7	<=0.03	88 – 124	≤ 6	≤ 0,50	≤ 0,23
B	201 – 1000	1,8 – 3,0	0,031 – 0,050	0.03 – 0.05 125 – 130	7 – 13	0,51 – 1,00	0,24 – 0,50
C	1001 – 2000	3,1 – 4,3	0,051 – 0,100	70 – 79 131 – 140	14 – 24	1,01 – 2,00	0,51 – 0,90
D	2001 – 3500	4,4 – 5,9	0,101 – 0,200	55 – 69 141 – 150	25 – 41	2,01 – 5,00	0,91 – 1,50
E	>3500	> 5,9	> 0,200	< 55 > 150	> 41	> 5,00	> 1,50

between the Moroccan and Quebec standards. These differences exhibit tougher quality thresholds in the IQBP for all water indicators in common between the two standards, especially for fecal coliform and total phosphorus.

One problem with this type of highly subjective water quality assessment rendered by both the IMBP and IQBP is that the final index does not take into consideration the uncertainty about acceptable threshold values for each indicator. In the next section an alternative to the IMBP

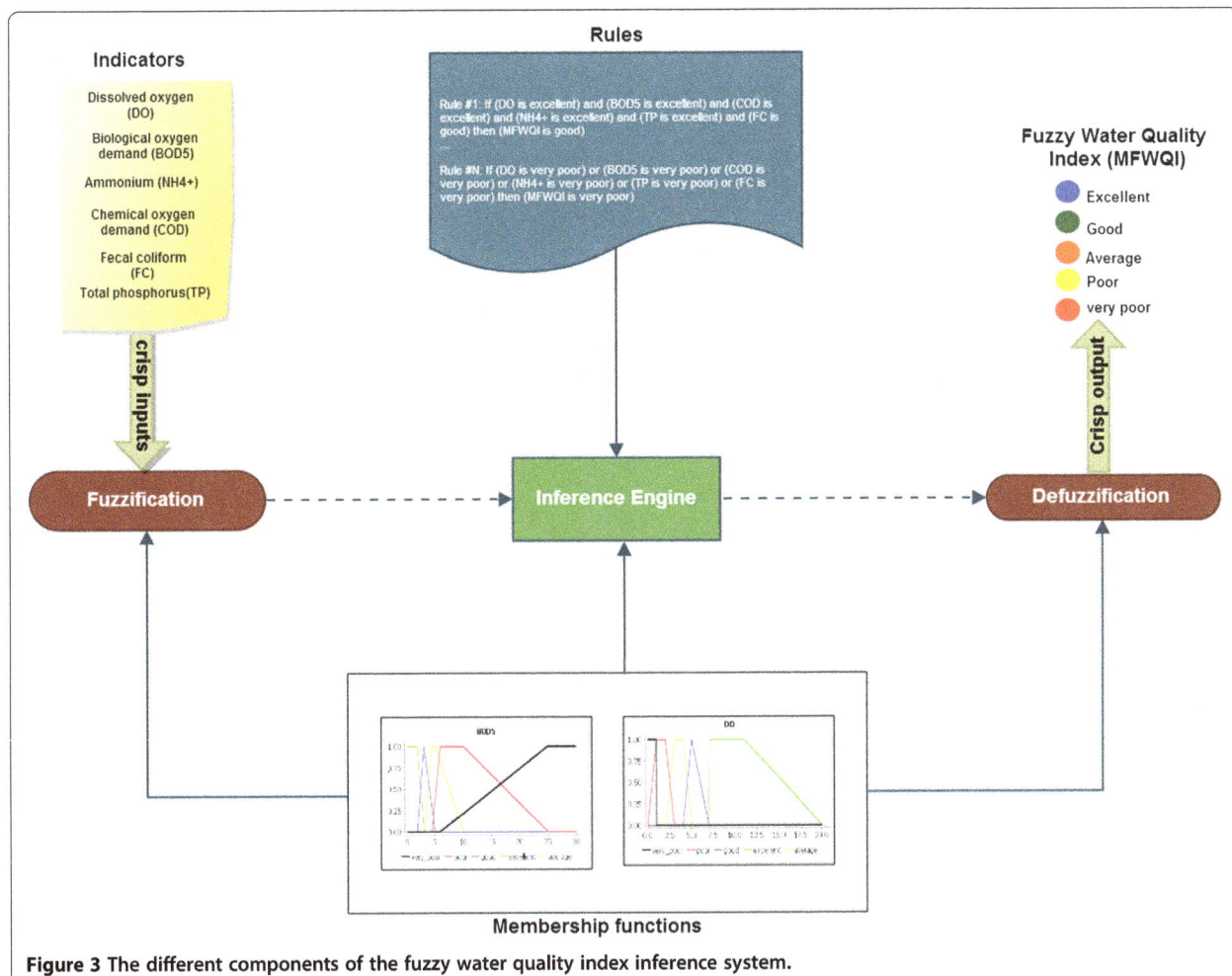

Figure 3 The different components of the fuzzy water quality index inference system.

index based on fuzzy logic is proposed with more relevance to the type of uncertainties involved in this particular problem.

Building a fuzzy water quality index for Morocco (MFWQI)
In order to combine the benefits from the two standards discussed above, fuzzy methodology is used to propose a new water quality index by conciliating quality thresholds from the Moroccan and Quebec legislations. Membership functions for the different water quality indicators were developed considering boundaries from both water regulatory bodies. A fuzzy inference system was designed and built to classify water quality with a membership grade. The components of the fuzzy inference system are depicted in Figure 3.

The method used to generate membership functions is very simple. Having two pieces of intervals about the same water quality class, these two pieces of information are fused. It can be concluded that the actual value belongs to the intersection of these intervals. If I and J respectively denote the quality ranges for a given class of water quality in the Moroccan standard and the Quebec one, then $I \cap J$ is considered as the interval where certainty is the highest, corresponding to a degree of membership equal to 1. Then, a linear interpolation is used for the remaining points not belonging to the intersection by linking the lower and upper bounds of both I and J to the intersection interval, which results in a trapezoidal shape. In cases where the intersection results in a single point, the shape is triangular.

For example, for FC to be considered as "good", FC measurements must fall within [201, 1000] and [20, 2000] for IQBP and IMBP respectively. For the purpose of our fuzzy index, a trapezoidal membership function for the term "good" corresponding to the indicator

FC is developed from these two intervals according to Eq. 6.

$$\mu_A(x) = \begin{cases} 0,\ x < 20\ or\ x > 2000 \\ \dfrac{(x-20)}{(200-20)},\ 20 \le x \le 200 \\ 1,\ 200 \le x \le 1000 \\ \dfrac{(2000-x)}{(2000-1000)},\ 1000 \le x \le 2000 \end{cases} \quad (6)$$

Similarly, BOD5 is considered "good" when BOD5 measurements fall within intervals [1.8, 3] and [3, 5] for IQBP and IMBP respectively. The intersection of these intervals results in a single point, hence the shape of the membership function "good" for the variable BOD5 is triangular as shown by Eq. 7.

$$\mu_A(x) = \begin{cases} 0,\ x < 1.8\ or\ x > 5 \\ \dfrac{(x-1.8)}{(3-1.8)},\ 1.8 \le x \le 3 \\ \dfrac{(5-x)}{(5-3)},\ 3 \le x \le 5 \end{cases} \quad (7)$$

Overall, five fuzzy sets, which are "excellent", "good", "average", "poor", and "very poor" have been considered for this study for both input indicators and for the output water quality index.

Depending on the overlap between quality thresholds in the two standards, trapezoidal and triangular membership functions were derived from the parameters as shown in Table 4. Membership curves for input indicators and MFWQI are shown in Figure 4.

Using the different fuzzy sets of the considered indicators, *if-then* rules were then generated automatically. Initially, since the inference methodology used relates the

Table 4 Parameters for membership functions of the different indicators used in the MFWQI

	DO				BOD5				COD			
	a	b	c	d	a	b	c	d	a	b	c	d
Excellent	7	7	11	20	0	0	1.8	3	0	0	20	30
Good	4	5	7	7	1.8	3	5	-	20	25	30	35
Average	2	3	4	5	3.1	4.3	5	10	25	35	40	-
Poor	0	1	2	3	4.4	5.9	10	25	40	40	80	80
Very poor	0	0	1	1	5.9	25	200	200	80	81	500	500
	NH4+				TP				FC			
	a	b	c	d	a	b	c	d	a	b	c	d
Excellent	0	0	0.1	0.5	0	0	0.03	0.1	0	0	20	200
Good	0.1	0.5	1	-	0.03	0.05	0.1	0.3	20	200	1000	2000
Average	0.5	1	2	2	0.05	0.1	0.3	0.5	1000	2000	20000	-
Poor	2	2	5	8	0.1	0.2	0.5	3	2000	3500	20000	50000
Very poor	5	8	50	250	0.2	3	20	20	3500	50000	1800000	1800000

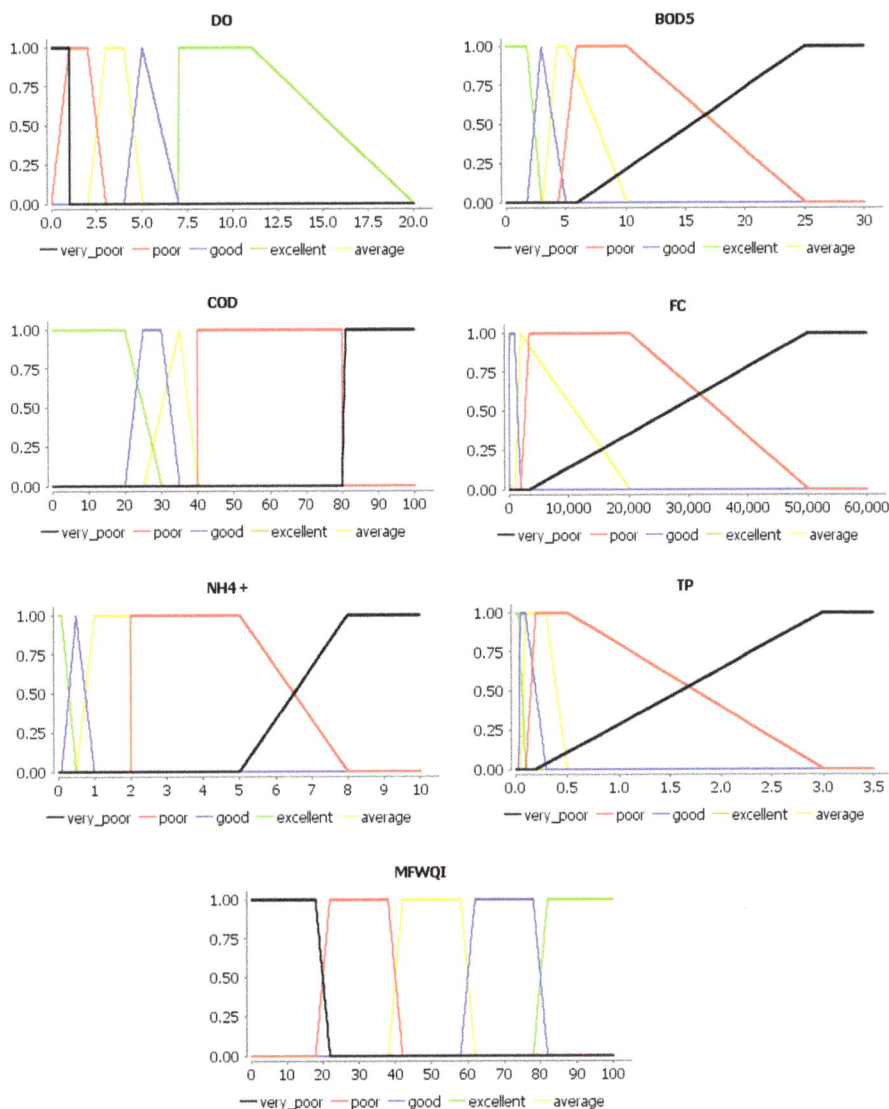

Figure 4 Membership functions for DO, BOD5, COD, FC, NH4+, TP and MFWQI.

relevant subsets of each input universal set to the subsets of the other system inputs through an intersection-rule configuration, a total number of $5^{^6}$ rules were generated, representing the model of the water quality assessment system from the set of 6 input indicators and their possible 5 classes. However, as the inference is based on the minimum sub-index, an optimization could be made using a disjunction of inputs by means of the "OR" operator: the MFWQI is considered "very poor" if one of the indicators is "very poor", and hence the rule used for all possible combinations in that case is Rule#3 as shown below. This optimization reduced the number of rules to 4097.

The examples below show three rules for "good", "poor", and "very poor" water quality respectively:

Rule #1: *If (DO is excellent) and (BOD5 is excellent) and (COD is excellent) and (NH4+ is excellent) and (TP is excellent) and (FC is good) then (MFWQI is good)*
Rule #2: *If (DO is excellent) and (BOD5 is good) and (COD is average) and (NH4+ is poor) and (TP is good) and (FC is excellent) then (MFWQI is poor)*
Rule #3: *If (DO is very poor) or (BOD5 is very poor) or (COD is very poor) or (NH4+ is very poor) or (TP is very poor) or (FC is very poor) then (MFWQI is very poor)*

In this work, the (Mamdani 1974) approach was used to build the MFWQI fuzzy inference engine. This approach is known for its simple structure and Max–Min inference. The implication method used is the "min" and

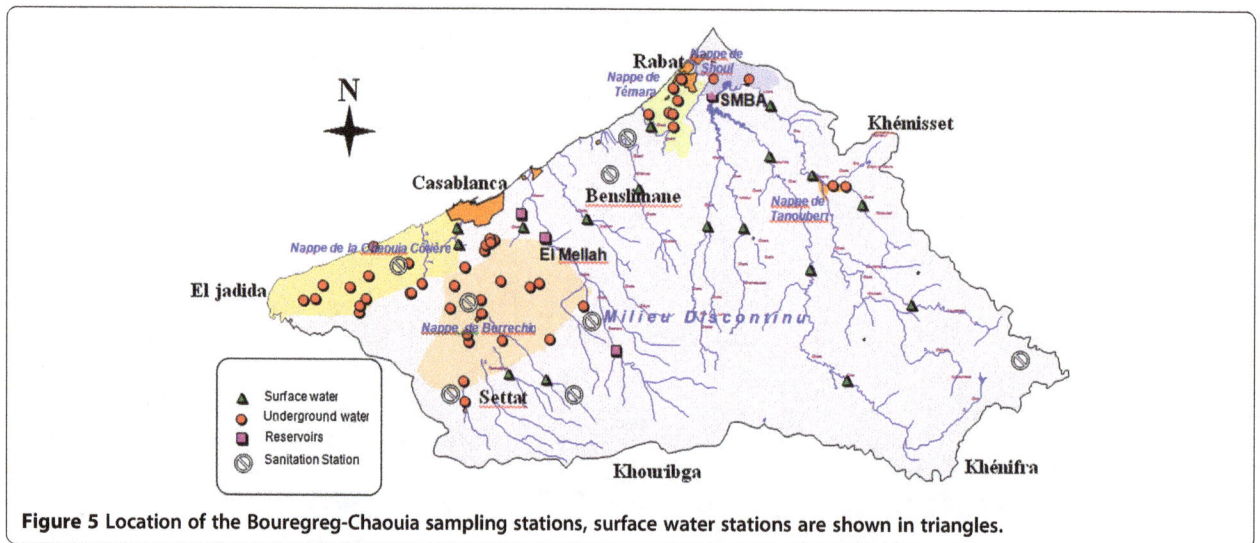

Figure 5 Location of the Bouregreg-Chaouia sampling stations, surface water stations are shown in triangles.

the aggregation method is "max". The defuzzification method used to determine the output is the center of gravity (COG) as expressed in Eq. (4). The computational tool used in modeling the overall system is the *Matlab Fuzzy Logic Toolbox package 7.6*.

Study area

In order to assess the proposed fuzzy index, a case study of water quality was performed using measured environmental data on each sampling site of the monitoring network of the Bouregreg-Chaouia basin collected during two primary campaigns in December-08 and April-09, and two more full campaigns that took place in January-09 and July-09 on its primary and secondary surface water networks. The total number of analyses performed over the study period for surface waters was up to 979. We also conducted a secondary case study on a less polluted area namely the basin Ziz-Rhéris during the period 2007–2008. All measurements were conducted according to standard methods.

Bouregreg-chaouia

The basin of Bouregreg-Chaouia (ABHBC 2000) extends over a surface of 20.470 km^2; that is to say nearly 3% of the national territory. Structurally, it is composed of the three following areas: 1) the basin of Bouregreg: which is the most important one; 2) the basin of Coastal rivers between Bouregreg and Oum Er Rbia. The main rivers are Owed Yquem, Cherrat, Nfefikh and Mellah, which flow into the Atlantic Ocean between Rabat and Casablanca; 3) and the Chaouia Plain.

The water quality monitoring network in the basin of Bouregreg-Chaouia was introduced in 1991. It tracks the quality status of surface water and groundwater in the region. It consists of 20 sampling stations, distributed as

follows: 1) the primary network, tracked four times a year, and which contains three hydrological stations: Tsalat Grou, Lalla Chafia on Oued Bouregreg and Ras El Oued on Fathia Aguenour; 2) the secondary network, tracked twice a year, consists of fourteen sampling points. The monitoring network is depicted on the map in Figure 5, where the surface water stations are shown in triangles.

Ziz-Rhéris

The basin Ziz-Rhéris corresponds to watersheds of the Ziz and Rhéris oueds. The unit extends over a surface of 24.900 km^2. The region is bounded on the north by the Moulouya Basin, to the north west by the basin of Oum-Rbia, to the west by the pool Draa, on the east by the watershed Guir and south by Algeria. This unit is located largely in the province of Errachidia, only the top Todgha upstream Tinjdad is part of the province of Ouarzazate.

Competing interests

The authors declare that they have no competing interests.

Authors' contributions

AM designed the study, conducted data analysis and wrote the manuscript. TR revised critically and helped editing and polishing the paper. All authors have read and approved the final manuscript.

Acknowledgement

The authors would like to express their deep gratitude to the environmentalists who provided water data used in this study besides reports about the state of water quality in the area of the Hydraulic Basin of Bouregreg-Chaouia and Ziz-Rhéis.

Author details

[1]School of Sciences and Engineering, Al Akhawayn University in Ifrane, PO.Box 2083, Ifrane 53000, Morocco. [2]School of Sciences and Engineering, Al Akhawayn University in Ifrane, PO.Box 1881, Ifrane 53000, Morocco. [3]Faculté des Sciences, Université Sidi Mohammed Ben Abdellah, Faculté des Sciences Dhar el Mehraz, Fès, Morocco.

References

ABHBC (2000) Agence du Bassin Hydraulique du Bouregreg et de la Chaouia., http://www.abhbc.com/index.php/bassin/la-zone-daction-de-lagence. Accessed 08 July 2012

Borja Á, Galparsoro I, Solaun O, Muxika I, Tello EM, Uriarte A, Valencia V (2006) The European Water Framework Directive and the DPSIR, a methodological approach to assess the risk of failing to achieve good ecological status. Estuar Coast Shelf Sci 66(1–2):84–96

Chang N-B, Chen HW, Ning SK (2001) Identification of river water quality using the Fuzzy Synthetic Evaluation approach. J Environ Manage 63(3):293–305

Gharibi H, Mahvi AH, Nabizadeh R, Arabalibeik H, Yunesian M, Sowlat MH (2012) A novel approach in water quality assessment based on fuzzy logic. J Environ Manage 112:87–95

Hébert S (1997) Développement d'un indice de la qualité bactériologique et physico-chimique de l'eau pour les rivières du Québec, Québec. Ministère de l'Environnement et de la Faune, Direction des écosystèmes aquatiques, Quebec

Icaga Y (2007) Fuzzy evaluation of water quality classification. Ecol Indic 7(3):710–718

Lermontov A, Yokoyama L, Lermontov M, Machado MAS (2009) River quality analysis using fuzzy water quality index: Ribeira do Iguape river watershed, Brazil. Ecol Indic 9(6):1188–1197

Linstone HA, Turoff M (1975) The Delphi method: techniques and applications. Addison-Wesley Pub. Co., Advanced Book Program, Boston, USA

Liou S-M, Lo S-L, Hu C-Y (2003) Application of two-stage fuzzy set theory to river quality evaluation in Taiwan. Water Res 37(6):1406–1416

Lu RS, Lo SL, Hu JY (1999) Analysis of reservoir water quality using fuzzy synthetic evaluation. Stoch Env Res Risk A 13(5):327–336

Mahapatra SS, Nanda SK, Panigrahy BK (2011) A Cascaded Fuzzy Inference System for Indian river water quality prediction. Adv Eng Softw 42(10):787–796

Mamdani EH (1974) Application of fuzzy algorithms for control of simple dynamic plant. Proc Inst Elec Eng 121:1585–1588

Nasiri F, Maqsood I, Huang G, Fuller N (2007) Water Quality Index: A Fuzzy River-Pollution Decision Support Expert System. J Water Resour Plann Manag 133(2):95–105, doi:10.1061/(ASCE)0733-9496(2007)133:2(95)

Nikoo M, Kerachian R, Malakpour-Estalaki S, Bashi-Azghadi S, Azimi-Ghadikolaee M (2011) A probabilistic water quality index for river water quality assessment: a case study. Environ Monit Assess 181(1–4):465–478, doi:10.1007/s10661-010-1842-4

Ocampo-Duque W, Ferré-Huguet N, Domingo JL, Schuhmacher M (2006) Assessing water quality in rivers with fuzzy inference systems: A case study. Environ Int 32(6):733–742

Ocampo-Duque W, Osorio C, Piamba C, Schuhmacher M, Domingo JL (2013) Water quality analysis in rivers with non-parametric probability distributions and fuzzy inference systems: Application to the Cauca River, Colombia. Environ Int 52:17–28

Ross TJ (1995) Fuzzy Logic with Engineering Applications. McGraw-Hill, New York, USA

Roveda SRMM, Bondança APM, Silva JGS, Roveda JAF, Rosa AH (2010) Development of a water quality index using a fuzzy logic: A case study for the Sorocaba river. In: 2010 IEEE International Conference on Fuzzy Systems (FUZZ)

Sadiq R, Tesfamariam S (2008) Developing environmental indices using fuzzy numbers ordered weighted averaging (FN-OWA) operators. Stoch Environ Res Risk Assess 22(4):495–505, 10.1007/s00477-007-0151-0

Shen G, Lu Y, Wang M, Sun Y (2005) Status and fuzzy comprehensive assessment of combined heavy metal and organo-chlorine pesticide pollution in the Taihu Lake region of China. J Environ Manage 76(4):355–362

Shepard RB (2005) Quantifying Environmental Impact Assessments Using Fuzzy Logic. Springer, USA

Silvert W (1997) Ecological impact classification with fuzzy sets. Ecol Model 96(1–3):1–10

Silvert W (2000) Fuzzy indices of environmental conditions. Ecol Model 130(1–3):111–119

Smith DG (1990) A better water quality indexing system for rivers and streams. Water Res 24(10):1237–1244

Wang Y, Sheng D, Wang D, Ma H, Wu J, Xu F (2014) Variable fuzzy set theory to assess water quality of the meiliang bay in taihu lake basin. Water Resour Manag 28(3):867–880

Zadeh LA (1965) Fuzzy sets. Inform Control 8:338–353

4

Performance of locally available bulking agents in Newfoundland and Labrador during bench-scale municipal solid waste composting

Khoshrooz Kazemi, Baiyu Zhang*, Leonard M Lye and Weiyun Lin

Abstract

Background: Newfoundland and Labrador (NL) has one of the highest waste disposal rates in Canada and it has 200 small communities without access to central composting facilities. During Municipal solid waste (MSW) composting, the selection of bulking agents is critical. Bench-scale composting systems plus locally available bulking agents are thus desired for economic and effective MSW management in NL communities. This study evaluated the performance of locally available bulking agents (i.e., NL sawdust and peat) during MSW composting in a bench-scale system. Physiochemical (temperature, oxygen uptake rate, pH, electrical conductivity, moisture and ash content, and C/N ratio) and biological (enzyme activities and germination index) parameters were monitored to evaluate compost maturity and stability.

Results: In peat composting, higher temperature for a longer duration was observed, indicating more effective pathogen removal and sterilization. High enzyme activities of dehydrogenase, β-glucosidase, and phosphodiesterase in the third week of composting imply high microbial activity and high decomposition rate. The low C/N ratio for compost product implies acceptable stability states. In sawdust composting, higher temperature and oxygen uptake rate (OUR) were observed in the third week of composting, and higher enzyme activities in the second week. Sawdust composting generated a higher germination index, indicating higher maturity.

Conclusions: Both sawdust and peat are effective bulking agents for the bench-scale composting. The choice of a bulking agent for a particular community depends on the availability of the agent and land in the region, convenience of transportation, price, and the expected quality of the compost product.

Keywords: Municipal solid waste (MSW); Bench-scale composting; Sawdust; Peat; Newfoundland and Labrador

Background

Population growth, aggregation of human settlements, higher living standards, and increased development and consumption of less biodegradable products have increased solid waste generation over the last 20 years (Adhikari et al. 2008; Asase et al. 2009). Municipal Solid Waste (MSW) management has thus become one of the biggest environmental concerns in recent decades (Iqbal et al. 2010). MSW contains high moisture content (60-70%) and large organic fraction (70-80%), posing adverse environmental impacts if it is not treated properly. Fortunately, the high organic fraction of MSW can be easily converted to energy sources through composting (Jolanun and Towprayoon 2010; Ponsá 2010). Therefore, composting has become an increasingly important strategy for the treatment of MSW. Composting is a biological process in which easily degradable organic matter is stabilized and converted into a humus-rich product by the action of microorganisms (Eiland et al. 2001). The advantages of composting are diverting organic matter from landfills, reducing waste volume, decreasing the potential odour, decreasing the moisture content of MSW, and amending soil/improving soil quality (Haug 1993; Cronje et al. 2003; Arslan et al. 2011; Hasan et al. 2012).

Some environmental conditions (moisture content, aeration rate, pH, and temperature) and substrate characteristics (C/N ratio, particle size, bulking agents, nutrients contents, and free air space) affect the composting

* Correspondence: bzhang@mun.ca
Faculty of Engineering and Applied Science, Memorial University of Newfoundland, St. John's, Newfoundland and Labrador A1B 3X5, Canada

process (Iqbal et al. 2010). Selection of a bulking agent which should be inexpensive and readily available in the vicinity of the composting region is very important because bulking agents can affect the condition of the starting composting mixtures, biodegradation kinetics and composting performance as well as the final compost quality (Blanco and Almendros 1995; Chang and Chen 2010; Jolanun and Towprayoon 2010). Bulking agents have different properties because of their carbon source, physical shape, particle size, water absorption capacity, and their bulking density (Iqbal et al. 2010). Bulking agents are usually fibrous and carbonadoes material with low moisture content; therefore they can absorb part of the leachate produced during decomposition to keep the moisture and sustain the microbial activity (Adhikari et al. 2008; Dias et al. 2010; Iqbal et al. 2010). The bulking agent provides structural support to prevent physical compaction, promotes porosity and air void, and improves the compost aeration and gas exchange (Adhikari et al. 2008; Yañez et al. 2009; Dias et al. 2010; Doublet et al. 2011). It can also act as a buffer against the organic acids in the early stages of composting and help maintain the mixture's pH within a range from 6–8 for proper microbial activity (Haug 1993), and adjust C/N ratio of the feedstock and encourage microbial activity without inhabitation (Jolanun and Towprayoon 2010). Numerous studies have used different bulking agents, which are mostly from agriculture byproducts. They include sawdust (Martin et al. 1993; Blanco and Almendros 1995; Banegas et al. 2007; Adhikari et al. 2008; Chang and Chen 2010; Yang et al. 2013), wheat straw (Blanco and Almendros 1995; Banegas et al. 2007), hay and pine wood shaving (Banegas et al. 2007), bagasse and paper (Adhikari et al. 2008), rice husk and rice barn (Chang and Chen 2010), wooden palette (Huet et al. 2012), cornstalks and spent mushroom substrate (Yang et al. 2013), wheat flour (Silva et al. 2014), peat (Mathur et al. 1986; Mathur et al. 1990; Martin et al. 1993; Vuorinen 2000; Nolan et al. 2011), and barley straw (Vuorinen 2000).

Sawdust is a by-product of cutting, grinding, drilling, and sanding of wood, and it is a very common and easily available bulking agent used in composting to provide the free air space, control moisture, and maintain the C/N ratio (Batham et al. 2013). Banegas et al. (2007) mixed aerobic and anaerobic sludge with sawdust in two ratios (1:1 and 1:3 v: v), and concluded that sawdust is a good bulking agent for sludge composting because of its dilution effect on the nutritional components of the compost. Iqbal et al. (2010) suggested that the effect of 40% addition of sawdust to MSW was best to optimize the moisture content to up to 60% in composting. Chang and Chen (2010) found more sawdust in the composting mixture resulted in the increase of the water absorption

capacity and the composting rate, shorter composting and acidification times, and lower final pH value.

Peat is an accumulation of partially decayed vegetation or organic matter which has been used as a bulking agent because it has high water absorption capacity, is rich enough in exchangeable H^+ ions to neutralize the ammonia and the cations released by decomposition prevents the loss of ammonia by remaining slightly acidic environment throughout the composting process. Peat has the capacity for adsorbing anions and retarding the leaching of NO_3^- and PO_4^{-3} when added to soil. It is fluffy to provide thermal insulation and replaceable air to prevent anaerobic production of malodours, and also has an exceptionally high capacity for enhancing soil organic matter (Mathur et al. 1990).

In Canada, Newfoundland and Labrador (NL) has the highest quantity of waste disposal per capita after Alberta. This amounts to about 429 kg of residential waste per capita (Statistics Canada 2008). NL comprises more than 200 small communities with population between 100 and 600. Most of these small communities are located in remote and isolated areas and cannot access large solid waste disposal sites or central organic processing facilities. Therefore, on-site composting facilities have been considered as a viable means to deal with organic wastes in the small communities. Although a lack of extensive agricultural production in the northern region of NL could limit the selection of bulking agents for composting, NL generally possesses extensive peat resources. In addition, the forestry industry in NL produces wastes organic materials in the form of sawdust, bark, and wood chips, which can be used as the bulking agent for MSW composting (Martin et al. 1993). The food waste constitutes approximately 40% of the MSW and it represents a significant proportion of organic material found in MSW. Diversion of food waste from landfill since it is the biggest organic stream in municipal solid waste is essential to reach high diversion target (Environment 2013). Therefore, detailed knowledge of the performance of the composting process with locally available bulking agents would allow the improvement of community-scale composts quality in the small communities of NL.

For compost quality assessment and practical use of composted materials in agriculture, maturity and stability indices are important (Mondini et al. 2004). Stability can be expressed by biological indicators such as the respiration index (i.e., oxygen uptake rate (OUR) or CO_2 evolution rate) and enzyme activity (Wu et al. 2000; Benito et al. 2003; Bernal et al. 2009). Important enzymes involved in the composting process include dehydrogenase activity for substrate oxidization by a reduction reaction, β – glucosidase activity for glucoside and amide hydrolysis, as well as phosphodiesterase activity for phosphate removal from organic compounds (Mondini et al. 2004). Maturity refers to the degree of decomposition where the compost does not pose any adverse effects on plants and growth of various crops (Zmora-Nahum et al. 2005; Castaldi et al. 2008). It is commonly reflected using the germination index (GI). There are currently limited studies on the effect of bulking agents (i.e., peat and sawdust) on the maturity and stability indices such as enzyme activities and GI.

Therefore, in this study, the performance of locally available bulking agents on the bench-scale MSW composting in NL was examined. Meanwhile, a comprehensive investigation of parameters indicating compost maturity and stability and monitoring composting process was conducted. The OUR and enzyme activities were selected to reflect compost stability, and GI was investigated to evaluate compost maturity.

Results and discussion
Temperature and OUR
The changes in composting temperature and OUR for FP (food waste + peat) and FS (food waste + sawdust) are shown in Figure 1. The temperature of the composting reactor indicates the breakdown of the organic matter and the quality of the compost, since the rise of temperature is the result of decomposition of readily available organic matter and nitrogen compounds by microorganisms (Ros et al. 2006; Lee et al. 2009). Temperature is one of the important indices to evaluate compost efficiency (Lee et al. 2009) because it affects the biological reaction rate, the population dynamic of microbes, and the physiochemical characteristics of the compost (Hue and Liu 1995). Godden et al. (1983) suggested three distinct stages during composting, including the mesophilic (below 40°C), thermophilic (above 40°C), and cooling (ambient temperature) stage. As the FP composting proceeded, the temperature of the decomposing waste rose rapidly and reached to a maximum temperature of 68°C after 2 days.

It is known that the highest thermophilic activity in the composting system was maintained at a temperature between 52 and 60°C (Liang et al. 2003; Kalamdhad et al. 2009). The high temperature ensured the elimination of all pathogens; only 3 days at 55°C was sufficient for elimination of pathogens (Rasapoor et al. 2009). Although the temperature of the FS compost showed an increase to 52°C on the third day, the high temperature period on compost was not sufficient to ensure the hygiene safety of the end product. Longer high temperature period was observed in FP composting, indicating effective pathogen removal and sterilization. The microbial activity and the organic matter breakdown rate decreased when the organic matter became more stabilized and consequently the temperature dropped for almost two weeks in both compost to the ambient temperature (Ros et al. 2006). Microbial respiration has been used to measure the microbial activity during composting. It has also been used to assess the evolution of the composting process and maturity of the final product (Ros et al. 2006). High OUR was recorded for FP during the first 5 days of composting and then it decreased sharply. High OUR indicates that organic matter are available for microorganisms to be degraded, and therefore the material is not stabilized yet. Low OUR indicates organic matter are more stabilized and most of the organic matter has been decomposed by microorganisms (Said-Pullicino et al. 2007). Increase of OUR for FS was smoother and reached the highest value at the end of the first week. Although the maximum OUR for FP was almost double the value of that for FS, the duration of the high OUR was much longer for FS than for FP. The OUR eventually decreased and appear to reach a steady state.

pH and electrical conductivity (EC)
The pH values for FS and FP ranged from 4.6 to 8.68 during composting. The pH value of the compost is one of the important factors to evaluate compost stability and maturity due to its influence on the physical-chemical and microbiological reactions in the compost (Banegas et al. 2007). The initial pH and the pH in the first week of FS and FP composting were slightly acidic as a result of organic acids such as acetic acid and butyric acid, partially contained in the food waste and partially produced by microorganism reactions (Smårs et al. 2002; Adhikari et al. 2009; Eklind and Kirchmann 2000). When microorganisms consume organic acids as a substrate, pH started to increase (Adhikari et al. 2009). The highest pH was observed after 8 days for FS and after 16 days for FP compost. This delay for FP compost could be due to the loss of ammonium through volatilization and nitrification, and accumulation of organic acid and CO_2 during decomposition of the simple organic matter like carbohydrates (Banegas et al. 2007; Chukwujindu et al. 2006; Kayikçioğlu and

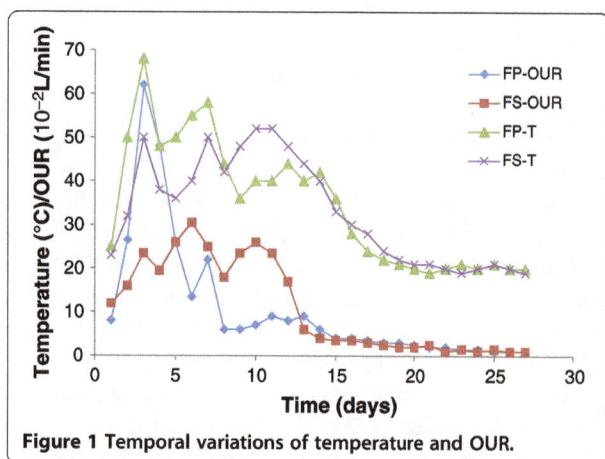

Figure 1 Temporal variations of temperature and OUR.

Okur 2011). Compost with low pH indicates lack of maturity due to the short composting time or occurrence of the anaerobic process (Iglesias Jiménez and Perez Garcia 1989). The final pH for FS and FP was above 8 and pH levels stayed almost steady by the end of composting.

Compost EC affects microbial population and organic matter transformation. High EC values of compost may have phytoxicity effects on the plant and negatively influence the plant growth and seed germination (Banegas et al. 2007; Kalamdhad et al. 2009; Arslan et al. 2011). Experimental results showed that EC values of FS compost increased earlier than FP compost (Figure 2). This increase could be due to the release of mineral cation concentration such as ammonium ions and phosphate which did not bind to the stable organic complex or went out of the system through leachate (Francou et al. 2005; Kalamdhad et al. 2009).

Moisture and ash content

Moisture and ash content variations are shown in Figure 3. As shown in the figure, moisture content showed descending trends in both compost. The combination of evaporation because of high temperature and aeration lead to the decrease of moisture content during composting, especially at high temperatures (Said-Pullicino et al. 2007; Lashermes et al. 2012). Moisture content for FS compost showed a slow declining trend by 10 days, which is an indication of decomposition of organic matter (Kalamdhad et al. 2009; Arslan et al. 2011). The temporary increasing trend observed for FS and FP compost was because temperature was not high enough to evaporate the water produced through microbial activity. The amount of ash increased consistently. The ash content increasing trend had a large slope at the thermophilic stage, and then the slope became smoother when the temperature dropped. During composting the organic matter was decomposed into volatile compounds, and consequently

the final compost has lower organic matter and higher ash content (Kalamdhad et al. 2009).

C/N ratio

Figure 4 shows the C/N ratio variation for both FS and FP compost. Both composts have close C/N ratio in the beginning (17). The initial C/N ratio has a strong influence on the performance of the composting process and the quality of the end product (Gao et al. 2010). An extremely high C/N ratio makes the composting process very slow as there is an excess of degradable substrate and lack of N for the microorganisms. On the other hand, a very low C/N ratio can lead to loss of N through NH_3 volatilization and generate potential odour problem (de Bertoldi Md, Vallini Ge, Pera A, 1983; Gao et al. 2010; Christensen 2011). For FS and FP composting, the initial C/N ratio were lower than the optimum value recommended for composting, i.e., 25 to 30 (Haug 1993). C/N ratio decreased for both composts during thermophilic phase. Decrease was very fast for FP whereas after the first week, the C/N ratio for FP dropped to 9 while it was 13 for FS. High microbial activity and high decomposition of organic matter after two weeks led to a C/N ratio decrease in both treatments. The C/N ratio stayed steady after two weeks by the end of the experiments for both FS and FP composts. The final value of C/N ratio of FP was low than that of FS.

GI

The maturity of the compost has been evaluated based on chemical parameters correlated with plant response (Bernal et al. 2009; Xiao et al. 2009). Seed germination test helps to evaluate the efficiency of the composting process for plant growth and seed germination (Banegas et al. 2007). As it is shown in Figure 5, GI is high at the beginning since the raw material is synthetic and non-toxic food waste. GI decreased as a result of formation of toxic compounds such as alcohols, phenolic compound, and organic acids during the thermophilic phase as a result of the composting process. This decrease was sharp for FS compost by the end of the first week and after that it started to increase quickly. It has been suggested that a GI over 80% indicates the absence of phytotoxicities in compost (Tiquia and Tam 1998; Zucconi et al. 1981). At the end of the composting, GI for FS was higher than 80%; but for FP, GI did not reach 40%, which can be associated with the stage of the composting. Higher degree of maturity was found for the FS compost.

Enzyme activities

Enzymes are responsible for the breakdown of several organic compounds characterised by complex structures, finally generating simple water-soluble compounds (Castaldi et al. 2008). Characterising and quantifying

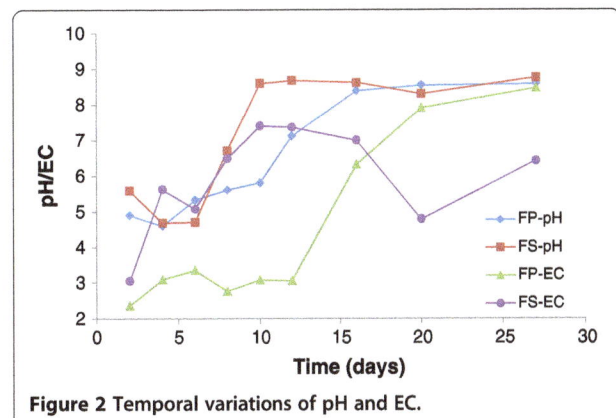

Figure 2 Temporal variations of pH and EC.

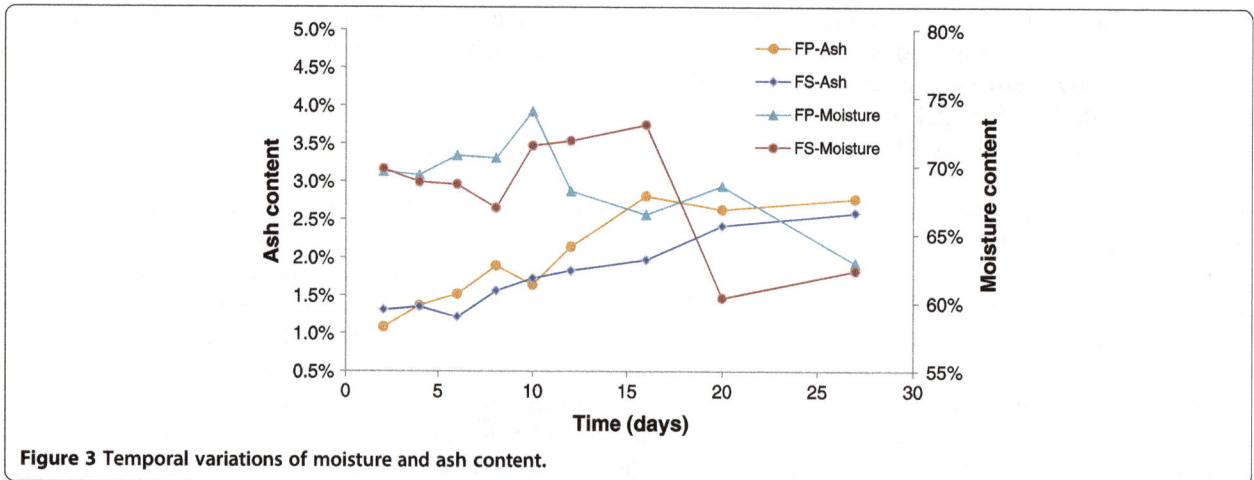

Figure 3 Temporal variations of moisture and ash content.

specific enzyme activities during composting could provide information of dynamics of the composting process. Enzyme activities can reflect the rate of transformation of organic residues and nitrogen, as well as the stability and maturity of end products (Mondini et al. 2004; Raut et al. 2008). Moreover, the determination of enzyme activity, in contrast to other analytical techniques used for compost stability evaluation, is easy, fast, and relatively inexpensive (Mondini et al. 2004). Garcia et al. (1993) confirmed that the hydrolytic enzymes were biomarkers of the state of the composting and evolution of the organic matter.

Dehydrogenase activity was 1,959 and 837 (µg TPF g dry matter $^{-1}$) on the second day for FP and FS, respectively. Dehydrogenase is an enzyme belonging to the oxidoreductase group which catalyzes the oxidation of organic substances (Kayikçioğlu and Okur 2011). Bernal et al. (2009) used dehydrogenase activity to monitor the composting process. They concluded that dehydrogenase is a useful parameter to follow the evolution of the biological activity of the composting process, since it correlates well with the temperature profile in the reactor. Dehydrogenase activity increases for FS and FP. FS reached the maximum

value, 19,106 (µg TPF g dry matter $^{-1}$) after 10 days corresponding to the peaks of temperature and OUR. The maximum value, 18,815 (µg TPF g dry matter $^{-1}$) for FP observed at 16 days at the end of the thermophilic phase or the beginning of the mesophilic stage is similar to the results of Kayikçioğlu and Okur (2011) and Bernal et al. (2009). Vargas-Garcia et al. (2010) stated that the higher dehydrogenase activity values were related to the higher microbial activity and large account of mesophilic and thermophilic bacteria and lower dehydrogenase activity values associated with the maturation phase. The longer period of high dehydrogenase activity was observed for FP compost. As shown in Figure 6, after 20 days the dehydrogenase activity decreased, which means that most of the organic matter has been degraded by the microorganism and converted to stable materials and consequently the respiratory process slowed down (Benitez et al. 1999; Benito et al. 2003; Tiquia 2005; Ros et al. 2006; Vargas-Garcia et al. 2010; Kayikçioğlu and Okur 2011). The cumulative dehydrogenase activity for FP (94, 899 µg TPF g dry matter $^{-1}$) was much higher than the cumulative dehydrogenase activity for FS (67, 924 µg TPF g dry matter $^{-1}$).

β-glucosidase is one of the key enzymes governing the C-cycle. It hydrolyses reducing terminations of b-D-glucose

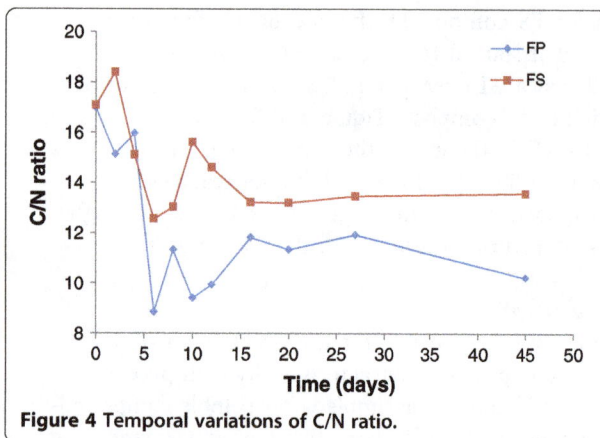

Figure 4 Temporal variations of C/N ratio.

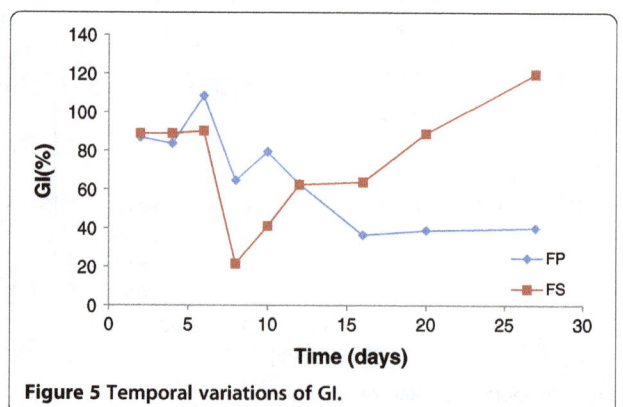

Figure 5 Temporal variations of GI.

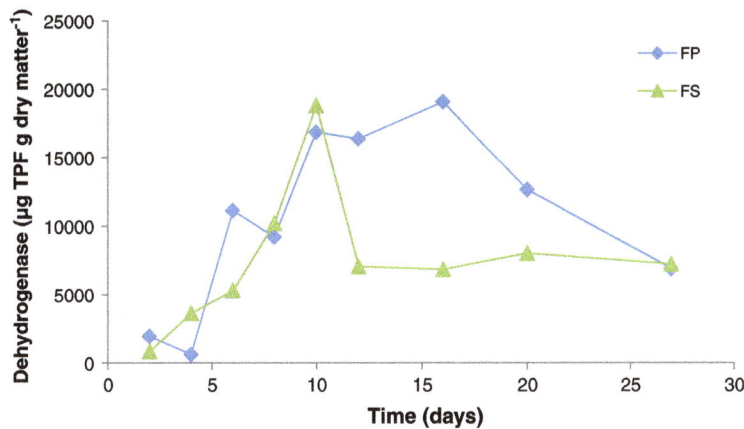

Figure 6 Temporal variations of dehydrogenase activity.

chains and form b-glucose. Its activity is therefore indicative of the presence of these terminations, which come from the labile organic matter (Vargas-Garcia et al. 2010; Kayikçioğlu and Okur 2011). The temporal variation of the β – glucosidase activity is shown in Figure 7. β – glucosidase activity was high at the beginning for both composts. At the end of the first week, β – glucosidase activity showed a peak, 11,980 (μg PNP g dry $matt \, r^{-1} h^{-1}$), and then dropped. The peak of β – glucosidase activity for FS was observed later than for FP after the second week but with almost the same value. β – glucosidase activity for both of the composts decreased by the end of composting and it was lower for the FS compost.

Phosphodiesterase (phosphoric diester hydrolases) hydrolyse one or two ester bonds in phosphodiester compounds including nucleases, which catalyze the hydrolysis of phosphodiester bonds of nucleic acids to produce nucleotide units or mononucleotides but not inorganic phosphates. Phosphodiesterase catalyzes phospholipids and nucleic acids degradation which are among the major sources of fresh organic P inputs (Nannipieri et al. 2011). In the beginning, phosphodiesterase activities were high in both composts. Phosphodiesterase activities showed the same trend for FS and FP in the first two weeks. The peak values observed at 8 days, 25,366 and 21,032 (μg PNP g dry matter$^{-1} h^{-1}$) for FP and FS, respectively. After 2 weeks, the phosphodiesterase activity dropped dramatically for FS compost and reached zero by the end of the experiment, whereas for FP compost, phosphodiesterase activity was 9,401 (μg PNP g dry matter$^{-1} h^{-1}$) at the end of experiment (Figure 8).

Conclusions

The results of different maturity and stability indices indicated the choice of bulking agents is important for composting performance and the quality of the end product.

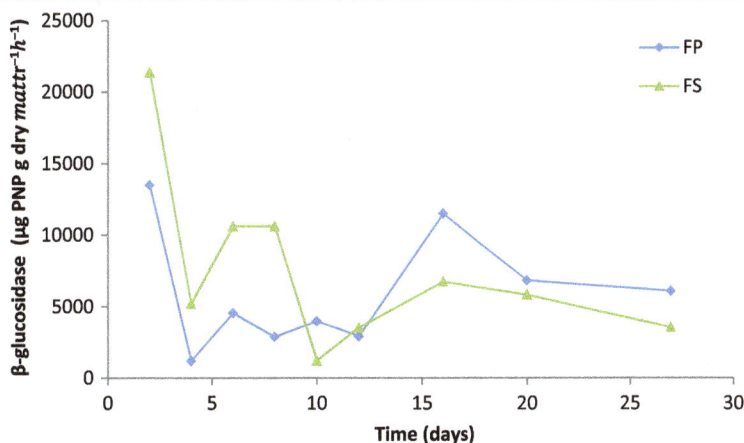

Figure 7 Temporal variations of β – glucosidase activity.

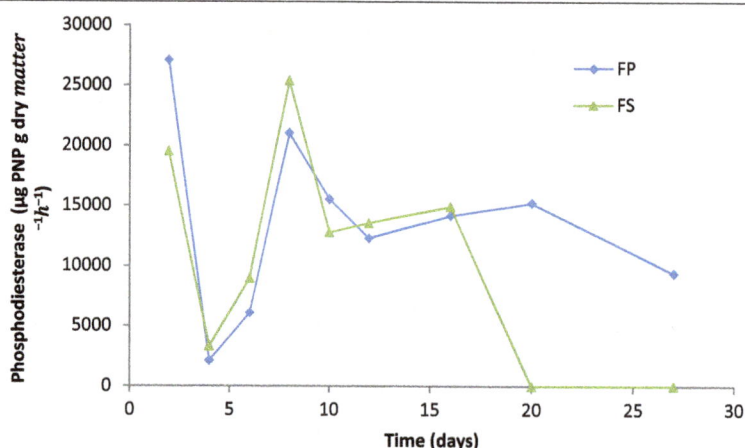

Figure 8 Temporal variations of phosphodiesterase activity.

Applying different bulking agents in composting influence temperature, OUR, GI, dehydrogenase activity and β-glucosidase activity. The final GI values for food waste composting with sawdust as a bulking agent was found to generate more mature compost with less phytotoxicity. The choice of bulking agent did not affect dehydrogenase and β – glucosidase activities values at the end of the composting for both treatments, but the final value for phosphodiesterase activity for FS was much lower than that for FP. High dehydrogenase and β-glucosidase activities during the third week of composting for FP indicate high microbial activities. To generate a high temperature and a longer duration of high temperature to kill pathogens and sterilize the compost, peat was considerably more effective. Both sawdust and peat are effective bulking agents for bench-scale composting. The choice depends on the availability of the bulking agent and land in the target community, the price of the bulking agent and its transportation, and the desired quality (e.g., higher maturity or stability) of the end compost.

Methods
Raw materials and experimental system
The synthetic MSW (food waste) consists of potato, carrot, meat, rice, cabbage, and soybean. The composition of the composting mixture is presented in Table 1. Food material was shredded with food processor to approximately 5 mm in diameter and was then mixed with locally available sawdust or peat (in a ratio of 10:1 by wet weight) with the moisture content adjusted to 70%. Two mixtures including FP (food waste + peat) and FS (food waste + sawdust) were composted in two identical lab-scale reactors for a month. Each composting reactor ($50 \times 20 \times 25$ cm) was made of acrylic sheets (Figure 9).

Six mixers were installed to enable homogenous materials. A perforated plate was installed over the bottom of reactor to distribute the injected air. The aeration rate was monitored by a flowmeter. The exhaust gas was discharged into a flask containing H_2SO_4 solution (1 M) to absorb NH_3, and then primarily monitored by gas monitoring system before released through ventilation system. The leachate outlet was used to collect the outcome leachate. A thermometer was used to monitor the temperature. The reactor was cover by heat insulating material to prevent the heat loss.

Sampling and analysis
Samples were collected randomly from 3–4 different points in the reactors after turning material, and then mixed together in a beaker on the 2nd, 4th, 6th, 8th, 10th, 12th, 16th, 20th, and 27th days. The effect of bulking agent on composting was evaluated through measuring pH, EC, C/N ratio, moisture content, ash content, dehydrogenase activity, phosphodiesterase activity, β – glucosidase activity, and GI. Temperature and OUR were recorded every

Table 1 Composition of composting mixtures (unit: kg)

	FP	FS
Meat	0.3	0.3
Rice	2.2	1.9
Carrot	2.2	2
Potato	1.1	1.1
Lettuce	0.2	0.2
Soybean	0.3	0.8
Peat	0.7	-
Sawdust	-	0.7

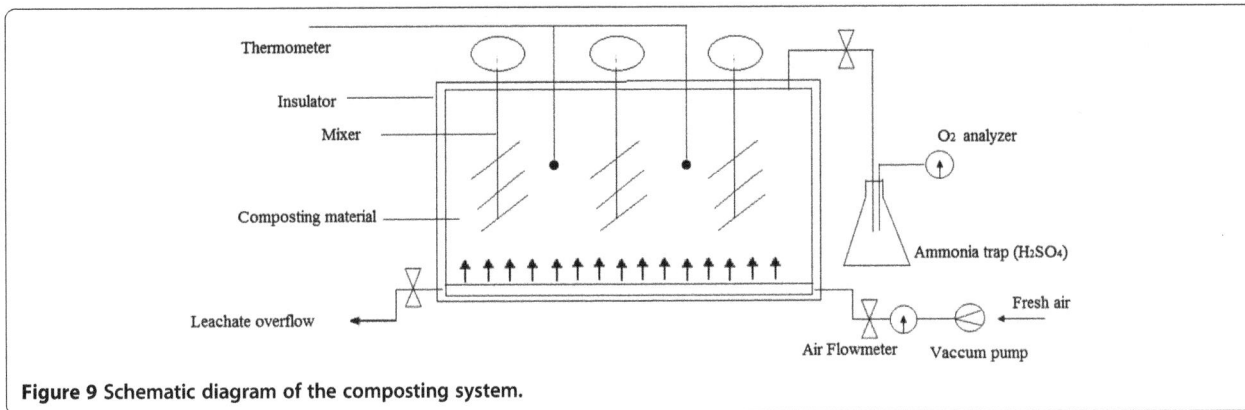

Figure 9 Schematic diagram of the composting system.

12 hours and all parameters were measured in duplicate. The average value for each duplicate measurement was used in figures and tables.

Temperature was recorded by bi-metal dial thermometer (H-B Instrument Company, PA). EC and pH were measured in 1:2 (w:v) aqueous extract by using a pH/Ion meter (Metller Toledo. EL20-Educational line pH, EL3-Educational line conductivity). The moisture content was determined by gravimetric loss on-ignition of 10 g sample at 105°C for 24 h, and the ash content of the dried samples after measuring moisture were determined by burning at 550°C in a muffle furnace (Blue M Electric Company, Blue Island, USA) for 4 h. The outlet oxygen concentration in the compost exhaust gas was monitored by passing the air through a M40 Multi-Gas Monitor (Industrial Scientific Corp., Oakdale, PA, USA). OUR was calculated through the following equation:

$$OUR = ((O_{2\ out}(\%) - O_{2\ in}(\%)) \times \text{airflow rate (L/ min)}$$
(1)

Where $O_{2\ out}$ (%) is the oxygen concentration in compost exhaust gas and $O_{2\ in}$ (%) is the oxygen concentration in the inlet air (20.9%) at airflow rate (0.5 L/min/kg) which is injected to the system.

For seed germination test, water was extracted from the samples by shaking fresh samples with double distilled water (DDW) at solids: DDW = 1:10 (w/v) for 1 h, then suspensions were centrifuged at 3,000 rpm for 20 min before filtering through Whatman No-1 filter paper. A filter paper was placed in the petri dish and almost 10 milliliter of water extract was introduced into the petri dish. Ten cucumber seed were placed on the filter paper. For control experiments, the DDW was used. The dishes were placed in the oven at 25°C in the darkness for 5 days. Test for each sample was run in triplicate. The GI was calculated according to Zucconi et al. (1981):

$$GI(\%) = (\text{Seed germination}$$
$$\times \text{Root length of the treatment}$$
$$\times 100)/(\text{Seed germination}$$
$$\times \text{Root length of the control})$$
(2)

The total carbon and nitrogen contents of the composting sample were determined by the Perkin Elmer 2400 Series II CHNS/O analyzer.

For dehydrogenase activity determination, a 5 g sample was suspended in 5 mL of 3% w/v 2, 3, 5-triphenyl-tetrazolium chloride (TTC) at 37°C for 24 h in the dark, and then 40 mL acetone was added and incubated at room temperature for 2 h in the dark. The suspension was filtered through a glass fiber filter and absorbance was measured at 546 nm (Thalmann 1968; Alef and Nannipieri 1995). Phosphodiesterase activity was measured using the method of Browman and Tabatabai (1978) and Tabatabai (1994). After the addition of a Tris buffere (pH 8) and Sodium bis-p-nitrophenyl phosphate (Sigma; for phosphodiesterase activity) to 1 g compost, samples were incubated for 1 h at 37°C. The p-nitrophenol released by phosphodiesterase activity was extracted and coloured with calcium chloride and determined spectrophotometrically at 400 nm. For β – glucosidase activity measurement, a 1 g sample was suspended in 0.25 mL toluene and 4 mL of MUB (Modified Universal Buffer, pH 6.0) plus 1 mL p-nitropenyl-β-D-glucopyranoside (Sigma; for glucosidase). After incubation for 1 h at 37°C, 1 mL of 0.5 M $CaCl_2$ and 4 mL Tris buffer (0.1 M, pH 12) were added and the suspension was filtered through a glass fiber filter. The release of p- nitrophenol was measured spectrophotometrically at 400 nm (Eivazi and Tabatabai 1988; Alef and Nannipieri 1995).

Temperature, OUR, moisture content, ash content, and C/N ratio were measured in duplicate. pH, EC, GI, and enzyme activities are tested in triplicate. The average value for each duplicate measurement was used in figures and tables.

Abbreviations

MSW: Municipal solid waste; NL: Newfoundland and Labrador; OUR: Oxygen uptake rate; EC: Electrical conductivity; FP: Food waste + Peat; FS: Food waste + sawdust; GI: Germination index.

Competing interests

The authors declare that they have no competing interests.

Authors' contributions

KK designed and conducted experiments, conducted data analysis, and drafted the manuscript. WL helped with laboratory experiment, manuscript drafting, editing, and refinement. BZ is the PI of the project, provided guidance and technical support throughout the study, and revised the manuscript. LL guided experimental design and revision of the manuscript. All authors read and approved the final manuscript.

Acknowledgments

This research was funded by the Harris Centre Memorial University of Newfoundland.

References

Adhikari BK, Barrington S, Martinez J, King S (2008) Characterization of food waste and bulking agents for composting. Waste Manage 28(5):795–804

Adhikari BK, Barrington S, Martinez J, King S (2009) Effectiveness of three bulking agents for food waste composting. Waste Manage 29(1):197–203

Alef K, Nannipieri P (1995) Methods in applied soil microbiology and biochemistry. Academic, London, UK

Arslan EI, Ünlü A, Topal M (2011) Determination of the effect of aeration rate on composting of vegetable–fruit wastes. CLEAN–Soil, Air, Water 39(11):1014–1021

Asase M, Yanful EK, Mensah M, Stanford J, Amponsah S (2009) Comparison of municipal solid waste management systems in Canada and Ghana: a case study of the cities of London, Ontario, and Kumasi, Ghana. Waste Manage 29(10):2779–2786

Banegas V, Moreno J, Moreno J, Garcia C, Leon G, Hernandez T (2007) Composting anaerobic and aerobic sewage sludges using two proportions of sawdust. Waste Manage 27(10):1317–1327

Batham M, Gupta R, Tiwari A (2013) Implementation of bulking agents in composting: a review. J Bioremediation Biodegradation 4(7):1–3

Benitez E, Nogales R, Elvira C, Masciandaro G, Ceccanti B (1999) Enzyme activities as indicators of the stabilization of sewage sludges composting with *Eisenia foetida*. Bioresour Technol 67(3):297–303

Benito M, Masaguer A, Moliner A, Arrigo N, Palma RM (2003) Chemical and microbiological parameters for the characterisation of the stability and maturity of pruning waste compost. Biol Fertility Soils 37(3):184–189

Bernal MP, Alburquerque J, Moral R (2009) Composting of animal manures and chemical criteria for compost maturity assessment. A review. Bioresour Technol 100(22):5444–5453

Blanco M-J, Almendros G (1995) Evaluation of parameters related to chemical and agrobiological qualities of wheat-straw composts including different additives. Bioresour Technol 51(2):125–134

Browman M, Tabatabai M (1978) Phosphodiesterase activity of soils. Soil Sci Soc Am J 42(2):284–290

Castaldi P, Garau G, Melis P (2008) Maturity assessment of compost from municipal solid waste through the study of enzyme activities and water-soluble fractions. Waste Manage 28(3):534–540

Chang JI, Chen Y (2010) Effects of bulking agents on food waste composting. Bioresour Technol 101(15):5917–5924

Christensen TH (2011) Solid Waste Technology & Management, vol 2, firstth edn. Wiley-Blackwell, West Sussex, UK

Chukwujindu M, Egun A, Emuh F, Isirimah N (2006) Compost maturity evaluation and its significance to agriculture. Pak J Biol Sci 9(15):125–131

Cronje A, Turner C, Williams A, Barker A, Guy S (2003) Composting under controlled conditions. Environ Technol 24(10):1221–1234

De Bertoldi M, Vallini G, Pera A (1983) The biology of composting: a review. Waste Manag Res 1(2):157–176

Dias BO, Silva CA, Higashikawa FS, Roig A, Sánchez-Monedero MA (2010) Use of biochar as bulking agent for the composting of poultry manure: Effect on organic matter degradation and humification. Bioresour Technol 101 (4):1239–1246

Doublet J, Francou C, Poitrenaud M, Houot S (2011) Influence of bulking agents on organic matter evolution during sewage sludge composting; consequences on compost organic matter stability and N availability. Bioresour Technol 102(2):1298–1307

Eiland F, Klamer M, Lind A-M, Leth M, Bååth E (2001) Influence of initial C/N ratio on chemical and microbial composition during long term composting of straw. Microb Ecol 41(3):272–280

Eivazi F, Tabatabai M (1988) Glucosidases and galactosidases in soils. Soil Biol Biochem 20(5):601–606

Eklind Y, Kirchmann H (2000) Composting and storage of organic household waste with different litter amendments. II: nitrogen turnover and losses. Bioresour Technol 74(2):125–133

Environment Canada (2013) Technical document on Municipal Solid waste Organics processing

Francou C, Poitrenaud M, Houot S (2005) Stabilization of organic matter during composting: Influence of process and feedstocks. Compost Sci Util 13(1):72–83

Gao M, Liang F, Yu A, Li B, Yang L (2010) Evaluation of stability and maturity during forced-aeration composting of chicken manure and sawdust at different C/N ratios. Chemosphere 78(5):614–619

Garcia C, Hernandez T, Costa C, Ceccanti B, Masciandaro G, Ciardi C (1993) A study of biochemical parameters of composted and fresh municipal wastes. Bioresour Technol 44(1):17–23

Godden B, Penninckx M, Piérard A, Lannoye R (1983) Evolution of enzyme activities and microbial populations during composting of cattle manure. Appl Microbiol Biotechnol 17(5):306–310

Hasan K, Sarkar G, Alamgir M, Bari QH, Haedrich G (2012) Study on the quality and stability of compost through a Demo Compost Plant. Waste Manage 32 (11):2046–2055

Haug RT (1993) The practical handbook of compost engineering. Lewis Publishers, Boca Raton, USA

Hue N, Liu J (1995) Predicting compost stability. Compost Sci Util 3(2):8–15

Huet J, Druilhe C, Tremier A, Benoist J-C, Debenest G (2012) The impact of compaction, moisture content, particle size and type of bulking agent on initial physical properties of sludge-bulking agent mixtures before composting. Bioresour Technol 114:428–436

Iglesias Jiménez E, Perez Garcia V (1989) Evaluation of city refuse compost maturity: a review. Biological wastes 27(2):115–142

Iqbal MK, Shafiq T, Ahmed K (2010) Characterization of bulking agents and its effects on physical properties of compost. Bioresour Technol 101(6):1913–1919

Jolanun B, Towprayoon S (2010) Novel bulking agent from clay residue for food waste composting. Bioresour Technol 101(12):4484–4490

Kalamdhad AS, Singh YK, Ali M, Khwairakpam M, Kazmi A (2009) Rotary drum composting of vegetable waste and tree leaves. Bioresour Technol 100 (24):6442–6450

Kayikçioğlu HH, Okur N (2011) Evolution of enzyme activities during composting of tobacco waste. Waste Manage Res 29(11):1124–1133

Lashermes G, Barriuso E, Le Villio-Poitrenaud M, Houot S (2012) Composting in small laboratory pilots: Performance and reproducibility. Waste Manage 32(2):271–277

Lee J, Rahman M, Ra C (2009) Dose effects of Mg and PO$_4$ sources on the composting of swine manure. J Hazard Mater 169(1):801–807

Liang C, Das K, McClendon R (2003) The influence of temperature and moisture contents regimes on the aerobic microbial activity of a biosolids composting blend. Bioresour Technol 86(2):131–137

Martin A, Evans J, Porter D, Patel T (1993) Comparative effects of peat and sawdust employed as bulking agents in composting. Bioresour Technol 44(1):65–69

Mathur S, Daigle J-Y, Lévesque M, Dinel H (1986) The feasibility of preparing high quality composts from fish scrap and peat with seaweeds or crab scrap. Biol Agric Horticulture 4(1):27–38

Mathur S, Patni N, Levesque M (1990) Static pile, passive aeration composting of manure slurries using peat as a bulking agent. Biological wastes 34(4):323–333

Mondini C, Fornasier F, Sinicco T (2004) Enzymatic activity as a parameter for the characterization of the composting process. Soil Biol Biochem 36(10):1587–1594

Nannipieri P, Giagnoni L, Landi L, Renella G (2011) Role of phosphatase enzymes in soil. In: Phosphorus in Action. Soil Biology 100:215–243

Nolan T, Troy SM, Healy MG, Kwapinski W, Leahy JJ, Lawlor PG (2011) Characterization of compost produced from separated pig manure and a variety of bulking agents at low initial C/N ratios. Bioresour Technol 102 (14):7131–7138

Ponsá S (2010) Different indices to express biodegradability in organic solid wastes, application to full scale solid waste treatment plant. Universitat Autonoma De Barcelona, Spain

Rasapoor M, Nasrabadi T, Kamali M, Hoveidi H (2009) The effects of aeration rate on generated compost quality, using aerated static pile method. Waste Manage 29(2):570–573

Raut M, Prince William S, Bhattacharyya J, Chakrabarti T, Devotta S (2008) Microbial dynamics and enzyme activities during rapid composting of municipal solid waste–a compost maturity analysis perspective. Bioresour Technol 99(14):6512–6519

Ros M, Garcia C, Hernández T (2006) A full-scale study of treatment of pig slurry by composting: Kinetic changes in chemical and microbial properties. Waste Manage 26(10):1108–1118

Said-Pullicino D, Erriquens FG, Gigliotti G (2007) Changes in the chemical characteristics of water-extractable organic matter during composting and their influence on compost stability and maturity. Bioresour Technol 98 (9):1822–1831

Silva MEF, De Lemos LT, Nunes OC, Cunha-Queda AC (2014) Influence of the composition of the initial mixtures on the chemical composition, physico-chemical properties and humic-like substances content of composts. Waste Manage 34(1):21–27

Smårs S, Gustafsson L, Beck-Friis B, Jönsson H (2002) Improvement of the composting time for household waste during an initial low pH phase by mesophilic temperature control. Bioresour Technol 84(3):237–241

Statistics Canada (2008) Human activity and the environment: annual statistics. Environment Accounts and Statistics Division. http://www.statcan.gc.ca/pub/16-201-x/2012000/t001-eng.htm. Accessed May 16, 2014

Tabatabai M (1994) Soil enzymes. Methods of soil analysis: Part 2—Microbiological and biochemical properties (methodsofsoilan2)., pp 775–833

Thalmann A (1968) Zur Methodik der bestimmung der dehydrogenaseaktivität im boden mittels triphenyltetrazoliumchlorid (TTC). Landwirtsch Forsch 21:249–258

Tiquia S (2005) Microbiological parameters as indicators of compost maturity. J Appl Microbiol 99(4):816–828

Tiquia S, Tam N (1998) Elimination of phytotoxicity during co-composting of spent pig-manure sawdust litter and pig sludge. Bioresour Technol 65(1):43–49

Vargas-Garcia M, Suárez-Estrella F, Lopez M, Moreno J (2010) Microbial population dynamics and enzyme activities in composting processes with different starting materials. Waste Manage 30(5):771–778

Vuorinen AH (2000) Effect of the bulking agent on acid and alkaline phosphomonoesterase and β-D-glucosidase activities during manure composting. Bioresour Technol 75(2):133–138

Wu L, Ma L, Martinez G (2000) Comparison of methods for evaluating stability and maturity of biosolids compost. J Environ Qual 29(2):424–429

Xiao Y, Zeng G-M, Yang Z-H, Shi W-J, Huang C, Fan C-Z, Xu Z-Y (2009) Continuous thermophilic composting (CTC) for rapid biodegradation and maturation of organic municipal solid waste. Bioresour Technol 100(20):4807–4813

Yañez R, Alonso J, Díaz M (2009) Influence of bulking agent on sewage sludge composting process. Bioresour Technol 100(23):5827–5833

Yang F, Li GX, Yang QY, Luo WH (2013) Effect of bulking agents on maturity and gaseous emissions during kitchen waste composting. Chemosphere 93(7):1393–1399

Zmora-Nahum S, Markovitch O, Tarchitzky J, Chen Y (2005) Dissolved organic carbon (DOC) as a parameter of compost maturity. Soil Biol Biochem 37(11):2109–2116

Zucconi F, Pera A, Forte M, De Bertoldi M (1981) Evaluating toxicity of immature compost. Biocycle 22(2):54–57

5

Detection and localization of harmful atmospheric releases via support vector machines

Ronald Taylor Locke[*] and Ioannis Ch Paschalidis

Abstract

Background: We present a Support Vector Machine (SVM) approach to the localization of hazardous particulate releases in an urban area using features constructed only from measurements obtained from a network of sensors.

Results: We find high levels of localization accuracy when a reasonable number of noisy sensors are deployed within the environment. We also compare SVM source localization performance to an existing stochastic localization technique over varying degrees of sensor noise and find it favorable for areas prone to urban canyon turbulence effects.

Conclusions: This approach is in contrast to earlier works which either use solutions to inverse dispersion problems for localization or apply maximum likelihood techniques. By using established SVM results, we also tackle the problems of release detection and optimal sensor placement.

Keywords: Environmental monitoring; Urban monitoring; Tracking; Simulation; Machine learning

Background

Modern cities face a multitude of atmospheric pollution threats from a plethora of sources, including large industrial plants in the outskirts of the city, smaller production or processing units in industrial parks or even within the city, warehouses and storage facilities, and large underground networks of gas pipes that are particularly leaky and can release methane.

In addition to the accidental release of some harmful pollutant in the atmosphere, an increasing concern are terrorist threats involving the deliberate release of some *Chemical, Biological, Radiological, or Nuclear (CBRN)* material.

With either accidental or deliberate releases, it is critical to not only provide a reliable indication of the presence of hazardous particles within the atmosphere (*release detection*) but to also indicate where the event originated (*release localization*). Understanding the source's point in time and space can guide emergency responders and

enable fast corrective actions to be more precisely targeted. In the case of a CBRN attack, for example, antidotal medications which are potentially already in short supply can be administered only to people who were exposed and in danger of succumbing to attack related illness or injury. Similarly, in the case of an accidental release, source localization can guide forensics and enable the prevention of future releases.

Prior work in the literature has tackled source localization by tracing, through dispersion modeling, observed particle concentrations back in time and space to the presumed origin (Atalla and Jeremic 2008; Ortner et al. 2007; Ortner and Nehorai 2008). The alternative presented in (Locke and Paschalidis 2012, 2013), as well as the alternative presented in this article, differ in that they depend only on *sensor measurements* and therefore do not require solution to complicated inverse dispersion problems. Additionally, and quite importantly, these source localization techniques provide insight to the related problem of placing sensors.

This work, as well as the work in (Locke and Paschalidis 2012, 2013), does require some a priori computation in lieu of the design of an analytical dispersion model.

*Correspondence: rtl@bu.edu
Department of Electrical & Computer Engineering, Boston University,
8 St Mary's St., 02215 Boston MA, USA

Specifically, knowledge of particulate dispersion behavior is needed in order to make sense of sequences of particulate concentration observations. Considering that large scale, physical simulant releases in urban areas is prohibitively costly in both time and money, we employ accurate numerical simulations of particulate dispersions in a Monte Carlo fashion to develop a mathematical characterization of sensor measurements under a variety of scenarios. A technique that we have found useful, and adapt for our purposes, which is described in greater detail in the Results and discussion Section, is the *Lattice Boltzmann Method (LBM)*. This numerical dispersion model is advantageous to the problem at hand because it handles complex geometries and changing phenomena typical of urban areas quite naturally. Also, LBM is easily parallelizable for use on a computational grid. This allows us to perform the numerous dispersion simulations required for enhanced accuracy of the presented localization methodology and enables large deployments of our methodology.

The major contributions of this article are:

1. a new, deterministic localization methodology that does not rely on solving any sophisticated inverse dispersion problem and is an alternative to the stochastic localization methodology presented in (Locke and Paschalidis 2013);
2. a novel sensor placement methodology that stems from a machine learning feature selection procedure; and
3. and a novel procedure inspired by machine learning techniques for the detection of hazardous atmospheric releases.

To our knowledge, no other machine learning hazardous release localization approaches have been presented in the relevant literature. The work contained in (Vujadinovic et al. 2008; Wawrzynczak et al. 2014) could be considered similar, however the localization processes presented therein employ Bayesian statistics as opposed to SVMs. A component that separates this work from (Vujadinovic et al. 2008; Wawrzynczak et al. 2014) is the development of feature vectors that could be applied to any machine learning technique. Also, our approach is not reliant on a forward propagation model.

The work presented here extends dispersion model-free localization into the realm of data-science and opens the door for application of continually evolving swath of machine learning algorithms to the problems of release detection and release localization.

While the focus of this work is urban environments, other environments, at all scales, are also amenable to the presented localization and detection techniques. For instance, downstream river contamination monitoring and underground chemical seepage are tangible problems. Indoor environments, such as nuclear power plants and chemical processing plants, also present potential monitoring applications.

Related work

Existing source localization approaches (Atalla and Jeremic 2008; Ortner et al. 2007; Ortner and Nehorai 2008) observe particulate concentrations and solve the inverse problem of tracing dispersion backward in time and space to the source of the release. Limitations of this methodology stem from the irregular and dynamic phenomena typically found in urban areas. For instance, buildings within a city tend to have irregular geometries and a plethora of external surface textures. Additionally, the micro-climate effects of urban canyon turbulence under generally uncertain weather conditions are challenging to model. All of these characteristics of urban environments work in unison to make source localization via solving inverse dispersion problems difficult. In (Ortner et al. 2007) the presence of challenging geographies and wind turbulence is accommodated by incorporating Monte Carlo simulation of fluid dispersion. However, all of these works suffer from the difficulty of determining, without a detection process, the point in time in which the release began. Barring this information, these inverse problem approaches can lead to erroneous localizations.

The work presented here provides a deterministic accompaniment to the stochastic localization approach presented in (Locke and Paschalidis 2013). There, considering that under a release, large (small) particulate concentrations observed at one time instance by a sensor are likely to be followed by similarly large (small) particulate concentrations, environmental sensor observations are modeled as a first-order Markov chain. This construct is made tenable by first encoding real-valued sensor concentration observations into a finite set of concentration states. Through the use of Monte Carlo simulation, marginal and transition probabilities of the concentration states are derived empirically to construct probability laws for concentration state evolution under a plethora of release scenarios. Localization is then performed through hypothesis tests that compare current empirical concentration state distributions to the previously derived probability laws. This approach is completely sensor measurement based and therefore does not fall victim to the inherent challenges of producing and then solving an inverse dispersion model.

The SVM-based localization technique presented in this article is similar to the stochastic approach in (Locke and Paschalidis 2013) in that it relies on previously conducted Monte Carlo simulation of release scenarios to build *training sets*. It is also entirely sensor measurement based, which brings the added benefit of not relying on solving an inverse problem. It differs, however, in that the training data is used to build *deterministic* decision boundaries that indicate the location of a detected release.

Results and discussion

To demonstrate the performance of the sensor placement approach and the localization methodology, we simulated several point release scenarios in an illustrative environment. We simulated CBRN releases in the *Quick Urban & Industrial Complex (QUIC)* Dispersion Modeling System (Paryak and Brown 2007) developed at the Los Alamos National Laboratory. QUIC first solves the fluid dynamics problem of determining local wind eddies throughout a modeled three-dimensional, outdoor setting using the methods of Röckle (1990). Using the fluid flow solution, QUIC simulates the travel of CBRN particulates via a Lagrangian random walk. Previously, the QUIC codes have been tested and validated for real-world situations (Paryak and Brown 2007).

Additionally, we simulated CBRN releases using the *Lattice Boltzmann Method (LBM)*. LBM evolved from the numerical fluid modeling technique *Lattice Gas Automata (LGA)*, in which parcels of air adhere to microscopic laws which dictate their movement. Macroscopic values of flow velocities and densities are then derived by the underlying microscopic properties propagated by the algorithm (Frisch et al. 1986). Unfortunately, LGA often falls victim to instability in the face of statistical noise

(Lallemand and Luo 2000). LBM extends LGA by considering air parcel movement more notionally by modeling microscopic air parcel velocities as distributions in the *Lattice Boltzmann Equation (LBE)*. It has been shown that under reasonable starting conditions LBM provides accurate approximations to fluid flows. The presented work continues the precedent set in(Locke and Paschalidis 2012) by using LBM in Monte Carlo simulation for analysis of CBRN events. It is not hard to imagine that it will be of use to future analysts as well. The macroscopic Navier-Stokes equations can be recovered from the microscopic LBE and LBM is easily adapted to parallel computation (Chen and Doolen 1998). This last characteristic is of particular interest, as large-scale real-world applications have the potential to require large amounts of computation.

Our modeled environment consists of geometries typical of dense urban areas, consisting of a city grid, four blocks-by-four blocks. Each block is 100 meters-by-100 meters with 10 meter-wide through-ways. Each block's height is drawn randomly from the uniform distribution ranging from 20 to 60 meters. Sensors are allowed to be placed at any intersection and five intersections are considered as potential release locations. The shape of the grid, as well as the location of simulated releases, is shown in Figure 1.

The agent concentration profiles of the two different dispersion models at a CBRN sensor located down wind of the point release are compared in Figure 2. These models

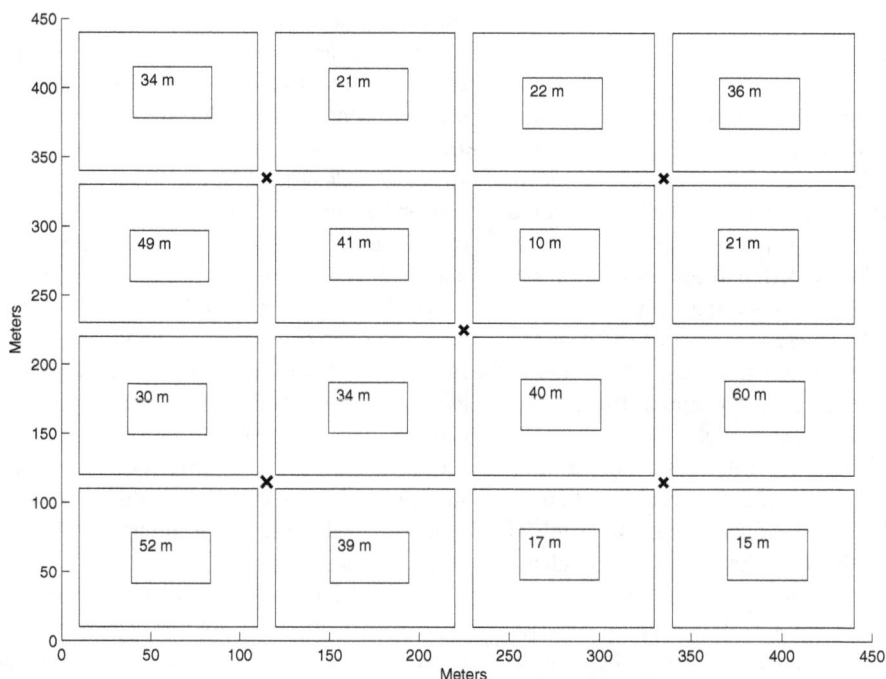

Figure 1 City model with CBRN release locations under consideration. Release locations are marked with an "x".

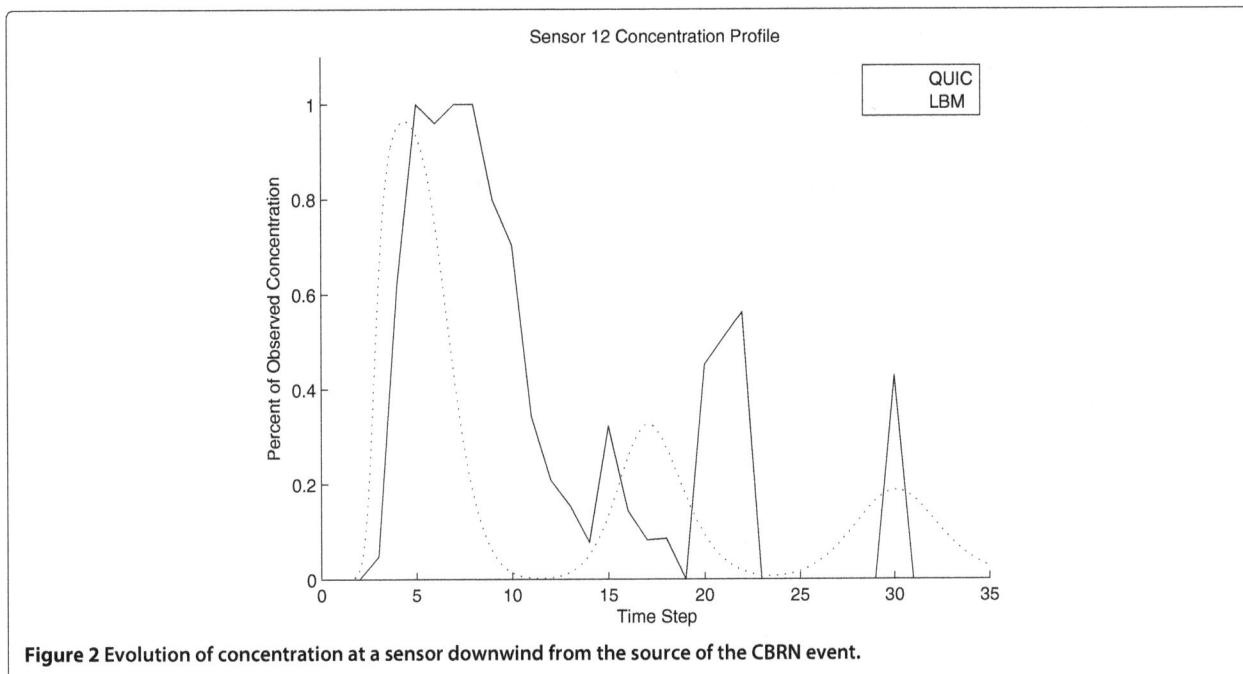

Figure 2 Evolution of concentration at a sensor downwind from the source of the CBRN event.

have different discretizations of the three-dimensional model and hence produce concentration values that differ in scale. This has been accommodated in Figure 2 by reporting the percentage of all observed concentrations reported by a single sensor downwind from a release. The LBM model produces a much smoother agent concentration evolution than the QUIC model. Thus, noise within the evaluation of the proposed methodologies when the LBM data are used is primarily due to the sensor false alarm model rather than the dispersion model.

QUIC is selected for testing purposes as it is representative of popular and traditional dispersion simulation approaches. LBM serves as a contrast to QUIC in modeling as well as in emerging atmospheric modeling trends that are intended to be scalable through parallelized computation. But it is important to note, the purpose of the presented work is hazardous atmospheric release detection and localization. The requisite data simulation process is simulator agnostic. We present results obtained by using training data generated from both simulators as a means to empirically demonstrate that assertion.

We consider 40 unique CBRN event scenarios, each containing a single point release of the same mass, spanning the five different release locations within the grid with wind blowing at 1 m/s or 5 m/s and originating from the four cardinal directions. A large set of training data was constructed via Monte Carlo simulation over each combination of wind direction, wind speed, and release location (data is available at ionia.bu.edu http://ionia.bu.edu/Research/Env_Loc_Data.html).

We use a real valued sensor model with additive white noise. We model these sensors as

$$\hat{C} = C + N(0, \sigma_\epsilon) \qquad (1)$$

where C is the actual concentration of particulate present at the sensor's location, $N(0, \sigma_\epsilon)$ denotes a normally distributed random variable with mean 0 and standard deviation σ_ϵ, and $\hat{C}$ denotes the sensor's reported concentration observation. First, in what will be referred to as the *mild noise case*, we use a value of σ_ϵ which produces an equivalent false alarm rate of 0.125. Also, in what will be referred to as the *large noise case*, we replace the normal random variate in (1) with $N(0, 10\sigma_\epsilon)$. The error term is intended to model both measurement error in the sensor and random perturbations in release concentration realizations. If a higher fidelity sensor is used, then the low-noise sensor data is a better model. Likewise, low-fidelity sensors correspond to the high-noise case.

SVM CBRN localization evaluation

We construct a test set for localization and placement performance evaluation for each release location by first selecting a wind direction and speed according to a "wind rose" which describes the likelihood of each unique wind speed and direction pair. Once a test simulation was run, we add sensor noise according to the sensor model in (1) for both the mild and large noise cases.

Since the sensor placement procedure is based entirely on the amount of sensors available for placement, we evaluate localization performance for various numbers of sensors, deployed according to our SVM feature selection adaptation to sensor placement. We train SVMs using both the maximum concentration feature space and mean concentration feature space outlined in Section Feature representation and test them using data generated via QUIC and LBM. As the number of sensors available placed within the urban grid increases, so does localization accuracy. Figure 3 depicts this result for all of the numerical examples.

Localization performed using the maximum concentration feature space and data generated using QUIC requires only four, in the case of mild-noise, or five, for the large-noise case, sensors to observe perfect localization on the data in the test set. When LBM data with large sensor noise is used for analysis, a performance plateau appears once five sensors are placed in the environment that is not overcome until 13 sensors are employed for localization. This is the result of many different sensor locations being selected an equal number of times in the iterative SVM feature selection process. The 13th-most commonly selected features provide the sensor location that produces the discernible information for a large portion of the data in the test set.

Features consisting of a time-series average of observed concentrations produce localization accuracy similar to when maximum concentration features are used. However, weaker performance on the QUIC generated data sets suggests the averaged concentration features have a greater sensitivity to noisy data.

The placement solutions produced adhere to an intuitive strategy. The features selected most frequently among all binary SVMs lie either directly on top of or adjacent to the five release locations. Thus, by the time five sensors are deployed, the city grid is covered by sensors that are not more than one block away from a release location. As is demonstrated in the bottom plot of Figure 3, placing five sensors according to this strategy is not always enough to produce acceptable localization accuracy.

One-class SVM CBRN detection evaluation

To evaluate our presented CBRN detection technique based on a one-class SVM novelty detector, we constructed a test set that contained sequences of sensor observations that are purely the product of noise as well as sequences of sensor observations from the test set used in localization evaluation. The sequences consisting of only sensor noise depict cases in which no CBRN release is present within the simulated environment and allow us to compute the detection methodology's probability of false alarm. The results appear in Figure 4 for the cases in which average concentration features and maximum concentration features are used. Sensor locations were determined according to the placement procedure found in the Methods Section. As shown, for both feature constructions and all simulated data sets, the probability of false alarm was consistently very low no matter how many sensors were deployed. Probability of detection, on the other hand, is only promising in cases in which low levels of measurement noise are considered. In general, when the features are constructed by using a sequence of average concentrations the detection procedure is more robust to large levels of measurement noise than when features are constructed from observed maximum concentrations.

By and large, in most real-world CBRN detection applications, the cost of a false alarm is prohibitively expensive. Emergency response when it is not necessary could mean that commercial or governmental buildings are closed to human access during critical times. Additionally,

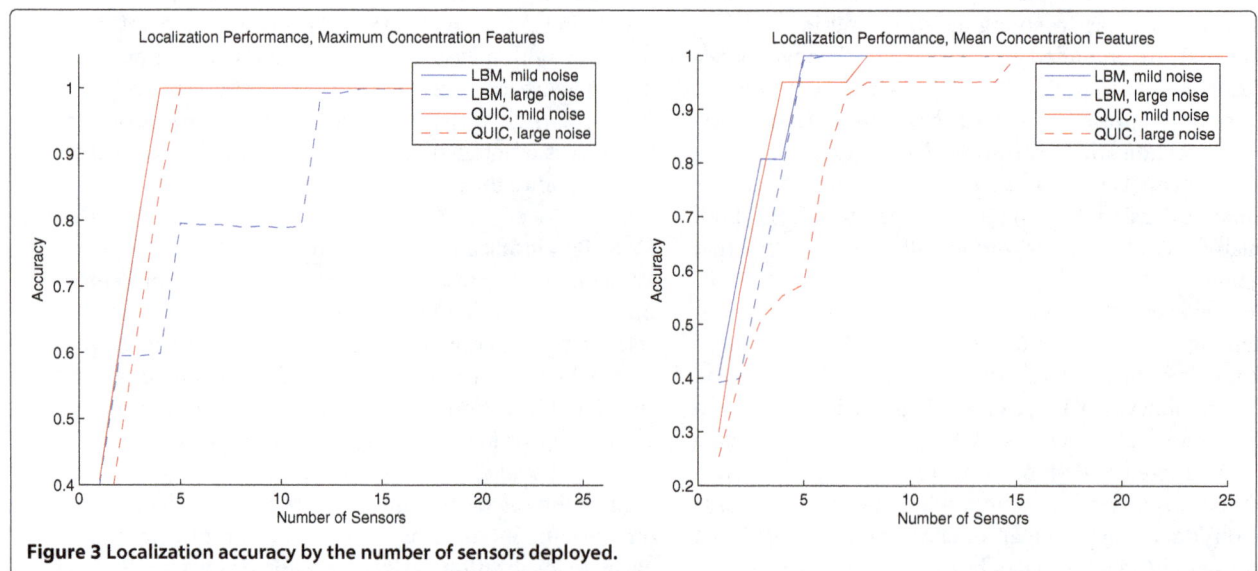

Figure 3 Localization accuracy by the number of sensors deployed.

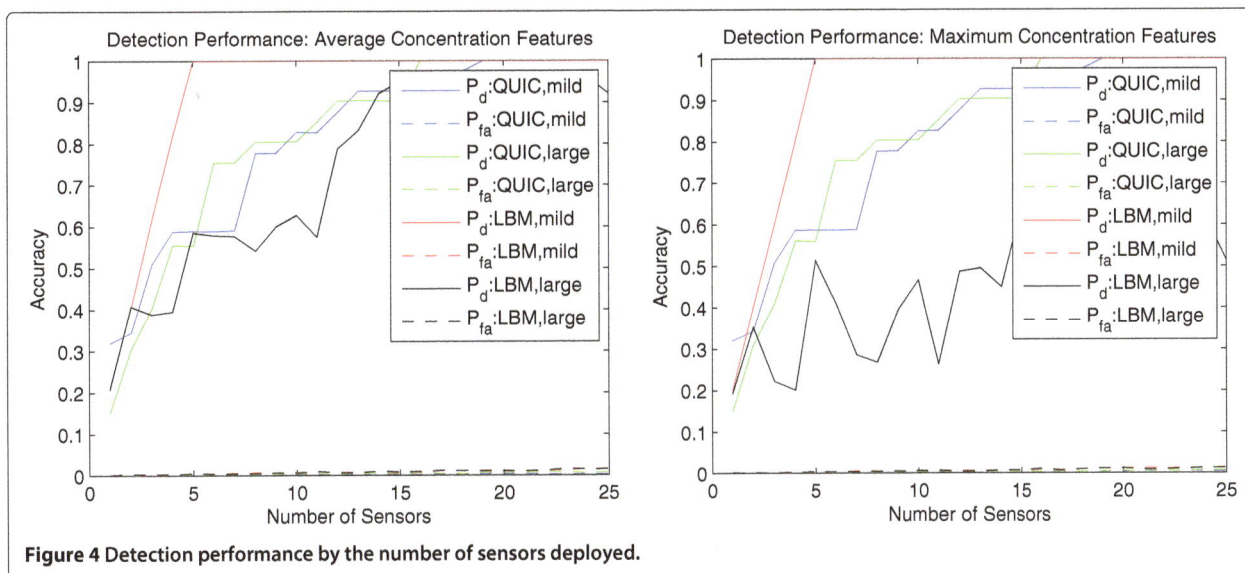

Figure 4 Detection performance by the number of sensors deployed.

otherwise healthy people may be administered antidotal medicines which could lead to hazardous and undue side-effects. These facts, coupled with the extreme rarity of a CBRN event, cause extremely low probabilities of false alarm to be the governing factor in performance analysis and implementation. In other applications, such as pollution monitoring, where false alarms do not accrue the same level of costs, the probabilities of detection reported here could be increased by increasing the allowable false alarm rate. This is accommodated in the presented methodology by a tunable parameter which defines the detection procedure's false alarm rate.

Comparison to stochastic CBRN localization

We conducted a comparison of the presented SVM CBRN localization methodology to the stochastic localization methodology presented in (Locke and Paschalidis 2013). Both schemes, under the right conditions and with the right sensor placements, can locate the origin of a CBRN release quite accurately. The proper comparison entails evaluation of the two methodologies when ideal conditions begin to break down. To aid in this comparison, we conduct localization performance of the stochastic and SVM localization techniques under varying degrees of sensor noise. For varying values of σ_ϵ, we construct training and test sets. The empirical probabilities of correct localization, as computed by localization performance on the test sets, appear in the Figure 5.

Clearly, as noise increases, both methodologies' localization accuracies deteriorate. However, it is evident that the SVM localization procedure is more robust to added noise than the stochastic localization procedure. To be fair, this may be a result of differences in fidelity. SVM features are based on real-valued concentration observations while the stochastic approaches rely on a discretization

of concentration measurements. Adding fidelity to the stochastic approaches by expanding the alphabet depicting concentration samples would close the gap shown in Figure 5, albeit at the cost of added computational overhead.

What this translates to in real-world application is a robustness to urban canyon turbulence. In cities where avenues and streets are dwarfed by the tall, densely-packed buildings that line them, turbulence from these urban walls could play a greater role in sporadic particle concentration samplings and should be considered when selecting a localization methodology.

In terms of computational workload, while both methodologies suffer from requiring copious amounts of simulated dispersion data from the gambit of release conditions expected within the environment under surveillance, the SVM training procedure requires much more work than that of the virtually nonexistent training required by the stochastic methodology. However, SVMs in general are an ongoing research topic in an already well developed community. It is likely that advances in SVM training procedures and the development of new "off the shelf" SVM software packages could deaden this computational constraint.

Conclusions

This work is a step away from the inherently challenging approach of solving inverse dispersion problems in highly dynamic urban environments. At the same time, it is a deterministic compliment to the previously established stochastic localization technique in (Locke and Paschalidis 2013).

Numerical evaluation of SVM CBRN localization shows promising accuracy in most situations with even a small number of sensors. We also found that this deterministic

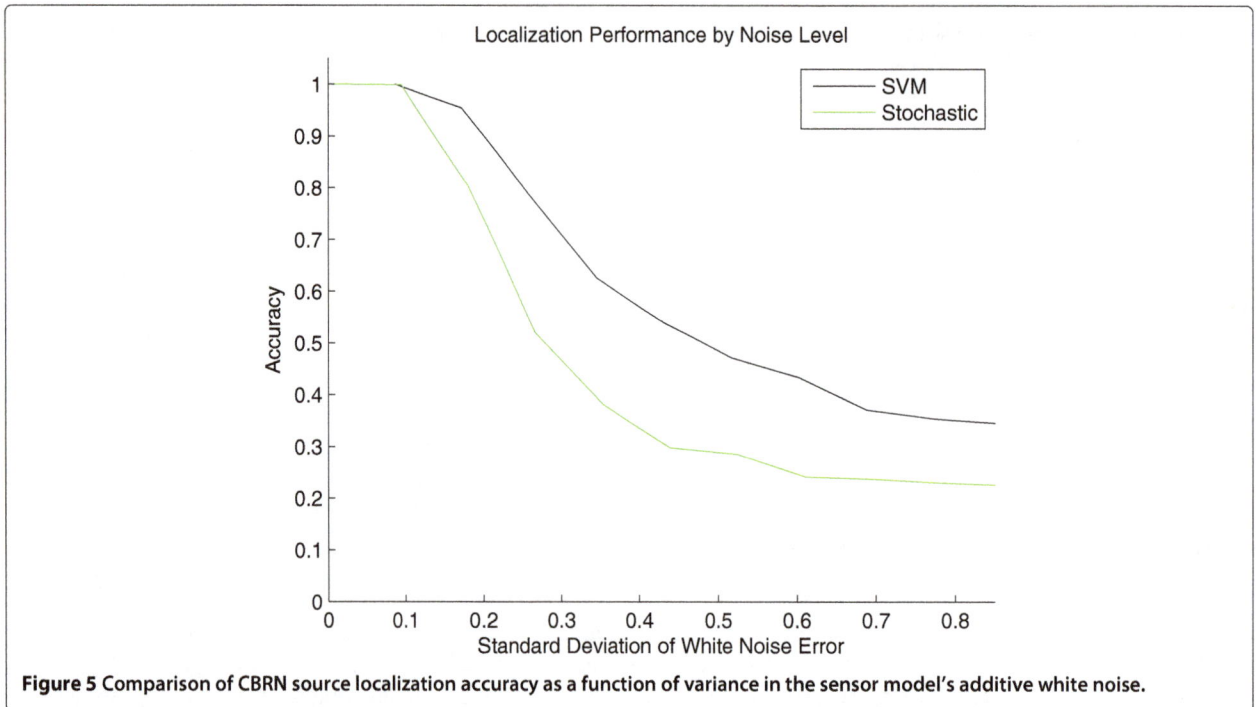

Figure 5 Comparison of CBRN source localization accuracy as a function of variance in the sensor model's additive white noise.

strategy is more robust to applications where either sensor measurement noise or chaotic urban canyon turbulence is to be expected as compared to stochastic CBRN localization.

The detection methodology presented is robust to multiple release cases since it is based on determining simply whether a hazardous element is present in the atmosphere. In theory, a multiple release event would be easier to detect than a single release event due to the increased levels of particulate. Localization, on the other hand, is not. To locate the origin of multiple simultaneous releases requires either a fusion of several single-source localizers or sufficient simulation of multiple release events to use the localization approach we have presented. Considering the impact on computational load to the latter option, it would be wise to employ a computational grid when obtaining localizer training data via simulation.

However, one-class SVM release detection performance still has some room for improvement. While dependable concentration observations, as represented by our LBM dispersion simulations with mild sensor noise in this evaluation, leads to ideal performance characteristics, the probability of detection at reasonable numbers of deployed sensors needs to be higher. In the event that false alarm performance demands are less stringent than those imposed on CBRN attacks, such as in pollution monitoring applications, probability of detection using a one-class SVM is likely to improve dramatically.

While the focus on this work has been on urban environments, this need not be the only application. Any

problems where detection and localization of the source of dispersed target particles are within the scope of application for our methods. These include finding a spurious pollution generating plant, downstream pollution monitoring, underground pollutant seepage tracing, and nuclear power plant monitoring, provided an accurate simulator is available for data set generation.

Methods

An SVM is a well established machine learning technique for binary classification problems (Cortes and Vapnik 1995). The premise is to construct the *decision function*

$$f(\mathbf{x}) = \sum_{i=1}^{m} y^i \alpha_i K(\mathbf{x}, \mathbf{x}^i) - b. \tag{2}$$

We classify a test data point $\mathbf{x} \in \mathbb{R}^n$, by assigning it a label equal to the value of $\mathrm{sgn}(f(\mathbf{x}))$. We review the aspects of SVMs pertinent to source localization in the remainder of this section.

The values y^i and $\mathbf{x}^i$, $i = 1, \ldots, m$, come from a *training set*. The parameters α_i, $i = 1, \ldots, m$, and b are found via a training procedure.

The function $K(\cdot, \cdot)$ denotes any of a set of *kernel functions*. Common choices include the Gaussian kernel, $K(\mathbf{u}, \mathbf{v}) = \exp\left(-\frac{\mathbf{u} - \mathbf{v}^T(\mathbf{u} - \mathbf{v})}{\sigma}\right)$, and the polynomial kernel of degree d, $K(\mathbf{u}, \mathbf{v}) = (\mathbf{u}^T \mathbf{v} + 1)^d$, where $(\cdot)^T$ denotes transpose.

By labeling a pattern $\mathbf{x}$ by the sign of the decision function (2), we are classifying it according to which side of a hyperplane it falls on. The use of kernel functions effectively augments the decision making space, thus allowing for accurate classification even in cases where two classes of data are not linearly separable.

Feature selection for SVMs

Feature selection can be seen as a technique for dimension reduction. The objective is to choose a subset of the existing features by excluding those that provide little benefit in differentiating between two classes of data. Features selected in this way (i.e., features that were not excluded) remain intact. This is in opposition to techniques like principal component analysis in which the resulting features are linear combinations of the original features.

A feature selection method designed specifically for SVMs is presented in (Weston et al. citeyearweston). There, the authors find a binary vector $\boldsymbol{\sigma} \in \{0,1\}^n$ whose positive elements indicate selection of k features. For a given $\boldsymbol{\sigma}$, a modified kernel,

$$K_\sigma(\mathbf{u}, \mathbf{v}) = K(\mathbf{u} \bullet \boldsymbol{\sigma}, \mathbf{v} \bullet \boldsymbol{\sigma}),$$

is used to build an SVM, where $\tilde{\mathbf{u}} = \mathbf{u} \bullet \boldsymbol{\sigma} \in \mathbb{R}^k$ is a vector whose elements are those of $\mathbf{u}$ corresponding to the elements of $\boldsymbol{\sigma}$ which are equal to one.

The *radius* of a modified kernel, R^2, is the radius of the smallest sphere that includes the training patterns in $\mathcal{X}$ after translation into the modified feature space. That is, R is the minimum non-negative value such that

$$\left[\Phi\left(\boldsymbol{\sigma} \bullet \mathbf{x}^i\right) - \mathbf{a}\right]^T \left[\Phi\left(\boldsymbol{\sigma} \bullet \mathbf{x}^i\right) - \mathbf{a}\right] \leq R^2, i = 1, \ldots, m,$$

where $\mathbf{a}$ denotes the center of the sphere. Lagrange duality produces a QP which defines the radius of a modified kernel,

$$\begin{aligned} \max_\beta \quad & \textstyle\sum_{i=1}^m \beta_i K_\sigma(\mathbf{x}^i, \mathbf{x}^i) - \sum_{i,j=1}^m \beta_i \beta_j K_\sigma(\mathbf{x}^i, \mathbf{x}^j) \\ \text{subject to} \quad & \textstyle\sum_{i=1}^m \beta_i = 1, \\ & \beta_i \geq 0, \quad i = 1, \ldots, m. \end{aligned}$$

(3)

A useful theorem, found in (Weston et al. 2000) but proven, in effect, in (Vapnik 1998) relates the likelihood of erroneous classification to the radius found by (3).

Error Rate Theorem. Let $E\left[P_{err}\right]$ denote the expected probability of erroneous classification of an SVM trained on a data set of m elements. If the data in $\mathcal{X} = \{\mathbf{x}^1, \ldots, \mathbf{x}^m\}$ with radius R are separable with a corresponding margin of ρ,

$$E\left[P_{err}\right] \leq \frac{1}{m} E\left[\frac{R^2}{\rho^2}\right] = \frac{1}{m} E\left[R^2 W^2(\boldsymbol{\alpha}^*)\right],$$

where $W^2(\boldsymbol{\alpha}^*)$ denotes the optimal value of the SVM soft margin objective function.

This theorem grants us an excellent metric for which different modifications of a kernel can be measured. An optimal $\boldsymbol{\sigma}$ is found as the minimizer of

$$R^2(\boldsymbol{\beta}^*; \boldsymbol{\sigma}) W^2(\boldsymbol{\alpha}^*, C; \boldsymbol{\sigma}),$$

(4)

subject to $\sum_{i=1}^n \sigma_i = k$, where $W^2(\boldsymbol{\alpha}^*, C; \boldsymbol{\sigma})$ represents the optimal objective function of the soft margin maximization problem with a kernel modified by $\boldsymbol{\sigma}$, and $R^2(\boldsymbol{\beta}^*; \boldsymbol{\sigma})$ is the radius of kernel $K_\sigma(\cdot, \cdot)$. By minimizing (4), features are selected according to excluding features that contribute least to discerning between two classes of data.

Ordinarily, optimization of the function (4) would prove to be a computationally intense search through the large number of possible feature combinations. Fortunately, derivative information for (4) is provided in (Weston et al. 2000). Relaxing $\boldsymbol{\sigma}$ to be a real-valued vector on the unit hyper-square, one can iteratively minimize (4) using an efficient gradient based solver. After each iteration the features whose corresponding elements of $\boldsymbol{\sigma}$ are sufficiently close to zero or one are excluded or included, respectively. This process continues until the desired number of features remain.

Source localization

Source localization is the problem of locating the origin of a hazardous atmospheric release in an urban environment with SVMs that operate solely on information obtained from a network of sensors deployed throughout the city under surveillance. These sensors are assumed to provide sequences of measured concentrations of a target particle. We assume for the time being that sensors have already been placed at K locations from a discrete set of locations $\mathcal{B} = \{B_1, \ldots, B_M\}$. We defer to the problem of selecting which locations from $\mathcal{B}$ will provide better localization performance to a later section. Our goal is then to, upon observing sequences of sampled particulate concentrations, determine which location from the set $\mathcal{L} = \{L_1, \ldots, L_N\}$ the particles originated.

Feature representation

Assuming concentration values to be sampled according to a fixed time interval, we represent concentration evolution observed by a sensor at location B_k as a sequence

$$c(k) = c_1^k, c_2^k, \ldots,$$

(5)

where c_t^k denotes the real-valued sampled concentration at location B_k at discrete time step t.

Several options are available to encode the sequences in (5) into feature vectors. For instance, the patterns could simply be composed of the first, say, n samples in $c(k)$, producing patterns of dimension nK of the form

$$\mathbf{x} = \left(c_1^{k_1}, \ldots, c_n^{k_1}, \ldots, c_1^{k_K}, \ldots, c_n^{k_K} \right),$$

where $k_1, \ldots, k_K$ depict the K location indices where sensors are deployed. A potentially smaller feature vector is the quantization of concentration observations made by taking aggregate mean concentrations over several consecutive elements of $c(k)$. If these means are taken over, say, m elements, the resulting patterns of dimension $\lceil \frac{n}{m} \rceil K$, where $\lceil \cdot \rceil$ denotes rounding up to the next integer, take the form

$$\mathbf{x} = \left(\frac{1}{m} \sum_{t=1}^{m} c_t^{k_1}, \ldots, \frac{1}{m} \sum_{t=m(n-1)+1}^{mn} c_t^{k_1}, \ldots, \right.$$
$$\left. \frac{1}{m} \sum_{t=1}^{m} c_t^{k_K}, \ldots, \frac{1}{m} \sum_{t=m(n-1)+1}^{mn} c_t^{k_K} \right).$$

In the simplest case, where $m = n$, these features are K dimensional and take the form

$$\mathbf{x} = \left(\frac{1}{n} \sum_{t=1}^{n} c_t^{k_1}, \ldots, \frac{1}{n} \sum_{t=1}^{n} c_t^{k_K} \right).$$

Another feature space we consider consists of the maximum value in the sequence $c(k)$ and its time step index. This feature contains only two elements per sensor, but still contains a differentiating feature (maximum concentration) as well as a temporal measure (time index), thus making it adequate for detecting differences in release scenarios. Patterns produced under this paradigm take the form

$$\mathbf{x} = \left(\max_t c(k_1), \arg \max_t c(k_1), \ldots, \right.$$
$$\left. \max_t c(k_K), \arg \max_t c(k_K) \right).$$

SVM localization

A question of how to adapt a binary classifier to select one of N release locations remains. For localization, we look to methods that repeatedly call upon the outcomes of binary SVM evaluations to select, based on the feature spaces described by any of the feature representations above, a release location. In effect, we are constructing an N-class classifier out of several binary classifiers. Popular approaches revolve around the idea of setting up some sort of tournament of several binary classifiers. Each match within the tournament prevents a test pattern from being labeled as a particular class. The class remaining at the conclusion of the tournament is assigned to the test pattern.

In (Hsu and Lin 2002), a comparison of the so-called "one-against-all", "one-against-one", and Directed Acyclic Graph (DAGSVM) methods for multi-class SVMs is conducted. In the "one-against-all" approach, N classifiers of the form (2) are found, where N is the number of classes. For classifier i, an SVM is trained with patterns belonging to class i labeled as 1 and those patterns belonging to all other classes as -1. When evaluating a test pattern $\mathbf{x}$, the class whose SVM achieves the maximum value of (2) is assigned.

In the "one-against-one" approach, $\frac{N(N-1)}{2}$ SVMs are trained. For each class combination (i, j), $i < j$, a binary SVM is trained on data belonging only to classes i and j. A test pattern $\mathbf{x}$ is assigned the class that was assigned the most number of times out of all of the binary classifications. DAGSVM is similar to the "one-against-one" approach, but instead of considering each of the $\frac{N(N-1)}{2}$ binary SVMs, it assigns labels by starting with the decision made on a particular class pair. Based on the results of the first classification, binary classification between the previously selected class and another particular class is performed. This process continues until no more prescribed comparisons remain.

Based on the results in (Hsu and Lin 2002) and ease of implementation, we use the "one-against-one" multi-class method for source localization throughout the following. From a large set of sensor concentration sequences $\mathcal{C}$, obtained from a particulate dispersion simulator, we form feature vectors according to one of the paradigms listed in Section Feature representation to produce a training set $\mathcal{X}$. Out of $\mathcal{X}$ we build $\frac{N(N-1)}{2}$ training sets denoted $\mathcal{X}_{1,2}, \mathcal{X}_{1,3}, \ldots, \mathcal{X}_{N-2,N}, \mathcal{X}_{N-1,N}$, where $\mathcal{X}_{i,j}$ represents the subset of $\mathcal{X}$ that consists only of training patterns obtained through simulation under release locations L_i and L_j. Then, for $i, j = 1, \ldots, N$ and $i < j$, we use $\mathcal{X}_{i,j}$ to train a binary classifier of the form (2), denoted as $f_{i,j}(\mathbf{x})$. For a new test vector $\mathbf{x}$, produced from newly observed concentration sequences, we select either location L_i or L_j according to the sign of $f_{i,j}(\mathbf{x})$ for all i and j, with $i < j$. The location out of $\mathcal{L}$ that was selected the most number of times out of all of the $\frac{N(N-1)}{2}$ selections is chosen as the release location.

Sensor placement

The question remains, if we are allowed to place sensors anywhere within the set of possible sensor locations $\mathcal{B}$, which sensor locations should be selected? A straightforward extension of the SVM localization method provides a guideline for placement of available sensors. Using the feature selection methods previously discussed, we can select K of the M sensor locations.

To illustrate this procedure, assume patterns are originally formed by M features of dimension d, with all potential sensor locations' associated features included.

That is, a pattern is originally of the form $\mathbf{x} = (\mathbf{x}_1, \ldots, \mathbf{x}_M)$ where $\mathbf{x}_i = (x_{i,1}, \ldots, x_{i,d})$ is a d-dimensional feature of one of the forms discussed in Section Feature representation or some other form. We select sensor locations by following the feature selection procedure of minimizing (4) with respect to $\boldsymbol{\sigma} = (\boldsymbol{\sigma}_1, \ldots, \boldsymbol{\sigma}_M)$, where $\boldsymbol{\sigma}_i \in \{0,1\}^d$ for $i = 1, \ldots, M$. More specifically, allowing the notation $\boldsymbol{\sigma}_i = (\sigma_{i,1}, \ldots, \sigma_{i,d})$, we minimize

$$R^2(\boldsymbol{\beta}^*; \boldsymbol{\sigma}) W^2(\boldsymbol{\alpha}^*, C; \boldsymbol{\sigma}) \qquad (6)$$

subject to the constraints

$$\sum_{i=1}^{M} \sum_{j=1}^{d} \sigma_{ij} = Kd,$$
$$\sigma_{i,1} = \cdots = \sigma_{i,d}, \quad i = 1, \ldots, M,$$
$$\boldsymbol{\sigma}_i \in \{0,1\}^d, \quad i = 1, \ldots, M.$$

Here, R^2 is the radius of the modified kernel $K_{\boldsymbol{\sigma}}(\cdot, \cdot)$ and is the optimal solution to

$$\min_{\boldsymbol{\beta}} \sum_{i=1}^{|\mathcal{X}|} \beta_i K_{\boldsymbol{\sigma}}(\mathbf{x}^i, \mathbf{x}^i) - \sum_{i,j=1}^{|\mathcal{X}|} \beta_i \beta_j K_{\boldsymbol{\sigma}}(\mathbf{x}^i, \mathbf{x}^j)$$
$$\text{subject to } \sum_{i=1}^{|\mathcal{X}|} \beta_i = 1, \qquad (7)$$
$$\beta_i \geq 0 \quad i = 1, \ldots, |\mathcal{X}|,$$

and W^2 is the optimal solution to

$$\max_{\boldsymbol{\alpha}} \sum_{i=1}^{|\mathcal{X}|} \alpha_i - \frac{1}{2} \sum_{i,j=1}^{|\mathcal{X}|} \alpha_i \alpha_j y^i y^j K_{\boldsymbol{\sigma}}(\mathbf{x}^i, \mathbf{x}^j)$$
$$\text{subject to } 0 \leq \alpha_i \leq C, \quad i = 1, \ldots, |\mathcal{X}|, \qquad (8)$$
$$\sum_{i=1}^{|\mathcal{X}|} \alpha_i y^i = 0,$$

for some prescribed C. In both (7) and (8) we use the convention introduced previously, where

$$K_{\boldsymbol{\sigma}}(\mathbf{u}, \mathbf{v}) = K(\mathbf{u} \bullet \boldsymbol{\sigma}, \mathbf{v} \bullet \boldsymbol{\sigma})$$

is the kernel modified by $\boldsymbol{\sigma}$, where $\mathbf{u} \bullet \boldsymbol{\sigma}$ returns a vector whose elements are those of $\mathbf{u}$ that correspond to the elements of $\boldsymbol{\sigma}$ that are equal to one. The minimizer $\boldsymbol{\sigma}^*$ of (6) becomes the vector indicating which K out of the M present d-dimensional features are to be used in localization. Thus, $\boldsymbol{\sigma}^*$ effectively selects which K locations sensors should be placed.

This process is replicated for each of the $\frac{N(N-1)}{2}$ binary SVMs used in the "one-against-one" localization technique. Sensors are placed at the K locations that are selected most frequently among these replications.

Release detection

An important component to locating the origin of a particulate release by concentration sampling is the amount of time between subsequent observations. It is hard to construct a feature space that does not rely on the starting time of the release. What is needed is a trigger mechanism that alarms immediately upon detecting a hazardous atmospheric dispersion.

Detection problems involving noisy observations have long been an area of research. When hazardous particulates are not present, sensor observations are purely due

to sensor noise. If we make the assumption that sensor noise is independent and identically distributed (iid) and behaves according to a known distribution, an elementary approach is to set, through analysis of the noise distribution and a tolerable false alarm rate, a concentration threshold. Whenever any sensor observes a value greater than this threshold, the time of the alarm can be used as the originating time of the release.

An SVM approach to release detection is the so-called "one-class SVM" method of novelty detection (Schölkopf et al. 2001). One-class SVM is similar to the binary SVM in its form and training, except that only representative training elements from a single set are available for analysis. The premise is, in the absence of one class's training patterns, to treat the origin of the higher dimensional space identified through the choice of a kernel function as the only member of the opposite class. Then, a hyperplane that separates all but a controlled number of training patterns from the origin effectively becomes an anomaly detector.

In the case of release detection, we train a one-class SVM using only features of the form described in Section Feature representation that represent sensor observations when no release is present. These patterns would therefore represent the perturbations in concentration observations that result from sensor noise or routine false alarming due to the presence of some non-target particulate. When any of the sensor observations are declared anomalous by the one-class SVM, we would presume that this is because a sensor's observations depart too largely from the usual behavior and declare a release.

Competing interests
The authors declare that they have no competing interests.

Authors' contributions
IP advised and directed the presented research as well as contributed to drafting of the manuscript. RTL developed the and implemented the methods, conducted the numerical experimentation, and led the drafting of the manuscript. All authors read and approved the final manuscript.

Acknowledgements
This work is sponsored by the United States Air Force under Air Force Contract #FA8721-05-C-0002. Opinions, interpretations, recommendations and conclusions are those of the authors and are not necessarily endorsed by the United States Government.
Research partially supported by the NSF under grants CNS-1239021 and IIS-1237022, by the ARO under grants W911NF-11-1-0227 and W911NF-12-1-0390, and by the ONR under grant N00014-10-1-0952.

References
Atalla A, Jeremic A (2008) Localization of chemical sources using stochastic differential equations. In: Acoustics, speech and signal processing, 2008. ICASSP 2008. IEEE international conference on. pp 2573–2576
Chen S, Doolen GD (1998) Lattice Boltzmann method for fluid flows. Ann Rev Fluid Mech 30(1):329–364
Cortes C, Vapnik V (1995) Support-vector networks. Mach Learn 20:273–297

Frisch U, Hasslacher B, Pomeau Y (1986) Lattice-gas automata for the
    Navier-Stokes equation. Phys Rev Lett 56:1505–1508
Hsu C-W, Lin C-J (2002) A comparison of methods for multi-class support
    vector machines. IEEE Trans Neural Netw 13:415–425
Lallemand P, Luo L-S (2000) Theory of the lattice Boltzmann method:
    Dispersion, dissipation, isotropy, galilean invariance, and stability. Phys Rev
    E 61:6546–6562
Locke RT, Paschalidis IC (2012) Stochastic localization of cbrn releases. In:
    Acoustics, speech and signal processing, 2012. ICASSP 2008. IEEE
    international conference on.
Locke, R T, Paschalidis IC (2013) Model-free stochastic localization of cbrn
    releases. Signal Process IEEE Trans 61(17):4246–4258
Ortner M, Nehorai A, Jeremic A (2007) Biochemical transport modeling and
    bayesian source estimation in realistic environments. Signal Process IEEE
    Trans 55(6):2520–2532
Ortner M, Nehorai A (2008) Biochemical transport modeling, estimation and
    detection in realistic environments. In: Acoustics, speech and signal
    processing, 2008. ICASSP 2008. IEEE international conference on.
    pp 5169–5172
Paryak ER, Brown MJ (2007) QUIC-URB v.1.1:theory and users guide. Technical
    report, Los Alamos National Laboratory. https://www.lanl.gov/projects/
    quic/open_files/QUICURB_UsersGuide.pdf
Röckle (1990) Bestimmung der strömungsverhlltnisse im bereich komplexer
    bebauungsstrukturen. PhD thesis, Technischen Hochschule, Darmstadt,
    Germany
Schölkopf B, Platt JC, Shawe-Taylor JC, Smola AJ, Williamson RC (2001)
    Estimating the support of a high-dimensional distribution. MIT Press,
    Cambridge, MA, USA. http://dx.doi.org/10.1162/089976601750264965
Vapnik VN (1998) Statistical learning theory. John Wiley & Sons, Inc., New York,
    NY, USA
Vujadinovic M, Rajkovic B, Grisc Z (2008) Locating a source of air pollution
    using inverse modelling and pre-computed scenarios. In: iEMSs 2008:
    International congress on environmental modelling and software
Wawrzynczak A, Jaroszynski M, Borysiewicz M (2014) Data-driven genetic
    algorithm in bayesian estimation of the abrupt atmospheric
    contamination source. In: Computer science and information systems
    (FedCSIS), 2014 IEEE federated conference on. pp 519–527
Weston J, Mukherjee S, Chapelle O, Pontil M, Poggio T, Vapnik V (2000) Feature
    selection for svms. In: Neural Information Processing Systems, 2000.
    pp 668–674

6

Risk assessment of heavy metal soil pollution through principal components analysis and false color composition in Hamadan Province, Iran

Alireza Soffianian[1*], Elham Sadat Madani[2] and Mahnaz Arabi[2,3]

Abstract

Background: In the process of decision making to combat heavy metals pollution, it is essential to have accurate quantitative information about heavy metals and their pollution hot-spots. The main purpose of this study was to determine spatial distribution of several elements (As, Sb, Cr, Cd, Ni, Co, Cu, Zn, Pb, Fe and V) on surface soils of Hamadan Province (Iran). It also sought to create a holistic view to determine the position, level, and anomaly of classified elements through principal components analysis (PCA), false color composition (FCC), inverse overlay method, and weighted linear combination. Finally, it tried to identify possible sources of pollution in the hotspots. Interpolation of heavy metal concentrations was performed using geostatistical methods and correlation analysis for locations. The most appropriate interpolation method was selected based on mean absolute error (MAE) and mean bias error (MBE) indices.

Results: According to Pearson's correlation analysis, the elements were categorized in four groups (Fe, V, and Co; Cu, Ni, and Cr; Pb, Zn, and As; Sb and Cd). For Fe, Zn, As, and Pb, the best method was disjunctive kriging. For Co, Sb, Ni, and Cr, ordinary kriging was the most appropriate. Radial basic functions was also the best method for Cd and Cu.

Conclusions: Overlaying of zoning maps of the elements and land use and geological layer maps showed that the distribution pattern of the studied elements did not fully conform to the existing land use pattern. Although the most influential factor on the concentration of elements in the studied soils was bedrocks, extensive use of chemical fertilizers should not be ignored. Moreover, urban pollution can also contribute to Pb contamination of soil.

Keywords: Kriging; Correlation analysis; Principal components analysis; False color composition; Inverse overlay method; Weighted linear combination

Background

Soil is a major natural resource whose properties and quality can be adversely affected by the over-concentration of agricultural and industrial activities. On the other hand, preserving soil quality and preventing its deterioration are fundamental to sustainable development. In recent decades, problems associated with increasing levels of heavy metals and their persistence in the environment have attracted the attention of researchers (Bowen 1979; Lindsay 1979; Lame and Leenaers 1997). Although low concentrations of these metals are naturally found in soils and

stones, human activities have elevated their release and propagation in the environment.

Determining the areas that are polluted either naturally or as a result of human activities is a means of evaluating the health of an ecosystem (Romic et al. 2007). Mining, industries, road transfer, waste burning, and agricultural use of fertilizers and chemicals are human activities that can lead to heavy metal contamination of the soil. On the other hand, natural factors contributing to heavy metal contamination of the soil include volcanoes, degassing of the Earth's crust, fires in forests, and chemical composition of parent materials (Lado et al. 2008).

A variety of sciences such as classical statistics, geostatistics, remote sensing, geographical information systems (GIS), soil science, and hydrology are employed to precisely

* Correspondence: soffianian@cc.iut.ac
[1]Department of Natural Resources, Isfahan University of Technology, Isfahan 84156-93111, Iran
Full list of author information is available at the end of the article

assess the heavy metal content of soils. In fact, soil pollution can currently be well determined using GIS and geostatistical methods. While classical statistical methods have been widely applied in previous studies on soil contamination, such methods are expensive and time-consuming and do not calculate estimation errors. Besides, preparing sufficient samples from the areas under-study is impossible. Hence, geostatistical methods have replaced classical statistics as they can accurately identify time and spatial changes of pollutants and calculate estimation error (Blom 1985; Bonham-carter et al. 1987).

The present research aimed to determine areas with heavy metal (Cu, Co, Ni, Cd, Cr, Sb, V, Fe, Pb, As, and Zn) contamination, to locate pollution hot-spots, and to recognize possible sources of contamination in surface soils of Hamadan Province (Iran) through principal component analysis (PCA) and false color composites (FCC).

Methods
Study area
Hamadan Province occupies an area of 19493 km^2 (from 33 degrees, 59 minutes to 35 degrees, 48 minutes north latitude and from 47 degrees, 34 minutes to 49 degrees, 36 minutes east longitude) (Figure 1). The agricultural and livestock development of this province (49.3% of its

lands are used for agricultural purposes) has turned Hamadan into an economic hub (Figure 2). The dominant geological structures of the area include alluvial terraces from the Quaternary period, orbitoline limestone, shale and marl from the late Cretaceous period, metamorphic sandstone from the Jurassic period, and andesitic lava and reef limestone from late Paleogene and early Neogene periods (geological map 1:1000000; Geological Survey of Iran). The area contains shallow to moderately-deep soils with small to medium-sized gravels and some amounts of calcareous material (2007).

Sampling
In order to perform systematic random sampling, the area was first divided to 5*5 km^2 networks. Afterward, based on the characteristics of the area and various types of land use and activities, networks of 2.5 × 2.5 and 10 × 10 km^2 were developed in areas with intensive and low-intensity land use, respectively. The points where the networks met were selected as sampling points. Finally, 286 soil samples were collected from the area, i.e. after determining sampling points, a 20 × 20 cm^2 macro-plate was drawn at each point. Then, three 3 × 3 cm^2 micro-plates were drawn inside it in the shape of a downward V. Five soil samples were taken from the four corners

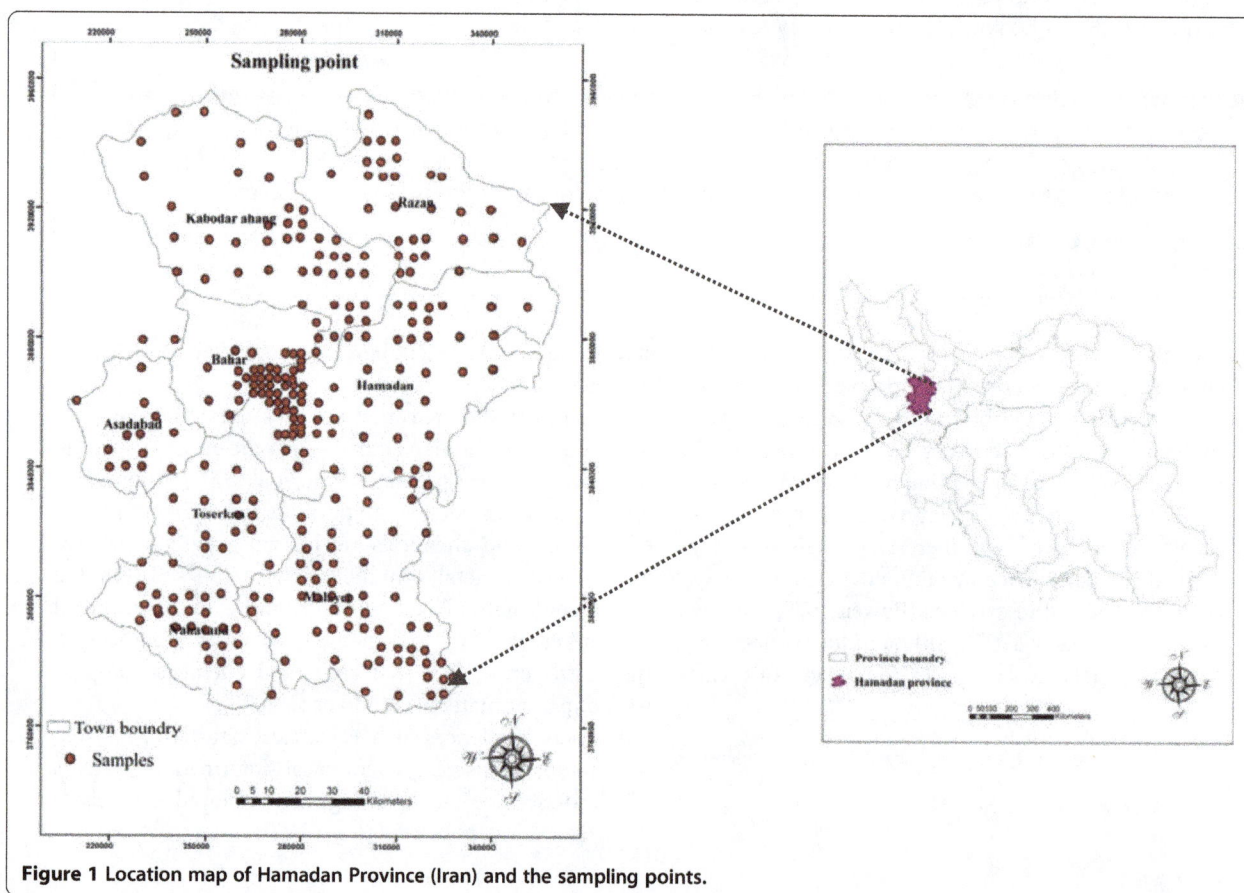

Figure 1 Location map of Hamadan Province (Iran) and the sampling points.

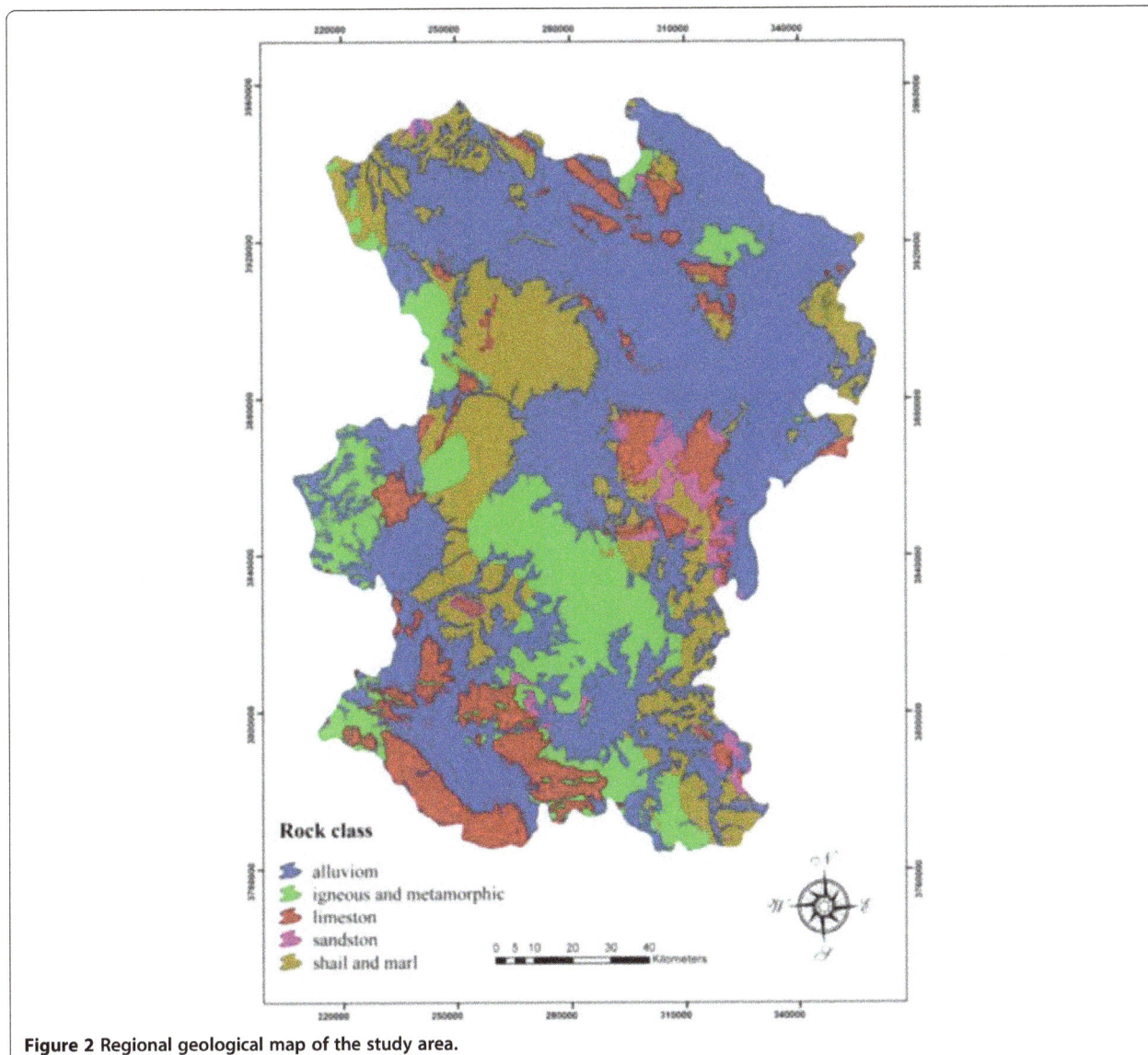

Figure 2 Regional geological map of the study area.

and the center of each micro-plate. The samples were all obtained at depths of 0–20 cm. The collected soils were eventually mixed and a final 2–3 kg sample was prepared.

Information about land use at the sampling point, type of agriculture (irrigated/rainfed), appearance of the land, kind of product, geographical characteristics, and the nearest village was recorded in a sampling form. Samples were taken from pristine areas as well as mountainous areas, deserts, salt marshes, lands under irrigated or rainfed agriculture, vineyards, gardens, and areas near villages (Figure 3).

Chemical analysis of soil

The prepared samples were sent to the laboratory and their Pb, Zn, As, Cr, Co, Ni, V, Fe, and Cu concentrations were measured via inductively coupled plasma-atomic emission spectrometry (ICP-AES). Cd content of the

samples was assessed using a graphite furnace and atomic absorption spectroscopy in the chemical laboratory of the Iranian Center for Advanced Mineral Processing Research.

Descriptive statistics

Kolmogorov-Smirnov test was applied to investigate the normal distribution of data at a confidence level of 95%. Moreover, the data was normalized with logarithmic transformation. In order to evaluate the effects of land use on the concentrations of heavy metals in soil, the map was first divided into three main categories of agricultural, urban-industrial, and non-agricultural use. Pearson's correlation coefficients were then calculate to identify the correlations between heavy metal concentrations and different land uses. A box plot and PCA were used to determine the outliers and classify the relations between the measured variables, respectively.

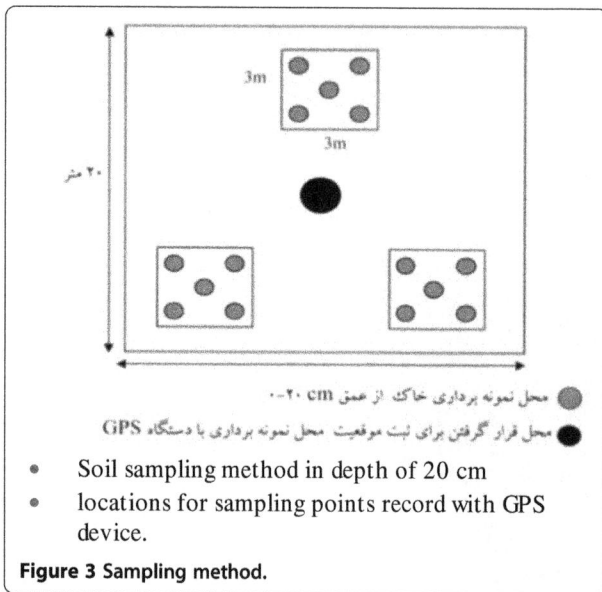

- Soil sampling method in depth of 20 cm
- locations for sampling points record with GPS device.

Figure 3 Sampling method.

Land use mapping

Considering the importance of land use map in analyzing heavy metal pollution, the map was prepared using Indian Remote Sensing (IRS-P6) Advanced Wide Field Sensor (AWiFS) images.

Data mapping

The current research employed kriging for data mapping.

False color composition (FCC)

Geographical information studies may utilize FCC to determine hot-spots of heavy metal pollution. This method can simultaneously compress data and interpret layers of information. Creating color composite images involved spatial correlation analysis of the heavy metals. After standardizing the zoning map of all elements' concentrations (0–255), the three elements in each group were colored as red, green, and blue. The FCC map was then developed accordingly.

Principal component analysis (PCA)

PCA is a statistical method to classify the relations between the measured variables, i.e. it linearly compresses a set of main data to an essentially smaller set of new, uncorrelated variables which represent nearly all information in the main data set. Meanwhile, understanding and working with a small set of uncorrelated variables is much easier than working with a large set of correlated variables (Eastman 2006). In the present study, PCA was carried out in SPSS for Windows 15.0 (SPSS Inc., Chicago, IL, USA). In addition, spatial correlation analysis was applied for each variable to select the most appropriate method of interpolation. Finally, the zoning map of the factors obtained from PCA was produced in ArcGIS 9.3 (Esri, CA, USA).

Inverse overlaying of elements

Since the images from the previous stage could not provide information about more than three elements in recognition of pollution hot-spots, the anomalies of elements (distribution of a combination of elements in each area) were determined using the zoning map of elements obtained from FCC, map overlay, and identification of heavy metal hot-spots. Overlaying of inverse layers facilitates analysis by creating an overall view on multiple elements (Suyash Kumar et al. 2007). It is, in fact, a different representation of elements for clearer observation of pollution hot-spots. In order to apply this method, the maps from FCC were first stretched between zero and 255. Afterward, pairs of layers (one as the top and one as the bottom map) were selected according to the correlations between their elements. In top and bottom layers, red to blue and blue to red (reverse) represented maximum to minimum amounts, respectively. Top layers were then demonstrated at a contrast of 100 and transparency of 60. Finally, similar to FCC, the anomaly of elements was interpreted based on the composition of the obtained colors.

Results
Filtration and normalization of primary data

Table 1 summarizes the density of heavy metals (mean As: 15.7993, Cd: 0.1535, Fe: 3.9472, Co: 18.9598, Cr: 96.8182, Cu: 36.4545, Ni: 69.0350, Pb: 31.8916: Sb: 2.9084, V: 109.4161, Zn: 80.0035). The box plot in Figure 4 was used to correct the outlier data about the concentration of heavy metals. The numbers of outlier and corrected data are given in Table 2. The results of Kolmogorov-Smirnov test showed the data for As, Sb, Cr, Co, and Pb not to be normal (Table 3). However, normal distribution of variables is essential in geostatistical studies. Moreover, too much stretching and crumpling can damage the structure of variogram and the results of kriging. Therefore, after ensuring the absence of negative data, logarithmic transformation was applied to normalize the concentrations of the mentioned elements (Figure 5). Pearson's correlation analysis was then performed again on the normalized data (Table 4).

Analysis of variance on the mean concentrations of heavy metals in soil did not prove any significant difference between the three different kinds of land use (agricultural, urban-industrial, and non-agricultural uses).

PCA

The appropriateness of data for PCA is assessed based on Kaiser-Mayer-Olkin (KMO) index, i.e. the test is recommended only if KMO > 0.70. Since KMO index was calculated as 0.84 in the current research, PCA was applied on

Table 1 Trace element data for the soil from Hamadan Province, Iran

Metal	N	Missing	Minimum	Maximum	Mean	Std. deviation	Europe mean
As	286	9	4.70	85.00	15.7993	8.94631	11.6
Cd	286	10	.10	0.88	0.1535	0.08408	0.284
Fe	286	9	1.80	6.00	3.9472	0.67775	2.17
Co	286	9	8.10	34.00	18.9598	3.89043	8.91
Cr	286	9	30.00	180.00	96.8182	26.37480	94.8
Cu	286	9	4.00	75.00	36.4545	10.09202	17.3
Ni	286	9	26.00	140.00	69.0350	20.43214	37.3
Pb	286	9	13.00	1800.00	31.8916	105.13443	32.6
Sb	286	9	0.50	28.00	2.9084	2.69343	1.04
V	286	9	50.00	190.00	109.4161	22.38291	68.1
Zn	286	9	35.00	200.00	80.0035	18.74679	60.9
Valid N (list wise)	286						

the primary data and 11 factors were found necessary for explaining 100% of variance. However, as a mere four out of the 11 factors could explain 71.36% of variance, the low importance of the other seven factors was deduced.

Factor loading was used to determine the relationships between the variables and each factor (Table 5). The first factor explained 39.25% of variance and its factor load was roughly the same for Fe, Zn, Co, Cr, Ni, and V. High correlation between these elements can indicate their common origin. The second factor explained 13.12% of the total variance of element distribution and suggested similarity between Cd and Cu (factor load: 0.85 for Cu and 0.65 for Cd). The third factor could only explain 9.55% of variance. According to the table of eigenvalues, Pb had a factor load of about 0.75 and was the only element in this factor. The fourth factor explained no more than 9.44% of variance and included As and Sb with factor loads of 0.49 and 0.89, respectively. The remaining factors did not have a significant role in increasing the total variance.

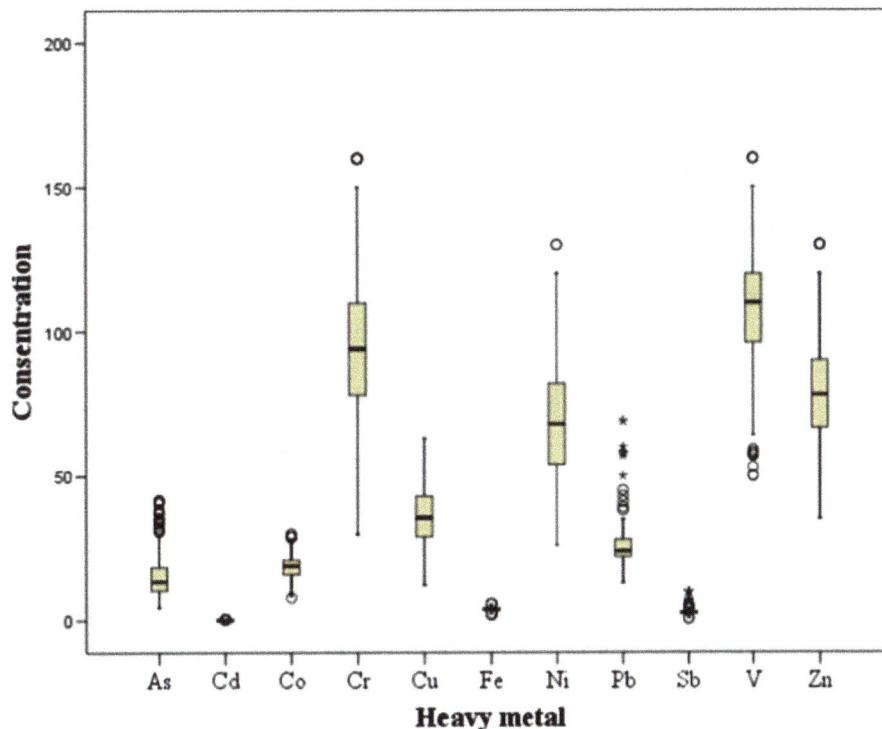

Figure 4 Box plot for outlier determination.

Table 2 The number of modified and element outlier data for studied heavy metals

Element	Cr	Ni	V	Zn	Cd	As	Co	Cu	Sb	F	Fe	Pb
The number of outliers	5	3	10	5	6	12	6	3	16	9	5	11
The number of modified data	2	1	2	3	3	4	2	3	5	1	2	6

Spatial distribution of components in PCA

In order to prepare the zoning maps of factors obtained from the PCA, spatial correlation analysis was conducted for each factor in SPSS (SPSS Inc., Chicago, IL, USA) and the most suitable method of interpolation was selected. Table 6 shows fitted models and the most appropriate methods of interpolation. Geostatistical analyses (Table 7) revealed the exponential model to be the best fitted model for the factors (Figure 6).

Geostatistical analysis

Kriging and radial basis function (RBF) were employed to investigate the spatial variations of heavy metals. All such analyses were performed in ArcGIS 9.3 (Esri, CA, USA). The precision of the methods was compared using the jack-knife technique and root mean square error (RMSE), mean bias error (MBE), and mean absolute error (MAE) which are all valid indices (Table 7). Based on the results, the most appropriate methods were disjunctive kriging for Fe, Zn, As, and Pb, ordinary kriging for Cr, Ni, Sb, and Co, and RBF for Cd and Cu (Figure 7).

FCC

According to Pearson's correlation analysis (Table 4), the elements were divided into four groups: Fe, V, and Co; Cu, Ni, and Cr; Pb and Zn; and Sb and Cd. Single-component images were standardized (0–255) to prepare the FCC map. Two-three elements were then combined based on the above-mentioned classifications and using the FCC method (Figure 8a and b). As the prepared images did not yield information about more than three elements in recognizing the hot-spots, after standardization of single-component images, the elements were categorized into two groups based on their correlation level.

Consequently, Fe, V, Co, Cu, Ni, and Cr were allocated to the first category and Pb, Zn, As, Sb, and Cd were placed in the second. The two groups were then combined using weighted linear combination (Figure 8c). This process simplified the analysis of hot-spot positions. In other words, to identify the anomalies of the elements (combined distribution of elements in each area), weighted linear combination and overlaying were applied on single-component images and the heavy metal hot-spots were recognized. Red and blue areas in Figure 8c have high and low anomalies, respectively.

Discussion

Correlation analysis

The results of spatial correlation analysis (Table 4) showed that the mean estimation error and RMSE for all variables were close to the ideal values (zero and one, respectively). This suggests high precision of our estimations. There was also a great spatial correlation between the concentrations of elements (especially As, Fe, Co, Ni, and Cr) in the samples. In other words, the concentrations of elements in closer samples were more similar to each other probably due to the effects of natural factors such as parent material, topography, and soil type. However, human factors (e.g. fertilizing) might have been responsible for weak spatial structure of Zn, V, and Pb.

MAE and MBE indices were calculated to assess the precision and deviation of interpolation models, respectively (Table 7). As these indices measure the difference between the measured and the estimated values, values closer to zero indicate higher spatial precision of the model and lower deviation, respectively (Hassani-pak 1998; Mohammadi 2006).

Table 3 One-sample Kolmogorov-Smirnov test results about the concentration (in mg/kg) of the studied heavy metals in Hamadan Province, Iran

		As	Cd	Fe	Co	Cr	Cu	Ni	Pb	Sb	V	Zn
N		286	286	286	286	286	286	286	286	286	286	286
Normal parameters (a,b)	Mean	15.49	.1457	3.9479	18.9458	96.6783	36.1612	69.0000	25.8392	2.702	109.241	79.5839
	Std. Deviation	7.34	.0720	.67360	3.84308	25.9820	9.75556	20.3185	7.22292	1.30851	21.8781	16.9531
Most extreme differences	Absolute	.15	.242	.070	.098	.096	.037	.078	.176	.180	.116	.052
	Positive	.15	.242	.049	.098	.096	.037	.078	.176	.180	.116	.052
	Negative	-.10	-.220	-.070	-.064	-.050	-.029	-.038	-.116	-.106	-.087	-.039
Kolmogorov- Smirnov Z		2.561	4.088	1.176	1.661	1.623	.624	1.324	2.978	3.052	1.954	.875
Asymp. Sig. (2-tailed)		.000	.054	.126	.008	.010	.831	.060	.000	.000	.064	.428

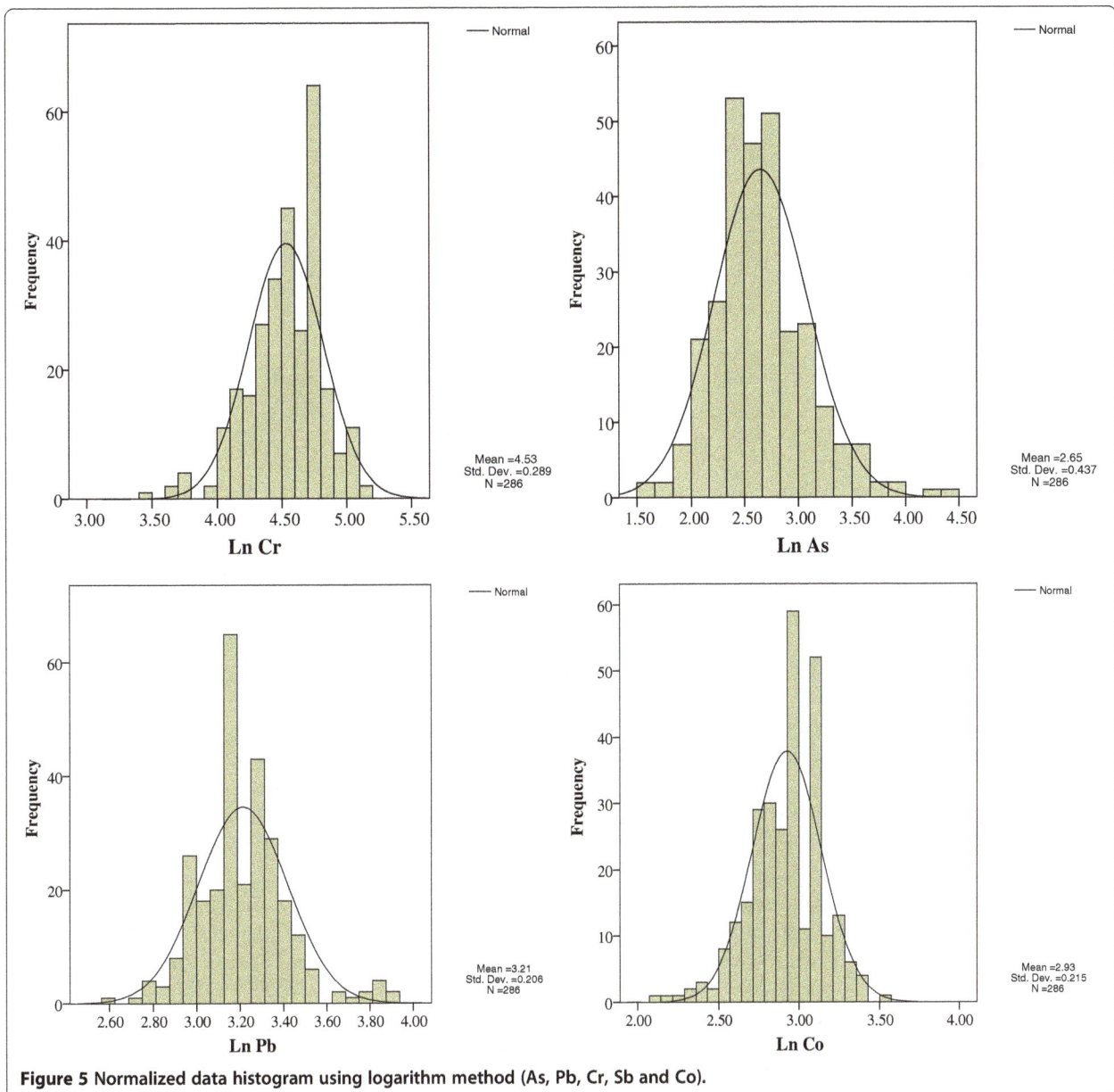

Figure 5 Normalized data histogram using logarithm method (As, Pb, Cr, Sb and Co).

Zoning maps

According to the distribution map of As (Figure 7), the element had the highest concentrations (20–50 mg/kg) in the northwest, two spots in the southwest and some spots in the center of the province. The geological structure of these areas comprised metamorphic rocks, shale, marl, and limestone.

The distribution maps showed the highest concentrations of Fe, Ni, Cr, and Co to be 4.0-5.9, 90–130, 140–160, and 20–30 mg/kg, respectively. These hot-spots were detected in the form of some red spots in the west, southwest, and northwest of the province. Furthermore, the bedrock was shale, igneous and alluvial rocks, and sandstone for Fe and Ni, limestone and igneous rocks for Cr, and shale-marl, sandstone, limestone, and metamorphic rocks for Co.

The highest concentration of Cd in the studied soils varied between 0.30 and 0.35 mg/kg (Figure 7) and was demonstrated as some spots throughout the province. It was found on geological structure of shale-marl, limestone, and alluvial rocks.

The distribution map of Zn suggested the maximum level of the element (100–130 mg/kg) to exist in the west, southeast, northeast, and partly in the center of the province on limestone, metamorphic rocks, and shale bedrocks.

The greatest concentration of Sb ranged from 5.0 to 9.8 mg/kg (Figure 7) and was seen in the form of two spots on igneous stones and shale bedrocks.

Table 4 Pearson's correlation coefficients between the heavy metals

	LnCo	LnSb	LnPb	LnCr	LnAs	Zn	V	Ni	Cu	Cd	Fe
LnCo	1										
LnSb	.616(**)	1									
LnPb	.108	.093	1								
LnCr	.159(**)	.085	.396(**)	1							
LnAs	.018	-.069	.098	.043	1						
Zn	.704(**)	.491(**)	.093	.168(**)	.031	1					
V	.860(**)	.553(**)	.160(**)	.224(**)	.023	.709(**)	1				
Ni	.020	-.014	.146(*)	.472(**)	.058	-.003	.072	1			
Cu	.082	-.042	.171(**)	.294(**)	.196(**)	.100	.109	.521(**)	1		
Cd	.005	.008	.072	.196(**)	-.098	.055	.064	.263(**)	.218(**)	1	
Fe	.899(**)	.608(**)	.116(*)	.129(*)	.006	.729(**)	.851(**)	.007	.062	.014	1

*Correlation is significant at the 0.05 level (two-tailed).
**Correlation is significant at the 0.01 level (two-tailed).

Overlaying of the distribution maps of As, Fe, Cd, Zn, Cr, Ni, and Sb, land use maps, and geological strata of the studied area revealed that the distribution pattern of the elements did not conform to the land use pattern of the area. On the other hand, since the concentration of each element is naturally high in its bedrock (Table 8) (De vos et al. 2005), the most important factor affecting the concentration of the mentioned elements in soil must have been the geological structure (bedrock).

As Figure 7 illustrates, areas with Pb concentration higher than 30 mg/kg were located in the southeast, center, and west of the province on igneous, metamorphic sandstone, shale, and limestone. Irrigated agriculture, pasture, and mining land use in these areas results in overuse of fertilizers and chemical herbicides in them. Moreover, sandstone and shale naturally have high concentrations of Pb (Table 8) (De vos et al. 2005).

Table 5 Load factors of the studied variables in Hamadan Province, Iran

Heavy metal	Component			
	1	2	3	4
As				0.897
Cd		0.848		
Fe	0.901			
Co	0.943			
Cr	0.898			
Cu		0.647		
Ni	0.833			
Pb			0.749	
Sb	0.496			0.491
V	0.882			
Zn	0.705			

Maximum Cu and V content of soil was 50–63 and 120–160 mg/kg, respectively (Figure 7). These values were detected in the west, southwest, and southeast of the province as well as some spots in the center. The bedrocks were shale and limestone for Cu and shale, igneous rocks, sandstone, and limestone for V. The concentration of these two elements is generally high in shale (Table 8). Meanwhile, overlaying of their distribution maps and the area's land use map showed excessive use of fertilizers and chemical herbicides in the agricultural activities performed in these lands. Hence, although the high concentration of Cu and V can be mainly justified by natural factors such as shale, lime, and alluvium bedrock, the accumulation of these elements in agricultural lands with unwarranted use of chemical fertilizers (the mean annual use of urea, potash, and phosphate fertilizers is about 500–700, 200–230, and 300–558 kg per hectare, respectively) can be expected.

Comparisons between our findings and the mean values in Europe and the world (Table 9) led to the conclusion that excluding Cd (which had a concentration lower than that in Europe and the world) and Pb (whose level was lower than the mean level in Europe), all elements had higher values in the studied area compared to Europe and the world.

Facchinelli et al. (2001) concluded that while Cu, Ni, Cr, and Co concentrations are controlled by bedrock, and Pb and Zn content of soil is influenced by human factors. Similarly, Leo et al. (2007) found Cu, Ni, Zn and Cr levels and Zn and Cd levels to depend on bedrock and human factors, respectively. Mico et al. (2006) reported that the amounts Cu, Fe, Ni, Zn, Cr, and Co are affected by bedrocks, but Cd content changes based on human factors such as the use of phosphate fertilizers. Lado et al. (2008) assessed heavy metals in the soils of Europe and concluded that Cu, Pb, Cd, and Zn have

Table 6 Models fitted for factors obtained from principal component analysis and selecting the best interpolation method

Factors	Interpolation method	Model	Nugget	Partial sill	Sill	Mean range	Mean	Root mean square
Factor 1	Ordinary kriging	Exponential	0	1.009	1.009	21035.1	0.027	0.853
Factor 2	Radial basis functions	-	-	-	-	-	0.015	1.003
Factor 3	Radial basis functions	-	-	-	-	-	0.001	0.874
Factor 4	Radial basis functions	-	-	-	-	-	0.002	0.785

high correlations with agriculture and limestone, whereas Cr and Ni levels are related to bedrocks. Inácio et al. (2008) evaluated the soils of Portugal to prepare a geochemical atlas. They suggested revealed that while the concentrations of As and Cr are controlled by parent material, the amount of V depends on mining activities (precipitation of ores). Likewise, Jiachun et al. (2004) indicated As content to be controlled by bedrock but to also be relatively dependent on human resources. They found the concentration of Cd to be affected by not only natural factors but also human activities.

Factor analysis

As it can be seen in the zoning map of factor 1, correlation coefficients of variables in each sample ranged from −0.93 to +2.76. High positive correlations existed between the seven elements in the southeast, southwest, and west where igneous and shale bedrocks were present. In contrast, high negative correlations were seen between the seven elements in the east and northwest on shale bedrocks.

Zoning map of factor 2 shows the correlations between the variables of each sample to vary between −1.24 and +4.40. High positive correlations between Cu and Cd were detected as four blue spots in the north and center of the province on alluvium bedrock. High negative correlations were demonstrated as orange spots throughout the province on shale, igneous, metamorphic, and limestone bedrock.

According to the zoning map of factor 3, correlations of variables in each sample took values between −1.31 and +4.20. High positive correlations were observed between Pb and F on shale and lime bedrocks in the southeast of the province. On the other hand, high negative correlations between the two elements were clear in the west and south (on alluvium bedrock) where agriculture is the most common land use. Other areas had values between these figures.

For factor 4, correlations between variables ranged from −0.84 to +4.14. High positive correlations between Sb and As were observed on shale, lime, igneous, and metamorphic bedrocks in the southeast, northwest, and partly center of the province. High negative correlations occurred on alluvium bedrock in the east of the province where the land is dominantly used for agricultural

purposes. Values in other areas lay between these two figures.

Correlations between these metals can indicate their common source, i.e. bedrock. Factor analysis maps can generally yield useful results about the source of heavy metals. In fact, negative and positive correlations mainly suggest the effects of human activities and geological sources, respectively.

In a study on heavy metal (Zn, Cd, As, Pb, Ni, Cu, and Hg) pollution and landscape patterns, Pin Lin et al. (2002) categorized the metals using factor analysis and then prepared the zoning maps of four factors. They concluded that the first (Ni, Cd, and Cr) and second factors (Pb, Zn, and Cu) were transferred to soil as a direct result of human activities. Similarly, studies in Italy (Facchinelli et al. 2001), northern Spain (Gallego et al. 2002), and other parts of Spain (Rodriguez Martin et al. 2006) classified Cr, Co, and Ni in the same factor and found them to be controlled by parent material of soil.

Moller (2005) reported that Cu in urban areas is produced following human activities. Mico et al. (2006) and Franco-Uria et al. (2009) suggested Cd, Cu, Pb, and Zn concentrations to change mainly through human activities. Rodriguez Martin et al. (2006) mentioned transportation and traffic as the main factors contributing to high levels of Zn and Pb in the soils of Spain. Mico et al. (2006) claimed that Pb is transferred to agricultural soils of Spain via atmospheric deposition and use of chemical fertilizers. On the contrary, Fe, Cr, and Co levels are basically controlled by parent material (Franco-Uria et al. 2009).

FCC analysis

FCC colors the three elements in each group as red, green, and blue. For instance, in a composition of As, Zn, and Pb, blue, red, and green will represent high concentrations of As, Zn and Pb, respectively. High levels of all elements will result in white areas. Black, on the other hand, shows minimum concentrations of all three elements. Therefore, as Figure 8 suggests, these three element had maximum levels in the southeast, northwest, west, and southwest of the province. Cd, Fe, and Vas well as Cr, Ni, and Cu (Figure 8a) had the highest concentrations in the west, southwest, and southeast of the province. In Figure 8b, the green and red areas correspond high levels of Sb and Cd, respectively.

Table 7 Results of spatial correlation analysis and the fitted models

Element	No. sample	Interpolation method	Model	Nugget (C_0) (mg/kg)	Partial sill (C) (mg/kg)	Sill (C_0+C) (mg/kg)	Major range(km)	Mean	RMS	Root- mean standardized	Trend	MAE (mg/kg)	MBE (mg/kg)	Anisotropy Isotropy
As	286	Disjunctive kriging	Exponential	0.4	0.8	1.2	106	0.076	7.085	1.16	no	5.13	0.076	344.5
Zn	286	Disjunctive kriging	Exponential	0.36	0.54	0.9	29.39	−0.059	14.6	1.046	no	11.48	0.059	*
Cu	286	Radial basis function	-	-	-	-	-	.068	9.3	-	-	7.42	0.045	-
V	286	Disjunctive kriging	Exponential	0.31	0.6	0.91	16.919	−0.92	19.22	1.003	no	14.85	.094	*
Ni	286	Ordinary kriging	Spherical	177.29	437.7	614.99	178	.034	15.98	1.068	no	12.54	.34	*
Co	286	Ordinary kriging	Exponential	7.89	10.67	18.56	178.01	0.0005	3.21	1.025	no	2.53	0.005	*
Pb	286	Disjunctive kriging	Exponential	0.39	0.49	0.88	13.49	0.018	6.68	1.09	no	4.06	0.018	*
Cd	286	Radial basis function	-	-	-	-	-	0.002	0.097	-	-	0.67	0.002	-
Fe	286	Disjunctive kriging	Exponential	0.52	0.7	1.22	123.97	0.0005	0.58	1.05	no	0.46	0.00005	*
Cr	286	Ordinary kriging	Exponential	241	669	910	178	0.006	20	1.11	no	16.25	0.006	*
Sb	286	Ordinary kriging	Exponential	0.086	0.111	0.197	178	0.042	0.35	1.14	-	0.74	0.0048	325.5

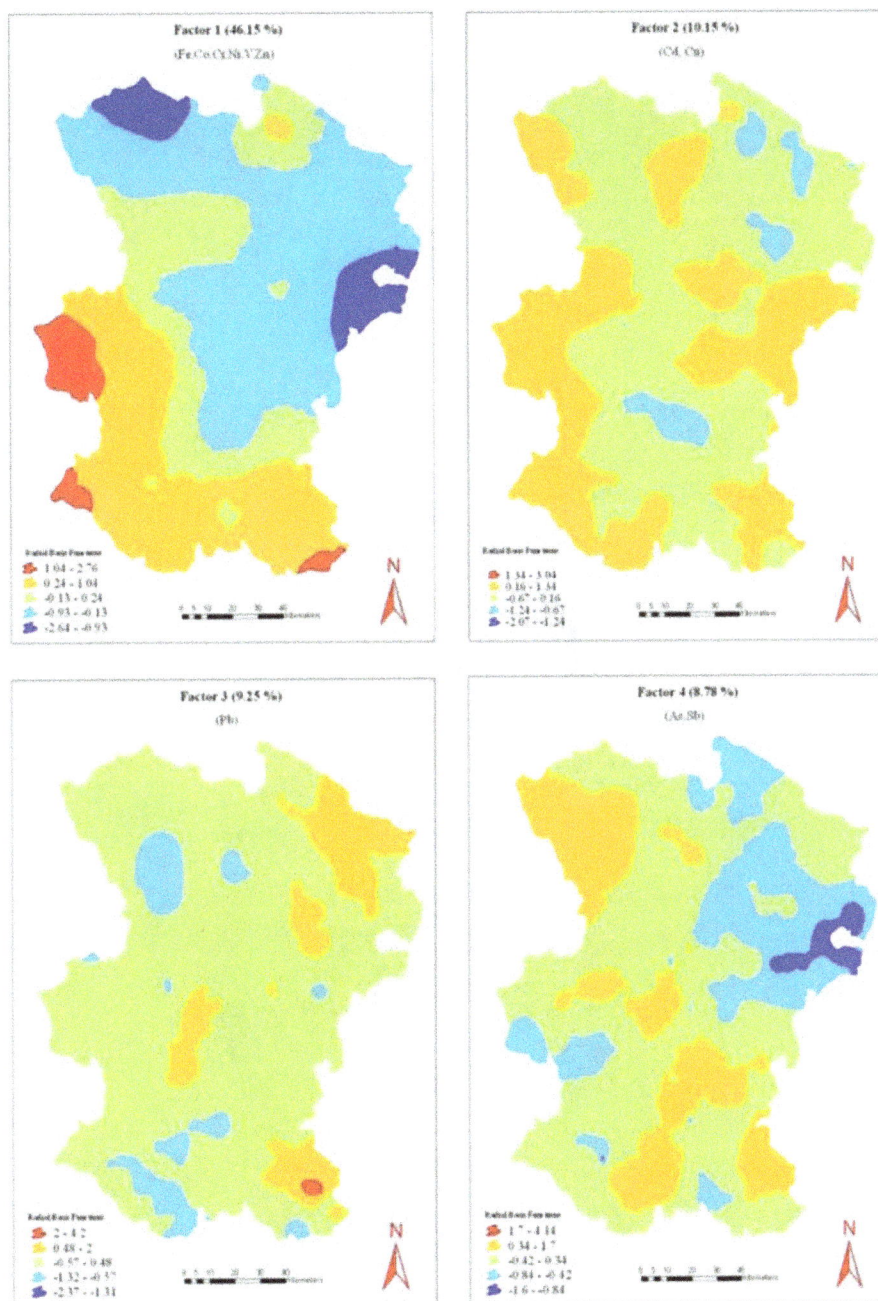

Figure 6 The first factor zoning maps (Zn, V, Ni, Cr, Co, Sb and Fe); The second factor (Cu and Cd) zoning maps; The third factor (Pb) zoning map; and the fourth factor (As and Sb).

In the maps produced by combining single-component images through weighted linear combination (Figure 8c), red areas (e.g. area number 1) illustrate high amounts of all five elements (As, Zn, Pb, Cd, and Sb). These areas were located on shale, alluvium, and igneous bedrocks and indicate the hot-spots of the mentioned elements. Areas 2 and 3 (in purple) have a layer with maximum content and a layer with minimum content. In other words, area 2 contained high concentrations of As, Pb, and Zn but low levels of Cd and Sb which is consistent with its shale, igneous, and alluvium bedrocks. However, the opposite was true in area 3. Blue areas suggested low values in both layers. In Figure 8c, the elements are at their maximum values in the west, southwest, and southeast. Based on our findings, FCC could clearly depict areas with high and relatively low concentrations of heavy metals.

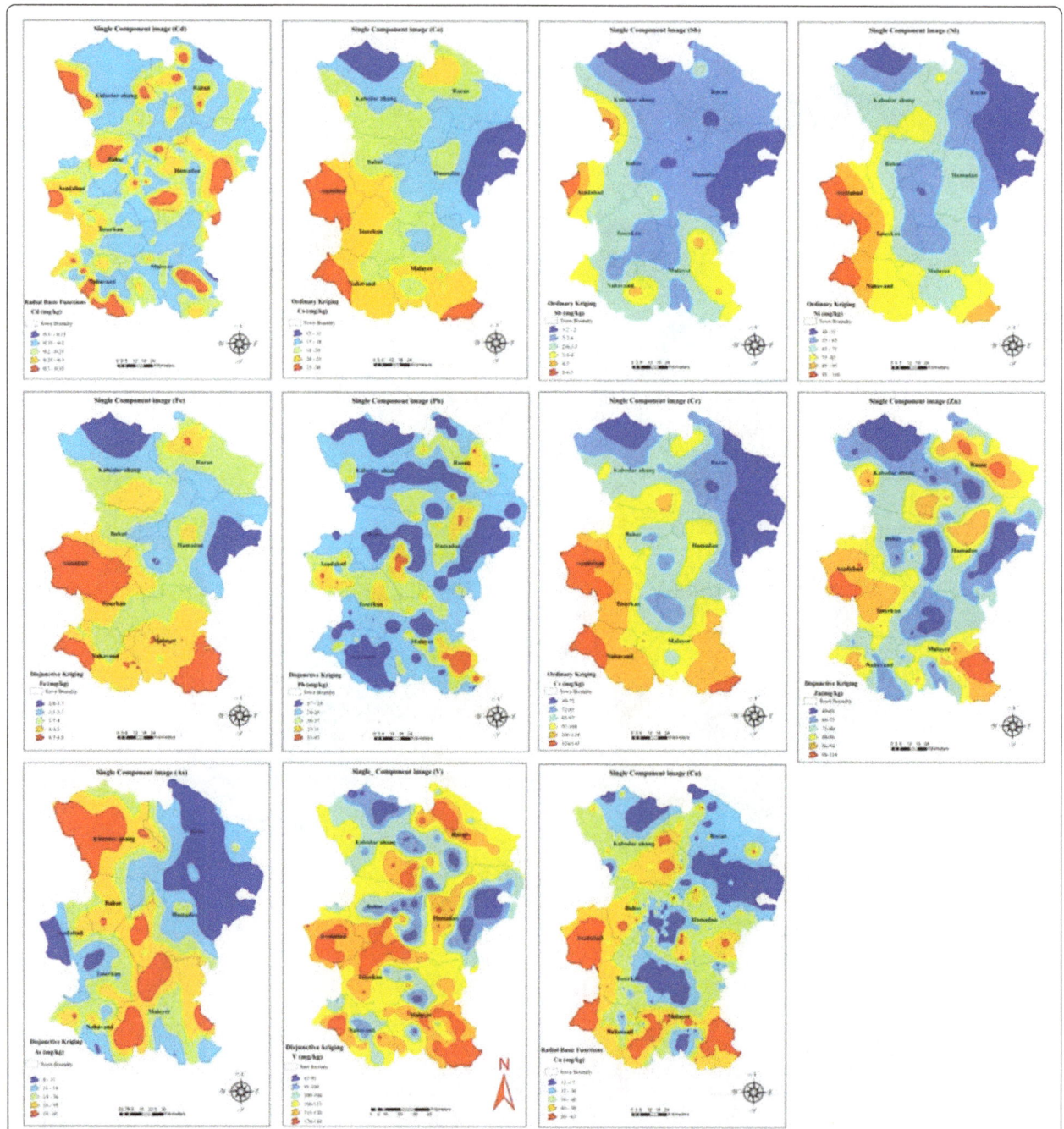

Figure 7 Single-component image of the study area (index values in PPw).

Conclusions

The current research aimed at locating the hot-spots of pollution with heavy metals. Its most significant findings are summarized below:

Comparison of the mean levels of heavy metals in Hamadan Province and other parts of Iran, suggested higher concentrations of As, Zn and V in the studied area than in Isfahan. The mean level of Zn was higher than that in Mashhad but lower than the value in Sepahanshahr.

The mean concentrations of Cd and Pb were lower than the levels in Isfahan and Sepahanshahr, respectively.

Comparisons between the obtained levels and the mean values in Europe and the world (Table 8) revealed that except for Cd and Pb, the concentrations of all other elements were higher in Hamadan Province than in Europe and the world. Overlaying of land use maps and those obtained from sampling revealed that higher concentrations of some elements in their bedrocks compared

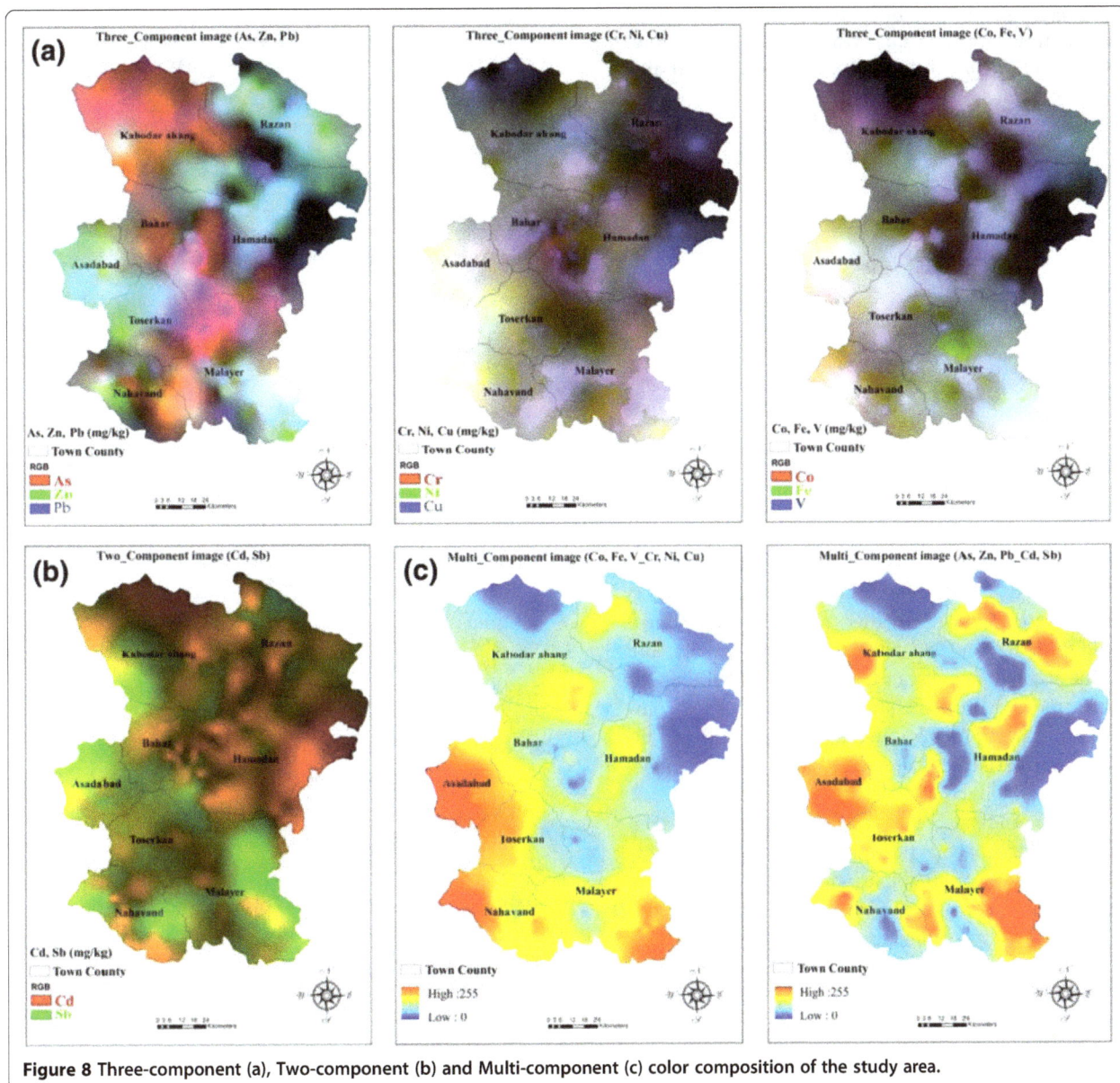

Figure 8 Three-component (a), Two-component (b) and Multi-component (c) color composition of the study area.

to standard levels existed in the agricultural lands where high amounts of chemical fertilizers are used.

Overlaying of zoning maps of the elements and land use and geological layer maps showed that the distribution pattern of the studied elements did not fully conform to the existing land use pattern. On the other hand, as the

Table 8 The mean concentrations of elements in different bedrocks

	Cu	Fe	Zn	Cr	Pb	Ni	V	Sb	Cd	Co
Ultramafic	40	9.6	50	1600	1	-				0.09
Sandstone	-	0.5		35	10	20		<1		
Shale	50	4.7	50-90	90	23	90	90-260	1.5		0.8
Limestone	-	0.98	50	11	-	<5		0.15		

level of each element is naturally high in its bedrock, it seems that the main factor affecting the concentrations of heavy metals in the studied soils must have been the geological structure (bedrock). However, excessive use of chemical fertilizers and industrial pollution (in case of Pb) should not be ignored.

According to factor analysis, the studied heavy metals lay in four factors. This strong correlation can be caused by a common source of transfer to soil through agricultural activities, atmosphere, or parent material of soil. Moreover, while the measured elements had maximum levels in shale, igneous stone, limestone, and sandstone bedrocks, their concentration was minimum in alluvium.

The zoning maps of pollution probability showed 99.65% of the area to have Cd values under the threshold.

Table 9 Comparison of the concentrations of elements in the studied area with the values from the region, Europe, and the world (Facchinelli et al., 2001; Franco-Uria et al., 2009)

	Europe mean	Area mean	World median	Area median
As	11.6	15.79	5	12.5
Cu	17.3	36.2	25	35.5
Fe	2.17	2.94	4	4
Cd	0.284	0.15	0.3	0.11
Zn	60.9	80.03	70	76
Cr	94.8	96.81	80	94
Co	8.91	18.95	10	19
Pb	32.6	25.66	17	24
Ni	37.3	69.03	50	68
V	68.1	109.41	90	110
Sb	1.04	2.9	0.5	2.5

Meanwhile, 100% of the area had Co and Sb concentrations higher than the threshold. Levels of As, Zn, Cu, V, Ni, Pb, Fe, and Cr were higher than the determined threshold in 80%, 90%, 98%, 99.7%, 70%, 98.8%, 85%, and 90% of the studied area. It can thus be concluded that the area had a more appropriate situation regarding Cd compared to other heavy metals. Greater management and filtering programs are essential to control the transfer of Co, Sb, As, Zn, V, Ni, Pb, Fe, and Cr.

Our findings indicated FCC to be an appropriate and fast method for analyzing heavy metal pollution hotspots. Weighted linear combination could also locate the hotspots by clear depiction of areas with high and relatively low levels of heavy metals.

Competing interests

The authors declare that they have no competing interests.

Authors' contributions

All authors read and approved the final manuscript.

Author details

[1]Department of Natural Resources, Isfahan University of Technology, Isfahan 84156-93111, Iran. [2]Environmental Sciences, Department of Natural Resources, Isfahan University of Technology, Isfahan 84156-93111, Iran. [3]Department of Agriculture, Payam-e-Noor University, Ardestan 83818-98951, Iran.

References

(2007) investigation of sludge disposal positioning in Hamadan province, meteorological reports, hydrology, geology, pedology, tectonic seismic prone and integrating to GIS. University of Shahid Beheshti research assistant

Blom HA (1985) Heavy metal contamination of soils. Agri Univ, Norway

Bonham-carter GF, Rogers PJ, Ellwood DJ (1987) Catchment basin analysis applied to surficial geochemical data, Cobequid highlands, Nova Scotia. J Geochem Explor 29:259–278

Bowen HJM (1979) The environmental chemistry of elements. Academic press, London, New York

De vos W, Batista MJ, Demetriades A, Duris MJ, Lexa J, Lis J, Sina K, O' c PJ (2005) Metallogenic mineral provinces and world class Ore deposits in Europe. In: Salminen R, Batista MJ, Bidovec M, Demetriades A, De Vivo B, De Vos W, Duris M, Gilucis A, Gregorauskiene V, Halamic J, Heitzmann P, Lima A, Jordan G, Klaver G, Klein P, Lis J, Locutura J, Marsina K, Mazreku A, O'Connor PJ, SÅ O, Ottesen RT, Petersell V, Plant JA, Reeder S, Salpeteur I, Sandström H, Siewers U, Steenfelt A, Tarvainen T (ed) Geochemical Atlas of Europe: part 1 background information, methodology and maps. Geological Survey of Finland

Eastman JR (2006) Idrisi Andes - tutorial, Clark Labs. Clark University, Worcester, MA

Facchinelli A, Sacchi E, Mallen L (2001) Multivariate statistical and GIS-based approach to identify heavy metal sources in soils. Environ Pollut 114:313–324

Franco-Uria A, Lopez-Mateo C, Roca E, Fernandez-Marcos ML (2009) Source identification of heavy metals in pastureland by multivariate analysis in NW Spain. J Hazard Mater 165:1008–1015

Gallego JLR, Ordonez A, Loredo I (2002) Investigation of trace. "Element sources from an industrialized area (A viles, northern Spain) using multivariate statistical methods". Environ 27:589–596

Hassani-pak A (1998) "geo statistic", the second edition. University of Tehran press, Iran, pp 286–314

Inácio M, Pereira V, Pinto M (2008) The soil geochemical Atlas of Portugal: overview and applications. J Geochem Explor 98:22–33

Jiachun S, Haizhen W, Jiaming X, Jianjun W, Xingmei L, Haiping Z (2004) Spatial distribution of heavy metal in soils: a case study of changing. China. Environ Goel 52:1–15

Lado LR, Hengel T, Reuter HI (2008) Heavy metals in European soils: a geostatistical analysis of the FOREGS geochemical database. Geoderma 148:189–199

Lame B, Leenaers L (1997) International ash working group. EEA, Brussels. 1999

Lindsay WL (1979) Chemical equilibria in soils. John Wiley & Sons, New York

Mico C, Recatala L, Peris M, Sanchez J (2006) Assessing heavvy metal sources in agricultural soils of an European Mediterranean area by multivariate analysis. Chemosphere 56:863–872. doi:810.1016/J.Chemosphere. 2006.1003.1016

Mohammadi J (2006) "Pedometry", the second volume, spatial statistic. pelk press, pp 262–279

Moller A (2005) Urban soil pollution in Damascus, Syria: concentrations and patterns of heavy metals in the soils of the Damascus Ghouta. Geoderma 124:63–71

Pin Lin Y, Teng TP, Chang TK (2002) Multivariat analysis of soil heavy metal pollution and landscape pattern in Changhua county in Taiwan. Landscape Urban Plan 62:19–35

Rodriguez Martin JA, Lopez Arias M, Grau Corbi JM (2006) Heavy metal contents in agricultural topsoils in the Ebro basin (Spain): application of multivariate geostatistical methods to study spatial variations. Environ Pollut 144:1001–1012

Romic M, Hengl T, Romic D, Husnjak S (2007) Representing soil pollution by heavy metals using continous limitation scores. Comput Geosci 33:1316–1326

Suyash Kumar KD, Shirke N, Pawar J (2007) GIS-based colour composites and relays to delineate heavy metal contamination zones in the shallow alluvial aquifers, Ankaleshwar industrial estate, south Gujarat, Ipdia. Environ Geol 54:117–129

7

Policy analysis of China inland nuclear power plants' plan changes: from suspension to expansion

Jinxin Zhu[1] and Gail Krantzberg[2*]

Abstracts

Background: China's inland nuclear power plants plan has been suspended until 2015 since Fukushima disaster. The policy on inland nuclear power plants becomes uncertain. This paper provides an overview of inland nuclear power plants the safety grantee, economic power on diminishing disparities between Western China and Eastern China, efforts on environmental improvement with reforming of energy restructure and essential public participation. The paper further discusses the government's current policy and successful experience from other countries.

Results: The paper gives the recommendations for promoting inland nuclear power plants' expansion, making economy develop healthily, improving environment and introducing public hearing into the nuclear power development.

Conclusions: With proper political guiding, China could obtain significant benefits from expanding nuclear power plants from coast to inland.

Keywords: Inland nuclear power plants; Public policy; West development strategy; Energy restructure

Background

With China's rapidly developing economy in these years, the problems of power supply restriction, an energy bottleneck, oil and coal supply tension, and so on are emphasized. At the same time, large-scale use of conventional energy sources leads to increased environmental pollution especially in air pollution and GHS emission. In order to raise the proportion of clean energy and catch up with the changing trends, Chinese government makes its mind to accelerate the development of nuclear power as soon as possible.

Ambitious goals for NPPs expansion have been set by China government. By 2020, China must reach a 40-GW nuclear power generation capacity, have 18 GW of additional nuclear power capacity under construction, and ensure that nuclear power takes approximately 4% of electricity generation (i.e., 260–280 billion kWh) (Hou, et al. 2011). However, since Fukushima disaster in Japan,

constructions of new nuclear power plants (NPPs)'s plan have been suspended from March 2011.

According to China's national nuclear plan, China will resume construction of new NPPs projects. Nevertheless, it had lowered its target for the construction of NPPs by 2015, notably by not building additional nuclear reactors at inland locations. Nuclear Power Safety Plan (2011-20) as well as the Mid-term and Long-term Development Plan for Nuclear Power (2011-20) was discussed and passed through executive meeting on Oct 24th, 2012. During the 2011-2015 periods, not any nuclear projects will be constructed in inland regions, but only a few projects in coastal areas that have gone through justification processes will be constructed (Xinhua 2012).

While acknowledging that nuclear power expansion plan in China presently is exhibiting symptoms of extreme stress from a combination of sources that include safety concern, economy concern, and environment concern. Moreover, public concern of NPPs' safety is not ignoring due to Fukushima incident. This paper is going to analyze factors related to China inland NPPs' strategies and discuss the current government policies, potential changes

* Correspondence: krantz@mcmaster.ca
[2]Engineering and Public Policy in the School of Engineering Practice, McMaster University, 1280 Main St., W Hamilton, ON L8S 4 L7, Canada
Full list of author information is available at the end of the article

to current policy from 2012 to 2015, the future of nuclear power after 2015, and the barriers to nuclear power development in China. Some recommendations are also provided.

Inland location NPPs from developed countries and China's high temperature Gas cool reactor technology (safety concern)

Other countries' primary considerations for NPPs construction are economic and social factors like local demand of electricity, and the situation of the primary energy and power distribution, there is not a distinction in safety between coastal and inland NPPs in the site selection. As a truth, countries like the Canada, France and United States all build a lot of NPPs in inland areas, and the amount of generating reactors and total installed capacity are much higher than coastal NPPs (Dynabond Powertech 2010). General information of percentage of inland reactors is shown in Table 1.

From regulations on NPPs of International Atomic Energy Agency and main nuclear reactors possessing countries, the security objectives and evaluations criteria on coastal nuclear power stations and inland NPPs are identical. There is no difference between regulations of inland NPPs and coastal NPPs' developments. No countries and organizations had ever post special requirements on inland nuclear power plants' safety (National Energy Administration 2012). These countries' operating experience of NPPs demonstrates that safety of inland NPPs can be fully guaranteed.

From Table 1, of the total 442 nuclear stations around the globe, 50 percent are located inland. In Canada, that number is at a striking 86 percent.

After Fukushima disaster, safety of traditional nuclear power plant technology is in fierce debate all over the country, construction work on a power plant using domestically developed, fourth generation technology has already began in Rongcheng City, Shandong Province in

2012. Because the project gained approved prior to the State Council's decision to impose a moratorium, it were unaffected by suspension of new NPPs project approvals VelkerMatt (2011). Fourth generation reactors represented by high temperate gas-cooled reactors are main trend of NPPs development in China.

China developed the fourth generation technology entirely. In 2003, China's first experimental modulized pebble bed high temperature gas-cooled reactor plant was designed, built by Tsinghua University and began to operate and achieved full capacity operating and supplying power with 10,000-kilowatt to the grid. In terms of its safety, the fourth generation is obviously superior to the former generations' technology, because it uses helium coolant's natural convection to control the reactor even in high temperatures (NGNP Industry Alliance 2012). The pressured water reactors units we see today are massive. However, the actual reactor itself is only about one thousandth the size of the total unit in reality. With the reactor made sufficiently large, it can radiate heat more easily, meaning no power and no water or other cooling fluid is required for high temperate gas-cooled reactor. If normal heat transport systems are not available, heat removal from the reactor occurs naturally and directly to the earth. Even if normal cooling systems are not functional, it takes days to reach limited temperatures (still under temperatures that could melt reactor core) with low energy density of the reactor core and the large heat capacity of the graphite structure. Therefore, reactors can be built and operated independent of water source.

Results and Discussion
Economy concern

For planning purposes, China's 31 provinces are divided into three macro-regions—eastern, central and western (see Figure 1). The developments of three regions are ranked with respect to the greatest in the first and the last in the third. Two kinds of disparities between Western China and Eastern China pervade China. The first one is regional disparities that focus on levels of economic development and it varies considerably from Western China to Eastern China. The second is referring to mismatch in energy resources supply and demand. Eastern China suffers severe shortages while Western China is blessed with significant surpluses (HuangTodd and ZhangDaniel 2011).

One important indicator, Vw coefficient, confirms the existence of regional disparities. The coefficient is used by Williamson (1965). With the regions being represented by China's provinces, it is a measure of variation in weighted GDP per capita the proxy for development—across a country's set of regions. A Vw of zero means the ideal outcome of an absence of any regional departures from a national average GDP; a coefficient of unity represents that the localization of wealth in a single region while all others were destitute. A Vw analysis fortheperiod 1952–2008

Table 1 General information of world's inland reactors (National Energy Administration 2012)

Nations	Number of inland reactors	Number of total reactors	Percentage of inland reactors
Global	220	442	49.8%
USA	64	104	61.5%
France	40	58	69.0%
Canada	12	14	85.7%
Germany	14	17	82.4%
Russia	18	31	58.1%
Ukraine	19	19	100%
China	0	11	0%
Others	53	199	26.6%

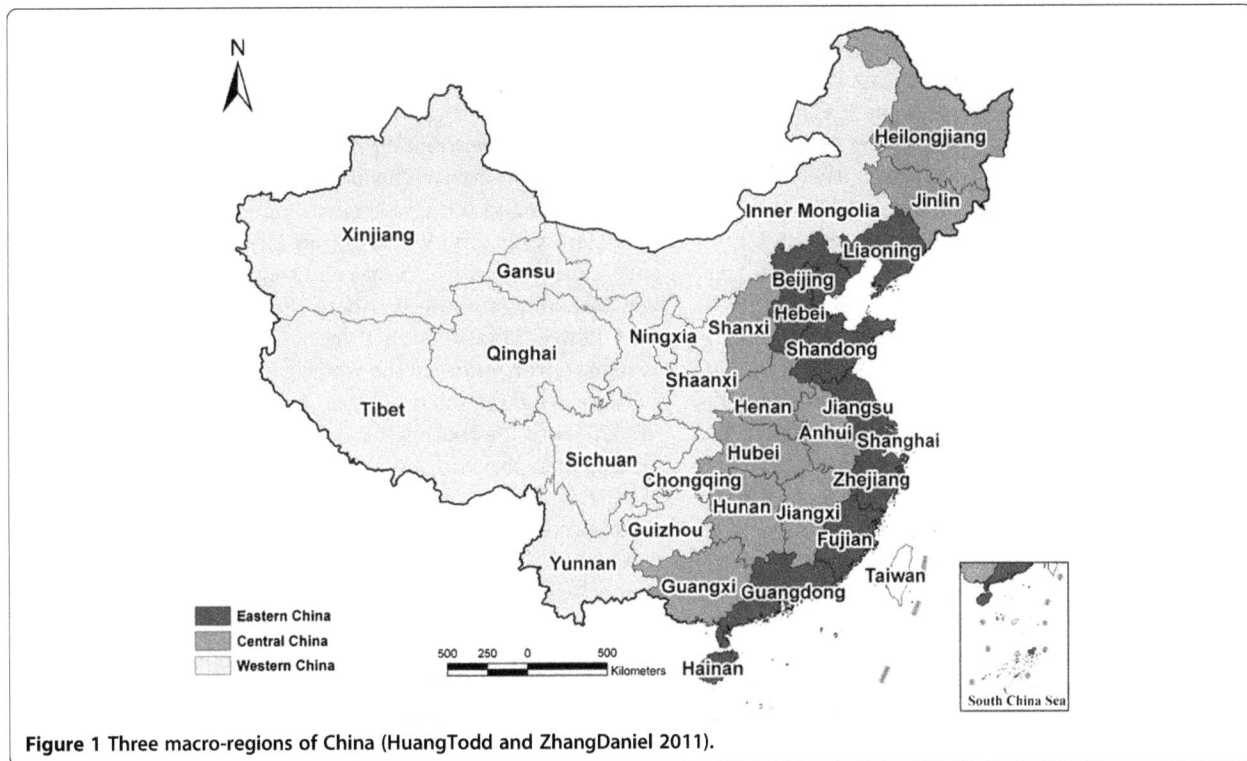

Figure 1 Three macro-regions of China (HuangTodd and ZhangDaniel 2011).

(with per capita GDP standardized on the former year, see Figure 2) demonstrates that Chinese regional disparities have increased shapely and consistently exceeded 0.7 since 1970 (HuangTodd and ZhangDaniel 2011). Little could be detected in the way of gap closing for much of that period, but some trend towards reduction in disparities has become discernible since the West Development Strategy initiated.

This disparities problem not only causes the economic issues to Western China, but also makes the environmental and social issues more severe in that region. Due to the underdeveloped economy, Western China is too dependent on mining energy or non-energy minerals and the over mining makes the fragile ecosystem even worse. Its education and welfare stay in quiet low level for decades because of lack of investment. These issues deter the region's

economic development in reverse. In national scale, disparities between Western China and Eastern China in China threat its well-being and integral development. To balance the development of Western China and Eastern China, China government established Western China Development Strategy in 1999 (HuangTodd and ZhangDaniel 2011). This strategy are mainly focusing on pulling demand and market of Western China, reforming the energy structure which is going to be discussed in environmental concern to conserve the ecosystem in this region, keep the society stable, and secure the borderland of China.

Nuclear plant and its construction will boost local economies because of its significant economic benefits. From direct and secondary effects, the average nuclear plant produces approximately \$470 million economic output in the local community and nearly \$40 million in

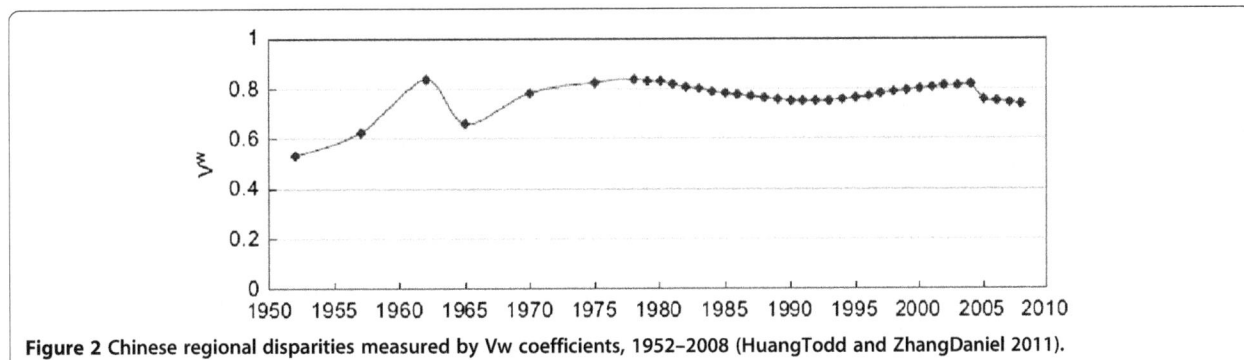

Figure 2 Chinese regional disparities measured by Vw coefficients, 1952–2008 (HuangTodd and ZhangDaniel 2011).

total labor income for each year (Nuclear Energy Institute of United States 2012). The direct effects reflect the labor, goods and services of plant's expenditures. The secondary effects include subsequent plant expenditures related to the presence of the plant and its employees that penetrate into the local economy. Analysis illustrates that $0.04 net benefits in the local community come from every dollar spent by the normal NPPs' construction. The NPPs generates nearly $16 million in local tax revenue for each year (Nuclear Energy Institute of United States 2012). The tax revenue could benefit education, transportation, and other local infrastructure. Nuclear power plants provide substantial economic benefits during their decades of operation. The taxes, direct and secondary expenditure from NPPs strengthen the economies of local communities and hosting a new nuclear plant will give a similar boost to any community. With different plant size, a new nuclear plant represents an investment ranging from $6 billion to $8 billion including interest (Nuclear Energy Institute of United States 2012). Moreover, construction of a new nuclear power plant will provide a substantial boost to manufacturers and suppliers of commodities such as steel, concrete and cement.

Current Chinese nuclear expansion plan that only covers the coastal regions is going to aggravate the disparities between these two regions. For NPPs' strong power on boosting economy, they can be a powerful tool to cooperate with Western China Development Strategy to realize its main purposes.

Restructure of energy (Environment Concern)

In China, coal covering 60% of the primary energy and 75% of the power generation leads to irrational energy structure. The coal based energy structure causes serious problems of environmental pollution and emission of greenhouse gases (GHG). China has become one of the most polluted countries in the world. For example, China emitted 20 Mt of SO_2 and 3 Gt of CO_2, both ranked the world No. 2 in 2006. Environmental pollutions cause the economic loss accounting for 3–7% of GDP in China. If these problems could not be solved properly and quickly, they will hinder the further development of the country. The China Energy Policy (2012) published by the State Council said that China decided to make up to 45 percent reduction of its carbon dioxide emissions by 2020 (Tiantian 2012). To achieve that goal, China has to shut down numerous coal plants and replace them with clean energy plants.

Simultaneously, compared the consumption of energy with east which is always in energy crisis, west only took less than 30% of total. As shown in Figure 3.

The West-East Power Transmission Project aims to turn the rich energy resources in West China into electricity and to transmit the electricity to East China, the economically dynamic and energy-deprived region of the country (Ouyang, and Changyu 2002). It is an important part of the West Development Strategy of the Chinese government, which will enhance the optimization of energy resources contributing to the healthy development of the national economy. This project will not only solve the energy crisis and the severe environmental contamination of Eastern China but will also transform the abundant resource of Western China into an economic advantage that will help mitigate poverty there. In this way, the West-East Power Transmission Project will promote the sustainable development of the whole country.

There are three main national power flows from Western China to the southeast coastal areas, in particular, the Guangdong province and Shanghai city referring to the West-East Power Transmission Project (Kang et al. 2012). As shown in Figure 4.

As a result of the rapid expansion of the West-East Power Transmission Project, air pollution is serious in the northern passageway region mainly where consists of coal-fired power plants, especially for Eastern China, and the main air pollutants are PM and SO_2 (Zhigang et al. 2005). Thus, the government and public have shown great concern about the proposed power plants in this region's impacts on environment. It is very important, therefore, to understand the impacts of building new clean power plants on improving the air quality in this region.

Nevertheless, Western China's energy shortfalls have surfaced with respect to remaining reserves. China's total remaining oil reserves 1.48 billion barrels by the end of 2009. The reserves/production (R/P) ratio of it was 10.7 years, which was far below the world's average oil R/P ratio of 45.7 years. In terms of remaining natural gas, it reserves totalled 2.46 trillion cubic meters with an R/P ratio of 28.8 years, which is lower than the world's average natural gas R/P ratio of 62.8 years. Remaining coal reserves 1.145 billion ton in total with an R/P ratio of 38 years which is far below the world's average of 119 years (British Petroleum (BP). 2011). Western China's energy shortfalls will not only affect local development, but also will threat the whole country's economy growing and environment improving. Needs for clean energy plants planned in these regions are urgent now.

Public concern

In addition to technical and economic factors, the public understanding and acceptance to the nuclear power play an important role in the development of nuclear power. The familiarity of public towards nuclear power that is terminology of public understanding is an important impact factor on public acceptance. With the increased familiarity with nuclear power, the proportion of people who consider NPPs' safety was corresponding increased. The percentages of public, scientists, energy scientists

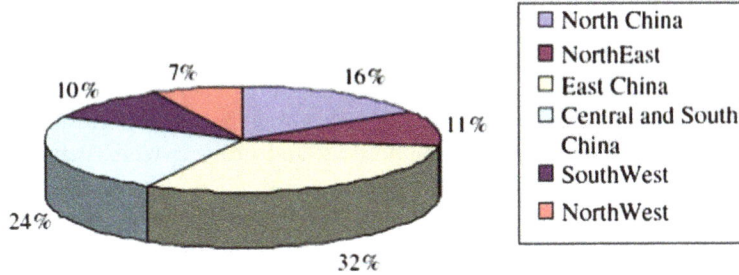

Figure 3 Electricity consumption by region (2000) (China Statistics Press 2002).

and nuclear experts who believed the NPPs' safety were respectively 40%, 60%, 76% and 99% (Barke RP HC Jenkins-Smith 1993). A study on the level of knowledge and attitude of nuclear power among Chinese residents around the Tianwan nuclear power plant within a radius of 30 km in Jiangsu of China shows the level of understanding and acceptance of nuclear power was generally not high (Ningle et al. 2012). Other 10 operating NPPs in China have the same situation.

Low level of understanding and acceptance of nuclear power from public will cause serious civic activism like local protest groups. Main issue prompting some local groups and communities to oppose nuclear projects is sensitivities over nuclear fuel leak risks. Local protest groups and adverse public opinion are emerged to challenge the legitimacy of proposed projects by a strong NIMBY (not-in-my-back-yard) element, especially in rural areas (International Atomic Energy Agency 2002). Although Chinese history is rife with prominent examples of civic activism, opposition groups have been unable to stop nuclear power

development. When a small group from local communities tried to protest licensing of the Daya Bay plant, officials' detained and arrested them. When more than one million people in Beijing signed a petition against nuclear power, Chiang Hsin-Hsiung, the Minister of Nuclear Industry, rebuffed the complainants by issuing a statement that the government "the unscientific objections from some people would not halt the Daya Bay project". When United Kingdom authorities and the Hong Kong Civic Association came out a series of reports questioning the safety of Daya Bay, the government answered "reports compiled by other countries do not constitute legal documents" and thus held no sway over the licensing process (Zhou SXZhang 2010). With no nuclear power mishaps happening, civic activism against nuclear power has waned since the late 1980s. However, the recent events at the Fukushima Daiichi nuclear power plant in Japan have resulted in a great focus on understanding both the safety of nuclear power and the public understanding and acceptance of nuclear power (FoE, F.o.t.E.,

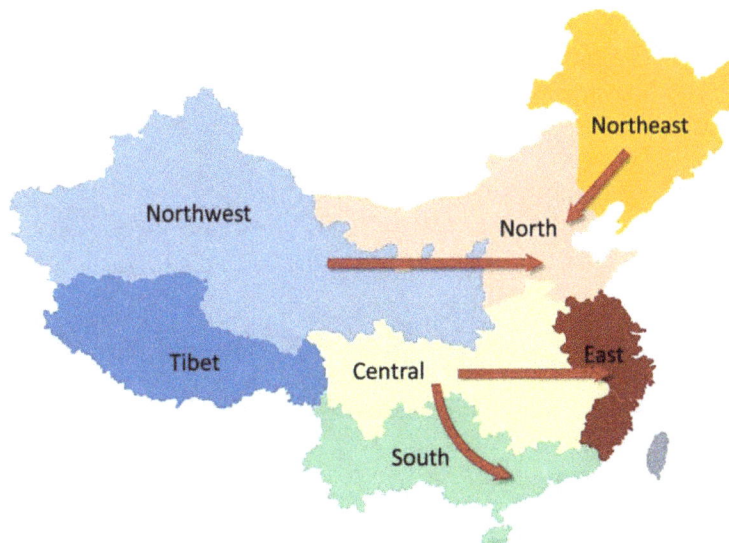

Figure 4 China's national power flow pattern (Ouyang and Changyu 2002).

GfKNOP 2011). Coercion policies of Chinese government like these, nowadays, lose their power on controlling of society and result in serious potential problem for social security.

On August and October 2012, plans to expand petrochemical plants in northern and eastern China have been shelved after days of protest (BBC NEWS 2012). The plants are producers of PX (a carcinogenic petrochemical used to create raw materials for the production of polyester film and fabric). Tens of thousands protested against chemical plants in Dalian city and Ningbo city. Environmental protests have been more common and influential in China. Many Chinese are becoming more environmentally aware and are deeply concerned about pollution and living quality surrounding them. China policy makers started to pay attention to public voice on NPPs. There are some changes made in its official documents. On March 2012, the National Nuclear Safety Administration

(NNSA) announced "if the environmental impact report does not have a public participation chapter, NNSA would not accept the report. According to the announcement of China's State Environmental Protection Department (SEPA), public participation will run through the whole process, and projects which have strong objections are likely to be stopped." (Dynabond Powertech 2010).

To deal with public issues, experience from other countries is quiet valuable for China. The UK Government's latest National Policy Statements for Energy Infrastructure recognised "there may be positive and negative effects from the construction of new nuclear plants and that organisations applying for site license should identify these effects in their license application." (DECC 2010). Although there are no specific comments on mitigating or avoiding these effects, the activity involved public participation is still widely adopted in nuclear power area in UK. The UK Health and Safety Executive has invited public

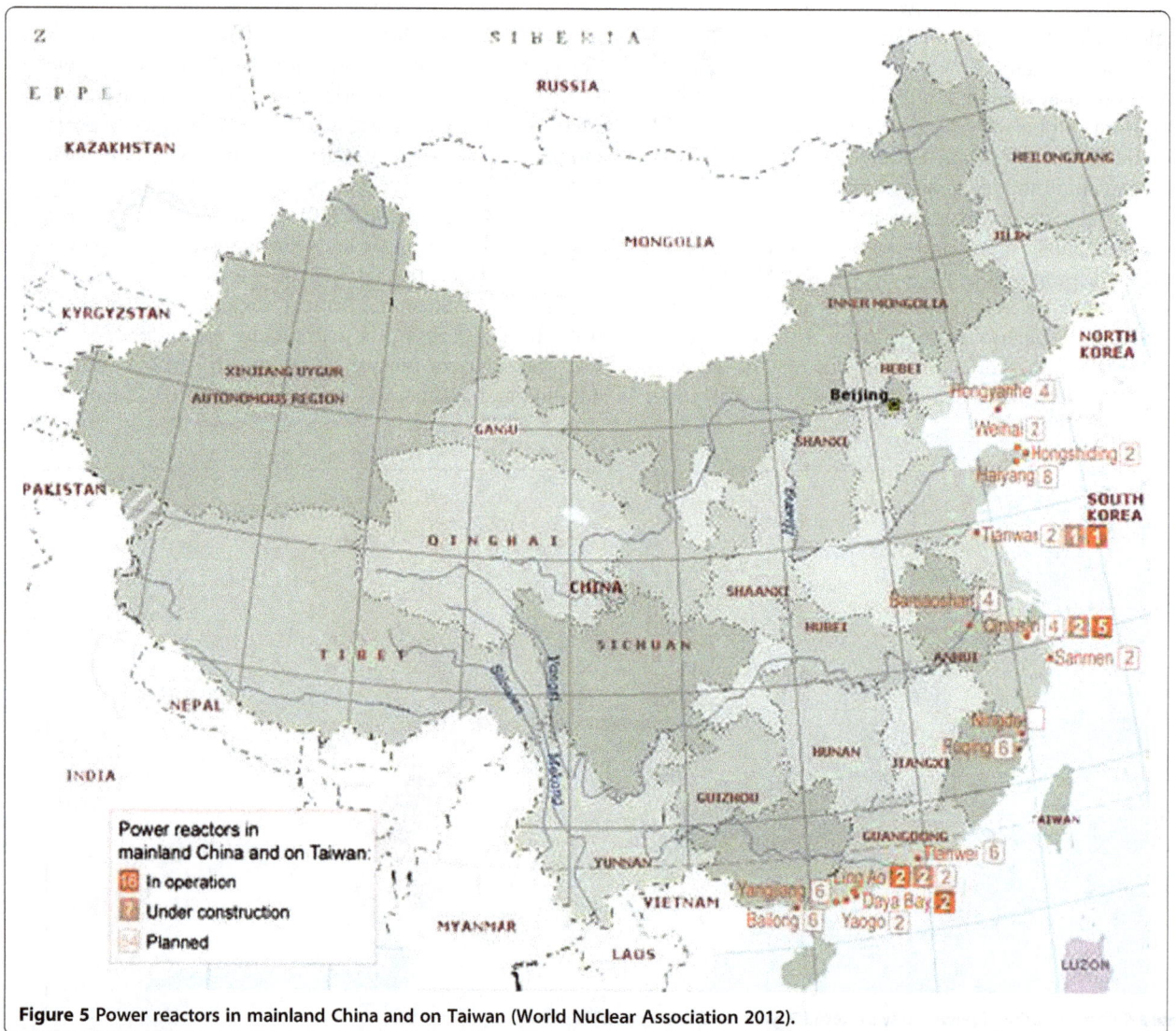

Figure 5 Power reactors in mainland China and on Taiwan (World Nuclear Association 2012).

input into its Generic Design Approval (GDA) process for the assessment of new nuclear reactor designs (HSE 2009). However, as the designing or licensing is almost finalised, any changes are might very costly and time consuming, and it is rare that such changes that are not critical for the safety will be made, even if they have potential ability to enhance the socio-economic benefit of the plant. Therefore, making changes earlier in the designing and licensing process would be less costly and more efficient.

Methods

After Fukushima disaster in Japan, China NPPs' expansion map changed as shown in Figure 5. Only coastal NPPs passed the new nuclear power plant project approvals. This is because dumping radioactive contaminations into sea has much lower risk and cost than dumping into river once the leaking happened in NPPs. However, only selecting locations for NPPs in coastal areas will not be able to meet the future needs of China's sustainable development. To maintain the energy supply for inland provinces which are suffering energy shortfalls and environmental contamination, NPPs are urgently need to be constructed and planned in these regions (Dynabond Powertech 2010). Aforementioned safety concern is the premise for satisfying this urgent need. The successful development of inland NPPs in several leading country such as USA, France and Canada shows that safety is not a main barrier for China to overcome to cover its inland province with nuclear power. Besides, the maturity of China self-developed high temperature gas cool reactor technology will make the siting process of NPPs independent from the limitation of closing to shore of river, lake or coast. Government has large flexibility to cover the benefits of NPPs expansion plan with inland provinces.

But the Fukushima disaster does reveal some designing flaws in NPPs like insufficient emergency producers design. So, revaluation and reexamination is necessary to the initial design of inland NPPs during the 3 years suspension from 2012 to 2015. However, if inland NPPs construction is put off easily by a slight risk, it would be a huge loss of great opportunities to achieve sustainable economy for China.

Disparity between west and east in China hinder the economy from developing healthily. Primary barrier is energy and resource misplacing as mentioned in previous section. West has main fossil fuel resources storage in China.

Misplacing and energy shortfalls make Western China's economy much more depressed and make disparity from Eastern China more severe. Following Western China Development Strategy, NPPs should be planned at inland locations. Not only boost local economy at west, but also improve the mobility of factors of production between west and east. The economic and trading cooperation and the mobilization of capital between Western China and eastern regions are the productivity of capital and could complement technological advances in Western China of the country. With these benefits, construction and planning of inland NPPs will contribute to diminish disparities and eliminate the risk of the economic gap between Western China and Eastern China being lager. Therefore, improve the society stability in Western China and remit the contradictions in whole country.

Western China-East Power Transmission Project has been proved to make a great contribution to environment improvement of China by reducing the transmission and burning of fossil fuel. However, coal fired plants from northern passageway of the project's environmental problems are still serious and these problem have the trend to

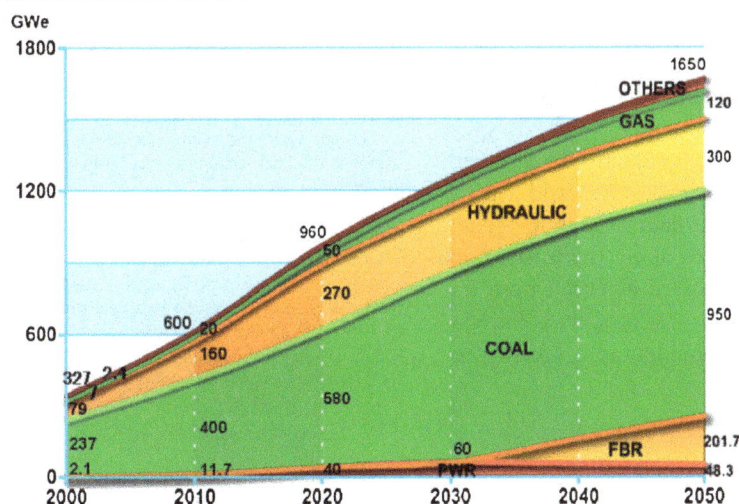

Figure 6 The electricity capacity development envisaged in China by 2050.

be worse due to the big energy gap for both west and east. To mediate the energy supplying intensity and improve the air-quality in western regions, connecting clean energy generators to electricity grids of northern passageway would be dispensable. In perspective of Chinese electricity capacity development, percentage of coal-fire electricity will decrease with respecting the growing percentage of nuclear power (see Figure 6). In terms of nuclear power's consistency supplying and low cost to generate characteristics, it is most stable and affordable power resource for west compared with solar and wind power. Policy makers should consider increasing the nuclear power's percentage of total energy transmission.

By involving the public at an earlier stage than is currently common, a nuclear plant can be built on a foundation of social consensus from most leading countries. Currently, interactions between the public and nuclear industry start with the licensing process and continue through to decommissioning in English. By extending this interaction to earlier stages, specifically during requirements analysis and plant design, it ought to be possible to increase public understanding of nuclear power and minimise the probability for disruption to new-build activities caused by misunderstanding and mistrust. It may not be possible to integrate all of the public views into a nuclear plant design, however the process of dialogue involved in such discussions can only reinforce current educational campaigns and help in building mutual trust. It is essential for keeping the social in Western China stable and the premise for promoting the inland NPPs' expansion.

For public participation, the operating and under constructing NPPs in eastern regions are all missing in their designing and licensing processes. The proposed NPPs projects in eastern regions are facing complicated public problems integrated with their affecting sources included the high population intensity, diverse education background and family income. In term of western regions' public problems, they are simpler than Eastern China's because communities in Western China has low population density, similar education background (average is quiet low) and not big difference in family income. To deal with public issues, west would be amore ideal test field. Having the experience and experimental data for public participation from west, next step is to apply them to future proposed projects especially located at east where has much more intricate situations. The process is like going from easy to hard. All stakeholders easy accept progressive reform.

Conclusions

Based on the analysis in this paper and experiences from other countries, several recommendations to promote

inland nuclear power development in China are given as follows:

- Inland Nuclear Power Plant plan should be written into Western China Development Strategy. After 2015, inland NPPs construction should be resumed. Under Western China-East Power Transmission Project, the electricity that is generated from the starting point of northern passageway grid must have no less than 5% from NPPs by 2020.

- During 3 years suspension from 2012 to 2015, we should attach great importance to the public acceptance of NPPs, enhance communication with the local government and local people by holding public hearing and meeting. In early designing stage, government is responsible to hold a public hearing for enhancing nuclear safety awareness, eliminating people's concerns, and creating a good social environment for the development and construction of nuclear power projects. In licensing stage, individuals who are directly affected by any licensing action involving a facility producing or utilizing nuclear materials have rights to ask for public hearing. Central government should give legislation support for all three hearings by establishing the formal regulations and local governments will be in charge of arranging the time and places to hold hearings.

- From 2015 on, all proposed NPPs projects will be evaluated their public participation chapter through early designing and licensing process by China's State Environmental Protection Department (SEPA). Reports without or forged public part will be rejected immediately. Public participation will run through the whole process and projects that have strong objections must be stopped.

Competing interests
The authors declare that they have no competing interests.

Authors' contributions
ZJ designed the study, collected and interpreted data, conducted data analysis and wrote the manuscript. KG revised the manuscript critically and approved the final version to be published. Both authors read and approved the final manuscripts.

Authors' information
ZJ's background is civil engineering. He received his master degree of engineering and public policy in Sept, 2013. Now, he is a PhD student under the supervision of Dr. Gordon Huang in University of Regina. He is performing as a research assistant in the Institute for Energy, Environment and Sustainable Communities.
KG is the Professor and Director of the Centre for Engineering and Public Policy in the School of Engineering Practice in McMaster University. KG was the direct supervisor of ZJ from 2012 to 2013.

Acknowledgements
We would like to thank the individuals who have given their time and expertise to in reviewing and offering advises as follows: Huijing Xu (Shanghai, China), Charles Jin (Hamilton, Canada) and Younggew Kim (Seoul, Korea).

Author details

[1]Institute for Energy, Environment and Sustainable Communities, University of Regina 3737 Wascana Parkway, Regina, SK S4S 0A2, Canada. [2]Engineering and Public Policy in the School of Engineering Practice, McMaster University, 1280 Main St., W Hamilton, ON L8S 4 L7, Canada.

References

Barke RP, Jenkins-Smith HC (1993) Politics and scientific expertise: scientists, risk perception, and nuclear waste policy. Risk Analysis, pp 425–439, McLean

BBC NEWS (2012) China Protesters Force Halt to Zhejiang Factory Plan. www.bbc.co.uk

British Petroleum (BP) (2011) Statistical review of world energy 2010. http://www.bp.com/statisticalreview

China Statistics Press (2002) China Statistical Yearbook 2002. www.stats.gov.cn

DECC (2010) Consultation on the revised draft NPSs for Energy Infrastructure. London

FoE, F.o.t.E.,GfKNOP (2011) Post Fukushima public perception of nuclear power. Energy of Japan, pp 42–50. Japan

Hou J, Zhongfu T, Jianhui W, Pinjie X (2011) Government policy and future projection. J Energy Eng 137:151–158

HSE (2009) New nuclear power stations, Generic Design Assessment, Safety assessment in an international context. Health and Safety Executive, Merseyside

HuangTodd Y, ZhangDaniel L (2011) Capitalizing on energy supply: Western China's opportunity for development. Resour Policy 36:227–237

International Atomic Energy Agency (2002) People's Republic of China. country profiles. pp 214–235, www.iaea.org

Kang C, Zhou T, Chen Q, Qianyao X, Xia Q, Ji Z (2012) Carbon emission flow in networks. Sci Rep:34–45, Beijing

National Energy Administration (2012) Discussion on Safty of China Inland NPPs. National Ennergy Administration Energy Science, Beijing

NGNP Industry Alliance (2012) The High Temperature Gas-Cooled Reactor (HTGR) - Safe, Clean and Sustainable Energy for the Future. High Tempreture Gas Rectors, pp 11–15

Ningle Y, Zhang Y, Wang J, Cao X, Fan X, Xiaosan X, Wang F (2012) Knowledge of and attitude to nuclear power among residents around Tianwan nuclear power plant in jiangsu of China. Int J Med Sci:361–369

Nuclear Energy Institute of United States (2012) Nuclear Power Plants Contributes Significantly to State and Local Economies. Reliable and Affordable Energy

Ouyang, and Changyu (2002) Implement the "Power Delivery from the West to East" strategy to promote the economy development of East and West China. Electric Power, pp 11–15, in Chinese. Beijing

Powertech D (2010) Study on China's inland NPPs. Dynabond Powertech Service

Tiantian B (2012) Nuclear coastline. Global Times

VelkerMatt (2011) Construction of China's 4G nuclear reactor to start soon. www.china.org.cn

World Nuclear Association (2012) Nuclear Power in China. www.world-nuclear.org/info/inf63.html

Williamson, Jeffrey G (1965) Regional Inequality and the Process of National Development: A Description of the Patterns. Economic Development and Cultural Change no. 4 (Julho):1–84

Xinhua (2012) No inland nuclear power plants to be built during 2011-15: statement. People's Daily Online, Beijing

Zhigang XH, Fahe C, Ning D, Yizhen C, Jindan L, Fu C, Simei L, Wenqing PJ (2005) Air quality impact of the coal-fired power plants in the northern passageway of the China west-east power transmission project. J Air Waste Manage Assoc 55(12):1816–1826

Zhou SXZhang (2010) Nuclear energy development in China: a study of opportunities and challenges. Energy. Energy 2010, in press. Shanghai

8

An inexact two-stage dynamic stochastic model for regional electricity and heat supply management with pollutants mitigation control

Wei Li[1*], Xiaoyu Liu[2], Guanzhong Sun[2], Ling Ji[3] and Guohe Huang[1]

Abstract

Background: Energy system management is an important tool for regional energy and environmental development, and many parameters and their interrelationships in energy-environmental management model appear complexity and uncertain. How to deal with these uncertainties and make a reasonable decision schemes are desired for managers.

Results: In this study, an inexact two-stage dynamic programming model is developed for regional electricity and heat supply management under considering the complexities and uncertainties in regional energy system. The model can reflect not only uncertainties expressed as probability distribution but also those being available as intervals. The developed model is applied to a case of planning regional electricity and heat supply as well as pollution emission reduction considered.

Conclusions: A number of scenarios corresponding to different pollutants emission reduction levels are examined; the results indicated that reasonable solutions have been generated under different pollutants reduction levels. They can be used for generating plans for energy resource/electricity/heat allocation and capacity expansion and help decision makers identify desired regional electricity and heat supply which need minimum cost under various standards of pollutants emission reduction control.

Keywords: Energy systems planning; Inexact two-stage stochastic programming; Electricity and heat supply; Pollutants mitigation control

Background

Currently, with the speedup of urbanization and the further economic development, the limitation of electricity resources to regional development becomes more and more obvious. On the one hand, globally, energy consumption grows more rapidly than the economy, meaning the energy intensity of economic activity has rosed. On the other hand, urban air quality is one of the hot topics, and some air pollutant emissions are from burning fossil fuels such as coal, natural gas which mainly generated in electricity and heat supply, especially, during long winter heating period which is the most serious air quality damage season. The goal of energy conservation

and pollutant emission reduction in the regional electricity and heat supply is in accordance with the stable and orderly development of the electricity and heat supply enterprises. Therefore, effective electricity and heat supply system planning method with pollutants emission reduction is desired urgently.

Previously, some deterministic models for energy management were developed. For example, Kwaczek et al. (1996) put forward an optimization model for comprehending economic impacts of various emission-reduction strategies on energy activities in Saskatchewan. Sailor (1997) conducted a comprehensive assessment of climate change influence on renewable energy demand and supply technologies in many places. Zhang et al. (2001) reviewed the relationship between global warming and energy system structural shift in power generation sector in south China. Heinrich et al. (2007) presented the South African ESI using a partial equilibrium E3 model approach, and

* Correspondence: weili1027@gmail.com

[1]MOE Key Laboratory of Regional Energy Systems Optimization, S&C Resources and Environmental Research Academy, North China Electric Power University, Beijing 102206, China

Full list of author information is available at the end of the article

extend the approach to include multiple objectives under selected future uncertainties. Klaassen and Riahi (2007) utilized the long-term MESSAGE (Model for Energy Supply Strategy Alternatives and their General Environmental Impact) to analyze energy planning management and climate change response. Chung et al. (2009) conducted a hybrid E-IO (Energy top-down approach) table which has higher classification sector resolutions to determine the strength of optimization model for making clear of economic impacts from various emission-reduction strategies on energy activities in Saskatchewan, Canada.

However, the above models emphasized on the planning of either electricity supply or heat supply, however, could hardly achieve energy efficiency. Though, some studies about combined heat and electricity generation have been done (Motevasel et al. 2011; Dong et al. 2012; Mehdi et al. 2013). However, there are defects: the uncertainties are not addressed in the previous studies which are critical for analyzing system reliability; on the other hand, some studies only pursue the maximum benefits in the operation of enterprises ignoring the strategic thinking of pollutants emission reduction control (Gustavsson and Madlener 2003; Cai et al. 2009).

Thus, in this study, an inexact two-stage dynamic programming (ITSDP) model is developed for regional electricity and heat supply management under uncertainties, aimed at the co-win between the enterprises profit and social benefits. Furthermore, this model performs satisfactory role on both cost reduction and pollutants emission reduction, with the ability to supply necessary information for decision makers, which could be applied to similar planning.

Methods

A general inexact two-stage stochastic programming can be formulated as follows:

$$\text{Min } f^{\pm} = \sum_{j=1}^{n_1} c_j^{\pm} x_j^{\pm} + \sum_{j=1}^{n_2} \sum_{h=1}^{v} p_{jh} d_j^{\pm} y_{jh}^{\pm} \tag{1a}$$

subject to

$$a_r^{\pm} x^{\pm} \le b_r^{\pm}, \quad r = 1, 2, ..., m_1 \tag{1b}$$

$$a_t^{\pm} x^{\pm} + a_t'^{\pm} y^{\pm} \ge \omega_h^{\pm}, \quad t = 1, 2, ..., m_2; h = 1, 2, ..., v \tag{1c}$$

$$x_j^{\pm} \ge 0, \quad j = 1, 2, ..., n_1 \tag{1d}$$

$$y_{jh}^{\pm} \ge 0, \quad j = 1, 2, ..., n_2; h = 1, 2, ..., v. \tag{1e}$$

where $a_r^{\pm} \in \{R^{\pm}\}^{m_1 \times n_1}$, $a_t^{\pm} \in \{R^{\pm}\}^{m_2 \times n_2}$, $b_r^{\pm} \in \{R^{\pm}\}^{m_1 \times 1}$, $c_j^{\pm} \in \{R^{\pm}\}^{1 \times n_1}$, $d_j^{\pm} \in \{R^{\pm}\}^{1 \times n_2}$, $x^{\pm} \in \{R^{\pm}\}^{n_1 \times 1}$, $y^{\pm} \in \{R^{\pm}\}^{n_2 \times 1}$ and $\{R^{\pm}\}$ denote a set of interval parameters and/or

variables. According to Huang et al. 2001, model (1) can be transformed into two deterministic submodels that correspond to the lower and upper bounds of the desired objective function. The objective function value corresponding to f^- is desired first because the objective is to minimize net system costs, and it can be formulated as follows (assume that $b_r^{\pm} \ge 0$):

$$\text{Min } f^- = \sum_{j=1}^{k_1} c_j^- x_j^- + \sum_{j=k_1+1}^{n_1} c_j^- x_j^+ + \sum_{j=1}^{k_2} \sum_{h=1}^{v} p_{jh} d_j^- y_{jh}^-$$
$$+ \sum_{j=k_2+1}^{n_2} \sum_{h=1}^{v} p_{jh} d_j^- y_{jh}^+ \tag{2a}$$

subject to:

$$\sum_{j=1}^{k_1} |a_{rj}|^+ \text{sign}\left(a_{rj}^+\right) x_j^- + \sum_{j=k_1+1}^{n_1} |a_{rj}|^- \text{sign}\left(a_{rj}^-\right) x_j^+ \le b_r^+, \forall r \tag{2b}$$

$$\sum_{j=1}^{k_1} |a_{tj}|^+ \text{sign}\left(a_{tj}^+\right) x_j^- + \sum_{j=k_1+1}^{n_1} |a_{tj}|^- \text{sign}\left(a_{tj}^-\right) x_j^+$$
$$+ \sum_{j=1}^{k_2} |a_{tj}'|^+ \text{sign}\left(a_{tj}'^+\right) y_{jh}^-$$
$$+ \sum_{j=k_2+1}^{n_2} |a_{tj}'|^- \text{sign}\left(a_{tj}'^-\right) y_{jh}^+ \ge \omega_h^-, \forall t, h \tag{2c}$$

$$x_j^- \ge 0, j = 1, 2, ..., k_1 \tag{2d}$$

$$x_j^+ \ge 0, j = k_1 + 1, k_1 + 2, ..., n_1 \tag{2e}$$

$$y_{jh}^- \ge 0, \forall h; j = 1, 2, ..., k_2 \tag{2f}$$

$$y_{jh}^+ \ge 0, \forall h; j = k_2 + 1, k_2 + 2, ..., n_2 \tag{2g}$$

where $x_j^{\pm}$, $j = 1, 2,..., k_1$, are positive coefficients; $x_j^{\pm}$, $j = k_1 + 1, k_1 + 2,..., n_1$ are negative variables; $y_{jh}^{\pm}$, $j = 1, 2,..., k_2$ and $h = 1, 2,..., v$, are random variables with positive coefficients; $y_{jh}^{\pm}, j = k_2 + 1, k_2 + 2,..., n_2$ and $h = 1, 2,..., v$, are random variables with negative coefficients. Solutions of $x_{j \text{ opt}}^-$ ($j = 1, 2,..., k_1$), $x_{j \text{ opt}}^+$ ($j = k_1 + 1, k_1 + 2,..., n_1$), $y_{jh \text{ opt}}^-$ ($j = 1, 2,..., k_2$), and $y_{jh \text{ opt}}^+$ ($j = k_2 + 1, k_2 + 2,..., n_2$) can be obtained through submodel (2). Based on the above solutions, the submodel f^+ can be formulated as follows:

$$\text{Min } f^+ = \sum_{j=1}^{k_1} c_j^+ x_j^+ + \sum_{j=k_1+1}^{n_1} c_j^+ x_j^- + \sum_{j=1}^{k_2} \sum_{h=1}^{v} p_{jh} d_j^+ y_{jh}^+$$
$$+ \sum_{j=k_2+1}^{n_2} \sum_{h=1}^{v} p_{jh} d_j^+ y_{jh}^- \tag{3a}$$

subject to:

$$\sum_{j=1}^{k_1} \left|a_{rj}\right|^- \text{sign}\left(a_{rj}^-\right)x_j^+ + \sum_{j=k_1+1}^{n_1} \left|a_{rj}\right|^+ \text{sign}\left(a_{rj}^+\right)x_j^- \leq b_r^-, \forall r$$

(3b)

$$\sum_{j=1}^{k_1} \left|a_{tj}\right|^- \text{sign}\left(a_{tj}^-\right)x_j^+ + \sum_{j=k_1+1}^{n_1} \left|a_{tj}\right|^+ \text{sign}\left(a_{tj}^+\right)x_j^-$$
$$+\sum_{j=1}^{k_2} \left|a_{tj}'\right|^- \text{sign}\left(a_{tj}'^-\right)y_{jh}^+$$
$$+\sum_{j=k_2+1}^{n_2} \left|a_{tj}'\right|^+ \text{sign}\left(a_{tj}'^+\right)y_{jh}^- \geq \omega_h^+, \forall t,h$$

(3c)

$$x_j^+ \geq x_{j\ \text{opt}}^-, j = 1, 2, ..., k_1$$ (3d)

$$0 \leq x_j^- \leq x_{j\ \text{opt}}^+, j = k_1+1, k_1+2, ..., n_1$$ (3e)

$$y_{jh}^+ \geq y_{jh\ \text{opt}}^-, \forall h; j = 1, 2, ..., k_2$$ (3f)

$$0 \leq y_{jh}^- \leq y_{jh\ \text{opt}}^+, \forall h; j = k_2+1, k_2+2, ..., n_2$$ (3g)

Solutions of $x_{j\ \text{opt}}^+$ ($j = 1, 2,..., k_1$), $x_{j\ \text{opt}}^-$ ($j = k_1 + 1, k_1 + 2,..., n_1$), $y_{jh\ \text{opt}}^+$ ($j = 1, 2,..., k_2$), and $y_{jh\ \text{opt}}^-$ ($j = k_2 + 1, k_2 + 2,..., n_2$) can be obtained through submodel (3). Through integrating solutions of submodels (2) and (3), the solution for model (1) can be obtained.

Model formulation

Typically, electricity and heat supply system often contains some components such as energy supply/demand, processing and transformation technologies, and electricity and heat generation (Heinrich et al. 2007). These components generally involve an array of economic activities, energy consumption and pollutants discharge. Electricity and heat supply options are usually classified as fossil energy and renewable resources. Each of those has its own industry representing the characteristics of the technologies.

Consider a case wherein a regional electricity and heat supply manager is responsible to allocate electricity and heat flows from the enterprise to users all the year (12 months). The manager can formulate the problem as minimizing the expected value of net system cost in the region during one year. Based on the local electricity and heat supply policies, a promised allowable supply quantity is defined. If this level for each user is not reached, it will result in the higher cost to the system, the system will then be subject to penalties of the system failure. In the mean time, the manager always seek a project which can assure the emissions of some environmental pollutants (e.g., nitrogen oxides (NO_x), sulfur

dioxide (SO_2), particulate matter (PM)), and greenhouse gas (GHG) could meet the regional environmental standard with sulfur dioxide emission fees deduced. As waste-generation and energy requirement amounts from the region are uncertain at the time when the planning decisions must be made, the problem under consideration can be formulated as a TSDP model as follows:

$$\min f^\pm = (1) + (2) + (3) + (4) + (5) + (6) + (7) + (8) + (9)$$

(4a)

$$(1) = \sum_{t=1}^{12} \sum_{i=1}^{I} ES_{it}^\pm * EP_{it}^\pm$$

(4b)

$$(2) = \sum_{t=1}^{12} \sum_{k=1}^{2} TC_{kt}^\pm * W_{kt}^\pm + \sum_{t=1}^{12} \sum_{k=1}^{3} \sum_{h=1}^{H} P_{ht}$$
$$*\left(TC_{kt}^\pm + PP_{kt}^\pm\right) * EQ_{kth}^\pm + TC_{3t}^\pm * \left(W_{3t}^\pm + SAE_t^\pm\right)$$

(4c)

$$(3) = \sum_{t=1}^{12} \sum_{j=1}^{J} TPH_{jt}^\pm * Y_{jt}^\pm + \sum_{t=1}^{12} \sum_{j=1}^{J} \sum_{L=1}^{l} PMlt$$
$$*\left(TPH_{jt}^\pm + PCH_{jt}^\pm\right) * HQ_{jtl}^\pm$$

(4d)

$$(4) = \sum_{t=1}^{12} \sum_{k=i}^{K} \sum_{h=1}^{H} \sum_{r=1}^{R} \left(W_{kt}^\pm + P_{ht} * EQ_{kth}^\pm\right) * PE_{rkt}^\pm * CE_{rkt}^\pm$$

(4e)

$$(5) = \sum_{t=1}^{12} \sum_{j=1}^{J} \sum_{l=1}^{L} \sum_{q=1}^{Q} \left(Y_{jt}^\pm + PM_{lt} * HQ_{jtl}^\pm\right) * PH_{qjt}^\pm * CH_{qjt}^\pm$$

(4f)

$$(6) = \sum_{t=1}^{12} \sum_{k=1}^{K} \sum_{m=1}^{M} \sum_{h=1}^{H} P_{ht} * JUE_{ktmh}^\pm * DIV_{kmt} * DCO_{kmt}^\pm$$

(4g)

$$(7) = \sum_{t=1}^{12} \sum_{j=1}^{J} \sum_{n=1}^{N} \sum_{l=1}^{L} PM_{lt} * JUH_{jtnl}^\pm * DVH_{jnt} * DCH_{jnt}$$

(4h)

$$(8) = \sum_{t=1}^{12} SAE_t^\pm \cdot CES_t^\pm$$

(4i)

$$(9) = \sum_{t=1}^{12} \sum_{k=1}^{K} \sum_{j=1}^{J} \left[PE_{1kt}^\pm * \left(1-\eta_{1kt}^\pm\right) + PH_{1jt}^\pm * \left(1-\mu_{1jt}^\pm\right)\right] * PDF_t^\pm$$

(4j)

where: i denotes the energy sources, $i = 1$ for coal, $i = 2$ for natural gas; k denotes the power generation technologies, $k = 1$ for coal-fired power, $k = 2$ for natural gas-fired power, $k = 3$ for wind power; t denotes the planning periods; r is the air pollutants, $r = 1$ for sulfur dioxide, $r = 2$ for nitrogen oxides, $r = 3$ for particulate matter; h is the demand level. j denotes the heat supply technologies, $j = 1$ for coal-fired heat, $j = 2$ for natural gas-fired heat, $j = 3$ for electric heat; l is capacity expansion size option for heat generation; $f^{\pm}$ = the net expected system total cost (million dollar); $EP_{it}^{\pm}$ = the supply of energy resources in month t (PJ); $ES_{it}^{\pm}$ = the supply cost of energy sources i in month t($million/PJ); $TC_{kt}^{\pm}$ = the variable cost for electricity generated by technology k in month t ($million/GWh); $W_{kt}^{\pm}$ = allowable power generation by technology k during month t (GWh); P_{ht} = probability of occurrence for scenario h during month t; PP_{kt} = penalty cost of excess electricity generated by technology k in month t ($million/GWh); $EQ_{kth}^{\pm}$ = the additional power generation because of the shortage of electricity generated by technology k in scenario h during month t (GWh); $TPH_{jt}^{\pm}$ = the variable cost for heat generated by technology j in month t ($million/GWh); $Y_{jt}^{\pm}$ = allowable heat generation by technology j during month t (PJ); PM_{lt} = probability of occurrence for scenario l during month t; PCH_{jt} = penalty cost of excess heat generated by technology j in month t ($million/PJ); $HQ_{jtl}^{\pm}$ = the additional heat generation because of the shortage of heat generated by technology k in scenario l during month t (PJ); $JUE_{ktmh}^{\pm}$ = binary variable for technology k with expansion option m in scenario h; DIV_{kmt} = capacity expansion size option m for power generation technology k in month t (GW); DCO_{kmt} = capacity cost of capacity expansion size m for power generation technology k in month t ($Million/GW); $JUH_{jml}^{\pm}$ = binary variable for technology j with expansion option n in scenario l during month t; DVH_{jnt} = capacity expansion size option n for power generation technology j in month t (PJ); DCH_{jnt} = capacity cost of capacity expansion size n for heat generation technology j in month t ($Million/PJ); $PE_{krt}^{\pm}$ = the emission intensity of pollutant r from power generation technology k in month t (kiloton/GWh); $CE_{krt}^{\pm}$ = the removal cost of pollutant r from power generation technology k in month t (dollar/kiloton); $PH_{jqt}^{\pm}$ = the emission intensity of pollutant q from power generation technology j in month t (kiloton/PJ);

$CH_{jrt}^{\pm}$ = the removal cost of pollutant r from power generation technology j in month t (dollar/kiloton); $CES_t^{\pm}$ = the cost of per unit wind power storage in month t (dollar/kWh);

$SAE_t^{\pm}$ = the stored wind power in month t (GW); $PDF_t^{\pm}$ = pollutants discharge fee for sulfur dioxide (dollar/tonnes); $\eta_{krt}^{\pm}$ = the removal efficiency of pollutant r from power generation technology k in month t; $\mu_{lqt}^{\pm}$ = the removal efficiency of pollutant q from heat generation technology j in month t.

Mass balance constraints

The mass balance constraints describe the balance of resource and energy flows in the system. They can be classified into three groups:

(1) balance for energy resource (4k) and (4l);
(2) balance for electricity generation (4m), (4n), (4o), (4p) and (4q);
(3) balance for heat generation (4r) and (4s).

These constraints are established to ensure that the input energy is greater than the output one.

$$\left(W_{kt}^{\pm} + EQ_{kth}^{\pm}\right) * EFE_{kt}^{\pm} + \left(Y_{jt}^{\pm} + HQ_{jtl}^{\pm}\right)$$
$$* EFH_{jt}^{\pm} + \sum_{s=1}^{S} ED_{its}^{\pm} \le T_{it}^{\pm}, \forall t,k,j,l,h,i \qquad (4k)$$

$$\left(W_{3t}^{\pm} + EQ_{3th}^{\pm} + SAE_t^{\pm}\right) * EFE_{3t}^{\pm} \le T_{it}^{\pm}, \forall t,k,h \qquad (4l)$$

$$\sum_{k=1}^{3}\left(W_{kt}^{\pm} + EQ_{kth}^{\pm}\right) + SAE_{t-1}^{\pm} \ge DTE_{th}^{\pm}, \forall t,h \qquad (4m)$$

$$\sum_{k=1}^{3} W_{kt}^{\pm} - DTE_{th}^{\pm} + SAE_{t-1}^{\pm} = SAE_t^{\pm}, \forall t,h \qquad (4n)$$

$$W_{kt}^{\pm} + EQ_{kth}^{\pm} \le ET_{kt}^{\pm}$$
$$* \left(RC_k + JUE_{ktmh}^{\pm} * DIV_{kmt}\right), \forall t,k,h,m \qquad (4o)$$

$$W_{3t}^{\pm} - \left[DTE_{th}^{\pm} - \left(W_{1t}^{\pm} + EQ_{kth}^{\pm}\right) - \left(W_{2t}^{\pm} + EQ_{kth}^{\pm}\right)\right] \qquad (4p)$$
$$+ SAE_{t-1}^{\pm} = SAE_t^{\pm}, \forall t \ge 2,k,h$$

$$W_{3t}^{\pm} + EQ_{3th}^{\pm} - \left[DTE_{th}^{\pm} - \left(W_{1t}^{\pm} + EQ_{1th}^{\pm}\right) - \left(W_{2t}^{\pm} + EQ_{2th}^{\pm}\right)\right]$$
$$+ SAEO^{\pm} = SAE_t^{\pm}, t = 1, \forall h$$
$$\qquad (4q)$$

$$\sum_{j=1}^{3}\left(Y_{jt}^{\pm} + HQ_{jtl}^{\pm}\right) \ge TDH_{tl}^{\pm}, \forall t,l \qquad (4r)$$

$$Y_{jt}^{\pm} + HQ_{jtl}^{\pm} \le HT_{jt}^{\pm}$$
$$* \left(RCH_j + JUH_{jtnl}^{\pm} * DVH_{jnt}\right), \forall t,j,n,l \qquad (4s)$$

Capacity constraints of technologies

For a single technology, it is assumed that its output or production should be less than the amount that total installed capacity. If this requirement is not satisfied, cost will be increased for additional capacities.

$$JUE_{ktmh}^{\pm} = \begin{cases} 1 \\ 0 \end{cases}, \forall k,t,m,h \qquad (4t)$$

$$JUH_{jtnl}^{\pm} = \begin{cases} 1 \\ 0 \end{cases}, \forall j,t,n,l \qquad (4u)$$

$$\sum_{m=1}^{M} JUE_{ktmh}^{\pm} \leq 1, \forall k, t, h \tag{4v}$$

$$\sum_{n=1}^{N} JUH_{jtnl}^{\pm} \leq 1, \forall j, t, l \tag{4w}$$

Environmental constraints

For an electricity and heat supply system planning, it is assumed that environmental requirement should be considered as an important constraint. Equation (4x) is the constraint of pollutants emission of power generation technologies; Equation (4y) is the constraint of pollutants emission of heat generation technologies. Equation (4z) is the constraint of GHG (Greenhouse Gas) emission of the electricity and heat supply system.

$$\sum_{k=1}^{3} \left(W_{kt}^{\pm} + EQ_{kth}^{\pm} \right) * PE_{krt}^{\pm} * \left(1 - \eta_{krt}^{\pm} \right) \leq TP_{rt}^{\pm}, \forall t, h, r \tag{4x}$$

$$\sum_{j=1}^{3} \left(Y_{jt}^{\pm} + HQ_{jtl}^{\pm} \right) * PH_{jqt}^{\pm} * \left(1 - \mu_{jqt}^{\pm} \right) \leq THP_{qt}^{\pm}, \forall t, l, q \tag{4y}$$

$$\sum_{s=1}^{S} \sum_{i=1}^{2} ED_{its}^{\pm} * INT_{s}^{\pm} + \sum_{k=1}^{3} \left(W_{kt}^{\pm} + EQ_{kth}^{\pm} \right) * COE_{kt}^{\pm}$$
$$+ \sum_{j=1}^{3} \left(Y_{jt}^{\pm} + HQ_{jtl}^{\pm} \right) * COH_{jt}^{\pm} \leq TC_{t}^{\pm}, \forall t, h, l \tag{4z}$$

where:

$EFE_{kt}^{\pm}$ = the conversion efficiency of power generation technology k in month t (PJ/GW);

$EFH_{jt}^{\pm}$ = the conversion efficiency of heat generation technology j in month t (PJ/PJ);

$ED_{its}^{\pm}$ = the demand of energy resource i in sector s during month t;

$T_{it}^{\pm}$ = the supply of energy i in month t;

$TDE_{th}^{\pm}$ = electricity demand in scenario h during month t (GWh);

$TDH_{tl}^{\pm}$ = heat demand in scenario l during month t (PJ);

$ET_{kt}^{\pm}$ = the working hours of power generation technology k in month t (hour);

$HT_{jt}^{\pm}$ = the working hours of heat generation technology j in month t (hour);

$RC_{k}^{\pm}$ = residual capacity of power generation technology k (GW);

$\eta_{krt}^{\pm}$ = residual capacity of heat generation technology j (PJ);

$TP_{rt}^{\pm}$ = the total allowable emissions of pollutant r in month t (kiloton);

$THP_{qt}^{\pm}$ = the total allowable emissions of pollutant q in month t (kiloton);

$INT_{s}^{\pm}$ = CO_2 emission intensity of sector s (kiloton/PJ);

$COE_{kt}^{\pm}$ = CO_2 emission intensity of power generation technology k in month t (kiloton/GWh);

$COH_{jt}^{\pm}$ = CO_2 emission intensity of power generation technology j in month t (kiloton/PJ);

$TC_{t}^{\pm}$ = the total allowable CO_2 emissions in scenario h during month t (kiloton).

In the ITSDP model, when the allowable amount of power generation $W_{kt}^{\pm}$ and heat generation $Y_{jt}^{\pm}$ are known, the above model can be transformed into two branches of deterministic sub-models, which related to the upper and lower bounds of the conceived objective-function value. The transformation is on the base of an interactive algorithm, which is different from traditional interval analysis as well as best/worst case analysis, and the previous methods for solving inexact linear programming problems cannot be used for granted directly (Li et al. 2006a, b; Klaassen and Riahi 2007). In this study, an optimized set would be made and correspond to minimize the system cost under the uncertain electricity and heat demands and supplies. Accordingly, let $W_{kt}^{\pm} = W_{kt}^{-} + \Delta W_{kt} \cdot u_{kt}$, $Y_{jt}^{\pm} = Y_{jt}^{-} + \Delta Y_{jt} \cdot v_{jt}$, where $\Delta W_{kt} = W_{kt}^{+} - W_{kt}^{-}$, $\Delta Y_{kt} = Y_{kt}^{+} - Y_{kt}^{-}$ $u_{kt} \in [0, 1]$, $v_{jt} \in [0, 1]$, u_{kt} and v_{jt} are decision variables that can identify an optimized set of target values $W_{kt}^{\pm}$ and $Y_{jt}^{\pm}$ so that the related policy analyses could be supported. For example, when $W_{kt}^{\pm}$ and $Y_{jt}^{\pm}$ reach their upper bounds (i.e., when $u_{kt} = 1$ and $v_{jt} = 1$), a relatively low cost would be procured when the electricity and heat demands are satisfied; a lower penalty might have to be paid if the promised electricity or heat is delivered. Thus, if $W_{kt}^{\pm}$ and $Y_{jt}^{\pm}$ approach their lower bounds (i.e., when $u_{kt} = 0$ and $v_{jt} = 0$), there might exist a higher cost as well as more risk of breaking the promise. Consequently, according to Huang and Loucks (2000) and Li et al. (2009), two transformed deterministic sub-models based on an interactive algorithm through introducing decision variables u_{kt} and v_{jt} could obtain the net system cost. Since the first objective is to minimize this cost, the sub-model relevant to lower-bound objective function value f^{-} is preferential. Therefore, we have model B:

$$\min f^{-} = (1) + (2) + (3) + (4) + (5) + (6) \\ + (7) + (8) + (9) \tag{5a}$$

$$(1) = \sum_{t=1}^{12} \sum_{i=1}^{I} ES_{it}^{-} * EP_{it}^{-} \tag{5b}$$

$$(2) = \sum_{t=1}^{12}\sum_{k=1}^{2} TC_{kt}^- * W_{kt} + \sum_{t=1}^{12}\sum_{k=1}^{3}\sum_{h=1}^{H} P_{ht} * \left(TC_{kt}^- + PP_{kt}^-\right)$$
$$* EQ_{kth}^- + TC_{3t}^- * \left(W_{3t} + SAE_t^-\right) \tag{5c}$$

$$(3) = \sum_{t=1}^{12}\sum_{j=1}^{J} TPH_{jt}^- * Y_{jt} + \sum_{t=1}^{12}\sum_{j=1}^{J}\sum_{L=1}^{l} PM_{lt}$$
$$* \left(TPH_{jt}^- + PCH_{jt}^-\right) * HQ_{jtl}^- \tag{5d}$$

$$(4) = \sum_{t=1}^{12}\sum_{k=i}^{K}\sum_{h=1}^{H}\sum_{r=1}^{R} \left(W_{kt} + P_{ht} * EQ_{kth}^-\right) * PE_{rkt}^- * CE_{rkt}^- \tag{5e}$$

$$(5) = \sum_{t=1}^{12}\sum_{j=1}^{J}\sum_{l=1}^{L}\sum_{q=1}^{Q} \left(Y_{jt} + PM_{lt} * HQ_{jtl}^-\right) * PH_{qjt}^- * CH_{qjt}^- \tag{5f}$$

$$(6) = \sum_{t=1}^{12}\sum_{k=1}^{K}\sum_{m=1}^{M}\sum_{h=1}^{H} P_{ht} * JUE_{ktmh}^- * DIV_{kmt} * DCO_{kmt} \tag{5g}$$

$$(7) = \sum_{t=1}^{12}\sum_{j=1}^{J}\sum_{n=1}^{N}\sum_{l=1}^{L} PM_{lt} * JUH_{jtnl}^- * DVH_{jnt} * DCH_{jnt} \tag{5h}$$

$$(8) = \sum_{t=1}^{12} SAE_t^- \cdot CES_t^- \tag{5i}$$

$$(9) = \sum_{t=1}^{12}\sum_{k=1}^{K}\sum_{j=1}^{J} \left[PE_{1kt}^- * \left(1-\eta_{1kt}^+\right) + PH_{1jt}^- * \left(1-\mu_{1jt}^+\right)\right] * PDF_t^- \tag{5j}$$

Subject to:

$$W_{kt} = W_{kt}^- + \left(W_{kt}^+ - W_{kt}^-\right)\cdot u_{kt}, \forall k, t \tag{5k}$$

$$0 \leq u_{kt} \leq 1, \forall k, t \tag{5l}$$

$$Y_{jt} = Y_{jt}^- + \left(Y_{jt}^+ - Y_{jt}^-\right)\cdot u_{jt}, \forall j, t \tag{5m}$$

$$0 \leq v_{jt} \leq 1, \forall j, t \tag{5n}$$

$$\left(W_{kt} + EQ_{kth}^-\right) * EFE_{kt}^+ + \left(Y_{jt} + HQ_{jtl}^-\right)$$
$$* EFH_{jt}^+ + \sum_{s=1}^{S} ED_{its}^- \leq T_{it}^-, \forall t, k, j, l, h, i \tag{5o}$$

$$\left(W_{3t} + EQ_{3th}^- + SAE_t^-\right) * EFE_{3t}^+ \leq T_{it}^-, \forall t, h, i \tag{5p}$$

$$\sum_{k=1}^{3} \left(W_{kt} + EQ_{kth}^-\right) + SAE_{t-1}^- \geq DTE_{th}^-, \forall t, h \tag{5q}$$

$$\sum_{k=1}^{3} W_{kt} - DTE_{th}^+ + SAE_{t-1}^- = SAE_t^-, \forall t, k, h \tag{5r}$$

$$W_{kt} + EQ_{kth}^- \leq ET_{kt}^-$$
$$* \left(RC_k + JUE_{ktmh}^- * DIV_{kmt}\right), \forall t, k, h, m \tag{5s}$$

$$W_{3t} - \left[DTE_{th}^+ - \left(W_{1t} + EQ_{kth}^-\right) - \left(W_{2t} + EQ_{kth}^-\right)\right]$$
$$+ SAE_{t-1}^- = SAE_t^-, \forall t \geq 2, k, h \tag{5t}$$

$$W_{3t} + EQ_{3th}^- - \left[DTE_{th}^+ - \left(W_{1t} + EQ_{1th}^-\right) - \left(W_{2t} + EQ_{2th}^-\right)\right]$$
$$+ SAEO^- = SAE_t^-, t = 1, \forall h \tag{5u}$$

$$\sum_{j=1}^{3} \left(Y_{jt} + HQ_{jtl}^-\right) \geq TDH_{tl}^-, \forall t, l \tag{5v}$$

$$Y_{jt} + HQ_{jtl}^- \leq HT_{jt}^-$$
$$* \left(RCH_j + JUH_{jtnl}^- * DVH_{jnt}\right), \forall t, j, n, l \tag{5w}$$

$$JUE_{ktmh}^- = \begin{cases} \leq 1 \\ \geq 0 \end{cases}, \forall k, t, m, h \tag{5x}$$

$$\sum_{m=1}^{3} JUE_{ktmh}^- \leq 1, \forall k, t, h \tag{5y}$$

$$JUH_{jtnl}^- = \begin{cases} \leq 1 \\ \geq 0 \end{cases}, \forall j, t, n, l \tag{5z}$$

$$\sum_{n=1}^{3} JUH_{jtnl}^- \leq 1, \forall j, t, l \tag{5aa}$$

$$\sum_{k=1}^{3} \left(W_{kt} + EQ_{kth}^-\right) * PE_{krt}^- * \left(1-\eta_{krt}^-\right) \leq TP_{rt}^-, \forall t, h, r \tag{5ab}$$

$$\sum_{j=1}^{3} \left(Y_{jt} + HQ_{jtl}^-\right) * PH_{jqt}^- * \left(1-\mu_{jqt}^-\right) \leq THP_{qt}^-, \forall t, l, q \tag{5ac}$$

$$\sum_{s=1}^{S}\sum_{i=1}^{2} ED_{its}^- * INT_s^+ + \sum_{k=1}^{3} \left(W_{kt} + EQ_{kth}^-\right) * COE_{kt}^+$$
$$+ \sum_{j=1}^{3} \left(Y_{jt} + HQ_{jtl}^-\right) * COH_{jt}^+ \leq TC_t^-, \forall t, h, l \tag{5ad}$$

Where T_{it}^-, EQ_{kth}^-, HQ_{jtl}^-, u_{kt} and v_{jt} are continuous decision variables, $JUE_{ktmh}^\pm$ and $JUH_{jtnl}^\pm$ are binary ones. Solution for f^- reach a extreme lower bound of system cost under uncertainties. Then, the optimized electricity and heat targets would be $W_{ktopt} = W_{kt}^- + \Delta W_{kt}\cdot u_{ktopt}$, $Y_{jtopt} =$

$Y_{jt}^- + \Delta Y_{jt} \cdot v_{jtopt}$. Conversely, the upper bound of the objective function value f^+ is:

$$\min f^+ = (1) + (2) + (3) + (4) + (5) + (6) \\ + (7) + (8) + (9) \tag{6a}$$

$$(1) = \sum_{t=1}^{12} \sum_{i=1}^{2} ES_{it}^+ * EP_{it}^+ \tag{6b}$$

$$(2) = \sum_{t=1}^{12} \sum_{k=1}^{3} TC_{kt}^+ * W_{kt} + \sum_{t=1}^{12} \sum_{k=1}^{3} \sum_{h=1}^{H} P_{ht} * (TC_{kt}^+ + PP_{kt}^+) \\ * EQ_{kth}^+ + TC_{3t}^+ * (W_{3t} + SAE_t^+) \tag{6c}$$

$$(3) = \sum_{t=1}^{12} \sum_{j=1}^{J} TPH_{jt}^+ * Y_{jt} + \sum_{t=1}^{12} \sum_{j=1}^{J} \sum_{L=1}^{l} PM_{lt} \\ * (TPH_{jt}^+ + PCH_{jt}^+) * HQ_{jtl}^+ \tag{6d}$$

$$(4) = \sum_{t=1}^{12} \sum_{k=i}^{K} \sum_{h=1}^{H} \sum_{r=1}^{R} (W_{kt} + P_{ht} * EQ_{kth}^+) * PE_{rkt}^+ * CE_{rkt}^+ \tag{6e}$$

$$(5) = \sum_{t=1}^{12} \sum_{j=1}^{J} \sum_{l=1}^{L} \sum_{q=1}^{Q} (Y_{jt} + PM_{lt} * HQ_{jtl}^+) * PH_{qjt}^+ * CH_{qjt}^+ \tag{6f}$$

$$(6) = \sum_{t=1}^{12} \sum_{k=1}^{K} \sum_{m=1}^{M} \sum_{h=1}^{H} P_{ht} * JUE_{ktmh}^+ * DIV_{kmt} * DCO_{kmt} \tag{6g}$$

$$(7) = \sum_{t=1}^{12} \sum_{j=1}^{J} \sum_{n=1}^{N} \sum_{l=1}^{L} PM_{lt} * JUH_{jtnl}^+ * DVH_{jnt} * DCH_{jnt} \tag{6h}$$

$$(8) = \sum_{t=1}^{12} SAE_t^+ \cdot CES_t^+ \tag{6i}$$

$$(9) = \sum_{t=1}^{12} \sum_{k=1}^{K} \sum_{j=1}^{J} \left[PE_{1kt}^+ * (1-\eta_{1kt}^-) + PH_{1jt}^+ * (1-\mu_{1jt}^-) \right] * PDF_t^+ \tag{6j}$$

subject to:

$$(W_{ktopt} + EQ_{kth}^+) * EFE_{kt}^- + (Y_{jtopt} + HQ_{jtl}^+) \\ * EFH_{jt}^- + \sum_{s=1}^{S} ED_{its}^+ \le T_{it}^+, \forall t,k,j,l,h,i \tag{6k}$$

$$(W_{3topt} + EQ_{3th}^+ + SAE_t^+) * EFE_{3t}^- \le T_{it}^+, \forall t,h,i \tag{6l}$$

$$\sum_{k=1}^{3} (W_{ktopt} + EQ_{kth}^+) + SAE_{t-1}^+ \ge DTE_{th}^+, \forall t,k,h \tag{6m}$$

$$\sum_{k=1}^{3} W_{ktopt} - DTE_{th}^+ + SAE_{t-1}^- = SAE_t^+, \forall t,k,h \tag{6n}$$

$$W_{ktopt} + EQ_{kth}^+ \le ET_{kt}^+ \\ * \left(RC_k + \sum_{t=1}^{12} \sum_{m=1}^{M} JUE_{ktmh}^+ * DIV_{kmt} \right), \forall t,k,h,m \tag{6o}$$

$$W_{3topt} - \left[DTE_{th}^- - (W_{1topt} + EQ_{kth}^+) - (W_{2topt} + EQ_{kth}^+) \right] \\ + SAE_{t-1}^+ = SAE_t^+, \forall t \ge 2, k, h \tag{6p}$$

$$W_{3topt} + EQ_{3th}^+ - \left[DTE_{th}^- - (W_{1topt} + EQ_{1th}^+) - (W_{2topt} + EQ_{2th}^+) \right] \\ + SAEO^+ = SAE_t^+, t = 1, \forall h \tag{6q}$$

$$\sum_{j=1}^{3} \left(Y_{jtopt} + HQ_{jtl}^+ \right) \ge TDH_{tl}^+, \forall t,l \tag{6r}$$

$$Y_{jtopt} + HQ_{jtl}^+ \le HT_{jt}^+ \\ * \left(RCH_j + JUH_{jtnl}^+ * DVH_{jnt} \right), \forall t,j,n,l \tag{6s}$$

$$JUE_{ktmh}^+ = \begin{cases} \le 1 \\ \ge 0 \end{cases}, \forall k,t,m,h \tag{6t}$$

$$\sum_{m=1}^{3} JUE_{ktmh}^+ \le 1, \forall k,t,h \tag{6u}$$

$$JUH_{jtnl}^+ = \begin{cases} \le 1 \\ \ge 0 \end{cases}, \forall j,t,n,l \tag{6v}$$

$$\sum_{n=1}^{3} JUH_{jtnl}^+ \le 1, \forall j,t,l \tag{6w}$$

$$\sum_{k=1}^{3} (W_{ktopt} + EQ_{kth}^+) * PE_{krt}^+ * (1-\eta_{krt}^-) \le TP_{rt}^+, \forall t,h,r \tag{6x}$$

$$\sum_{j=1}^{3} \left(Y_{jtopt} + HQ_{jtl}^+ \right) * PH_{jqt}^+ * (1-\mu_{jqt}^-) \le THP_{qt}^+, \forall t,l,q \tag{6y}$$

$$\sum_{s=1}^{S}\sum_{i=1}^{2}ED_{its}^{+} * INT_{s}^{-} + \sum_{k=1}^{3}\left(W_{ktopt} + EQ_{kth}^{+}\right) * COE_{kt}^{-}$$

$$+\sum_{j=1}^{3}\left(Y_{jtopt} + HQ_{jtl}^{+}\right) * COH_{jt}^{-} \leq TC_{t}^{+}, \forall t, h, l$$

(6z)

$$T_{it}^{+} \geq T_{itopt}^{-}, \forall i, t \tag{6aa}$$

$$EQ_{kth}^{+} \geq EQ_{kthopt}^{-}, \forall k, t, h \tag{6ab}$$

$$HQ_{jtl}^{+} \geq HQ_{jtlopt}^{-}, \forall j, t, l \tag{6ac}$$

$$JUE_{ktmh}^{\pm} \geq JUE_{ktmhopt}^{-}, \forall k, t, m, h \tag{6ad}$$

$$JUH_{jtnl}^{\pm} \geq JUH_{jtnlopt}^{-}, \forall j, t, n, l \tag{6ae}$$

where T_{it}^{+}, EQ_{kth}^{+} and HQ_{jtl}^{+} are continuous decision variables, and JUE_{ktmh}^{+} and JUH_{jtnl}^{+} are binary ones; T_{itopt}^{-}, EQ_{kthopt}^{-}, HQ_{jtlopt}^{-}, $JUE_{ktmhopt}^{-}$ and $JUH_{jtnlopt}^{-}$ are solutions of the model B. As a result, the solutions for electricity and heat supply model under the optimized targets could be obtained by incorporating the solutions of the above two sub-models.

Case study

The following regional electricity and heat supply management problem is used to demonstrate the applicability of the developed ITSDP model. In this study system, a decision maker is responsible for electricity and heat to multiple users through several technologies on an annual cycle based on different pollutants emission permissible levels. Since different environmental restrictions, the decision maker would take expected electricity and heat demand into account to make fossil and renewable resources utilization optimal, manage electricity and heat generation and plan the equipment expansion. Mostly, increasing electricity and heat demand could lead to facility expansion, fuel exploitation and energy imports. However, it is unpractical to achieve sustainable development with the ever-increasing economic and environmental costs, unlimited facility expansion and energy exploitation (Liu et al. 2000). Thus, the problem is how to incorporate different pollutants emission reduction control levels into regional electricity and heat planning. In this study system, planning period is on an annual cycle, with heating seasons and no heating seasons. Multiple energy resources/technologies need to be allocated to multiple end users. Traditional energy resources (e.g., coal, natural gas) with limited availabilities and renewable resources (e.g., wind) are utilized to meet the electricity and heat demand. More specifically, coal and natural gas are both used for electricity generation; wind is a supplement for electricity generation; the cogeneration system of electricity and heat is another

energy-efficient way. The electricity and heat demands during one year are affected by many uncertainties, such as the growing population, energy-transformation rate, electricity and heat prices and changing weather; all of those factors would produce many complicated uncertainties (Li et al. Li et al. 2006a, b; Liu 2007; Li et al. 2008). In addition, these uncertainties are intricate because of a series of imprecise information (e.g., social, economic, environmental, seasonal and geographic conditions, energy carrier characteristics). They could only be expressed as distribution information or intervals when most data can hardly be available determinately. Once these uncertainties are determined, system costs, efficiencies and capacities of each technology could be defined.

Under different environmental quality standards, decision maker is responsible for these followings (Lehtila and Pirila 1996):

(1) Assigning electricity load to four convention technologies (e.g., coal-fired power, natural-gas-fired power, wind power and cogeneration of power and heat) and heat load to three convention technologies (e.g., coal boiler heating generation, natural-gas boiler technology and cogeneration of power and heat);
(2) Planning the facility expansion of electricity and heat generation;
(3) Managing the fossil fuels purchases (including coal and natural gas).

If actual supply could not meet users' demands, decision maker would invest more funds on capacity expansion or

Table 1 Electricity demands under different probability distributions

Electricity demand (10^3GWh)			
Level	Low	Medium	High
Probability	0.2	0.6	0.2
t = 1	[486, 600]	[526.5, 650]	[567, 700]
t = 2	[445.5, 550]	[486, 600]	[526.5, 650]
t = 3	[445.5, 550]	[486, 600]	[526.5, 650]
t = 4	[486, 600]	[526.5, 650]	[567, 700]
t = 5	[486, 600]	[526.5, 650]	[567, 700]
t = 6	[526.5, 650]	[567, 700]	[607.5, 750]
t = 7	[567, 650]	[607.5, 700]	[648, 750]
t = 8	[567, 650]	[607.5, 700]	[648, 750]
t = 9	[500, 600]	[607.5, 650]	[648, 700]
t = 10	[445.5, 550]	[486, 600]	[526.5, 650]
t = 11	[445.5, 550]	[486, 600]	[526.5, 650]
t = 12	[469.8, 580]	[510.3, 630]	[550.8, 680]

Table 2 Heat demands under different probability distributions

Heat demand (TJ)

Level	Low	Medium	High
Probability	0.2	0.6	0.2
t = 1	[810, 1000]	[972, 1200]	[1053, 1300]
t = 2	[810, 1000]	[972, 1200]	[1053, 1300]
t = 3	[810, 1000]	[972, 1200]	[1053, 1300]
t = 4	/	/	/
t = 5	/	/	/
t = 6	/	/	/
t = 7	/	/	/
t = 8	/	/	/
t = 9	/	/	/
t = 10	[810, 1000]	[972, 1200]	[1053, 1300]
t = 11	[810, 1000]	[972, 1200]	[1053, 1300]
t = 12	[810, 1000]	[972, 1200]	[1053, 1300]

turn to purchase shortfall from other enterprises with higher costs; both of the programs would make the region economic interests punished.

In this study, sulfur dioxide, nitrogen oxide and particulate matters (PM10) are considered as main gaseous emission generated from fuel combustion in electricity and heat conversion; based on the gaseous pollutants emission intensity and specific demand of fuel in power and heat supply, the total amount of sulfur dioxide,

nitrogen oxide and particulate matters (PM10) emission could be calculated. The fuel demand has been predicted in each month, whether or not consider pollutants emission reduction, fuel supply would be determined to meet users demand. Therefore, under the scenarios of pollutants emission reduction, three electricity generation processes are forced to decarbonize in energy system. At the same time, pollutants emission intensity are affected by many uncertain factors (e.g., pollutants emission inventory, reduction control measures, related costs, weather situation), which can be presented as interval numbers without distribution information. The availabilities of electricity and heat demand are directly influenced by natural fluctuations, which can be expressed as probability distributions. Most of other parameters (energy demand, technological convention efficiency, and utilization factors) are indicated as intervals. Tables 1 and 2 show the available electricity demands and heat demands under different probability distributions; Table 3 presents some corresponding economic data; gaseous pollutants emission intensity is listed in Table 4. Besides, coal-fired power has a residual capacity of 0.9 GW, natural-gas-fired power has a residual capacity of 0.6 GW, CHP (Combined Heat and Power) has a residual capacity of 0.75 GW, wind power has a residual capacity of 0.27 GW; coal boiler heating generation has a residual capacity of 1.8 TJ, natural-gas boiler technology has a residual capacity of 1 TJ and cogeneration of heat and power has a residual of capacity of 1.5 TJ. The technical data and representative costs are investigated in government report

Table 3 Economic data

	t = 1	t = 3	t = 5	t = 7	t = 9	t = 11
Regular cost for electricity generated by each technology ($10³/GWh)						
Coal-fired	[4.88, 5.78]	[4.88, 5.78]	[4.88, 5.78]	[4.88, 5.78]	[4.88, 5.78]	[4.88, 5.78]
Natural gas-fired	[4.30, 5.20]	[4.30, 5.20]	[4.30, 5.20]	[4.30, 5.20]	[4.30, 5.20]	[4.30, 5.20]
Wind power	[5.23, 7.51]	[2.79, 3.29]	[2.79, 3.29]	[2.79, 3.29]	[2.79, 3.29]	[2.79, 3.29]
CHP	[2.79, 3.29]	[2.79, 3.29]	[2.79, 3.29]	[2.79, 3.29]	[2.79, 3.29]	[2.79, 3.29]
Regular cost for heat generated by each technology ($10³/TJ)						
Coal	[1.14, 1.61]	[1.14, 1.43]	[1.14, 1.43]	[1.14, 1.43]	[1.14, 1.43]	[1.14, 1.49]
Natural gas	[1.78, 1.93]	[1.44, 1,77]	[1.44, 1,77]	[1.44, 1,77]	[1.44, 1.77]	[1.51, 1,77]
CHP	[0.61, 0.91]	[0.61, 0.80]	[0.61, 0.80]	[0.61, 0.80]	[0.61, 0.80]	[0.61, 0.84]
Surplus cost for electricity generated by each technology ($10³/GWh)						
Coal-fired	[2.88, 3.35]	[2.88, 3.35]	[2.88, 3.35]	[2.88, 3.35]	[2.88, 3.35]	[2.88, 3.35]
Natural gas-fired	[2.46, 2.88]	[2.46, 2.88]	[2.46, 2.88]	[2.46, 2.88]	[2.46, 2.88]	[2.46, 2.88]
Wind power	[1.78, 2.33]	[1.78, 2.33]	[1.78, 2.33]	[1.78, 2.33]	[1.78, 2.33]	[1.78, 2.33]
CHP	[2.88, 3.35]	[2.88, 3.35]	[2.88, 3.35]	[2.88, 3.35]	[2.88, 3.35]	[2.88, 3.35]
Surplus cost for heat generated by each technology ($10³/PJ)						
Coal boiled	[0.56, 0.76]	[0.33, 0.53]	[0.11, 0.31]	[0.16, 0.36]	[0.36, 0.56]	[0.40, 0.60]
Natural gas boiled	[0.33, 0.53]	[0.25, 0.45]	[0.25, 0.45]	[0.25, 0.45]	[0.25, 0.45]	[0.33, 0.53]
CHP	[0.44, 0.64]	[0.28, 0.48]	[0.278, 0.48]	[0.28, 0.478]	[0.28, 0.48]	[0.37, 0.57]

Table 4 Parameters of contamination

	t = 1	t = 3	t = 5	t = 7	t = 9	t = 10	t = 11	t = 12
SO_2 emission intensity of each power generation technology (tones/GWh)								
Coal-fired	[7.2, 8]	[7.2, 8]	[7.2, 8]	[7.2, 8]	[7.2, 8]		[7.2, 8]	
NG-fired	[0.054, 0.06]	[0.054, 0.06]	[0.054, 0.06]	[0.054, 0.06]	[0.054, 0.06]		[0.054, 0.06]	
CHP	[7.38, 8.2]	[7.38, 8.2]	[7.38, 8.2]	[7.38, 8.2]	[7.38, 8.2]		[7.38, 8.2]	
Wind	/	/	/	/	/		/	
NO_X emission intensity of each power generation technology (tones/GWh)								
Coal-fired	[6.26, 6.95]	[6.26, 6.95]	[6.26, 6.95]	[6.26, 6.95]	[6.26, 6.95]		[6.255, 6.95]	
NG-fired	[0.78, 0.87]	[0.78, 0.87]	[0.78, 0.87]	[0.78, 0.87]	[0.78, 0.87]		[0.78, 0.87]	
CHP	[6.26, 6.95]	[6.26, 6.95]	[6.26, 6.95]	[6.26, 6.95]	[6.26, 6.95]		[6.255, 6.95]	
Wind	/	/	/	/	/		/	
PM_{10} emission intensity of each power generation technology (tones/GWh)								
Coal-fired	[2.9, 3.4]	[2.9, 3.4]	[2.9, 3.4]	[2.9, 3.4]	[2.9, 3.4]		[2.9, 3.4]	
NG-fired	[0.05, 0.075]	[0.05, 0.075]	[0.05, 0.075]	[0.05, 0.075]	[0.05, 0.075]		[0.05, 0.075]	
CHP	[2.9, 3.4]	[2.9, 3.4]	[2.9, 3.4]	[2.9, 3.4]	[2.9, 3.4]		[2.9, 3.4]	
Wind	/	/	/	/	/		/	
SO_2 emission intensity of each heat generation technology (tones/PJ)								
Coal boiled	[2.48, 2.92]	[2.48, 2.92]	[2.38, 2.92]	[2.48, 2.92]	[2.38, 2.92]		[2.48, 2.92]	
NG boiled	[0.016,0.019]	[0.016,0.017]	[0.016,0.014]	[0.016,0.011]	[0.016,0.014]		[0.016,0.019]	
CHP	[0.745,0.816]	[0.745,0.816]	[0.745,0.816]	[0.745,0.816]	[0.745,0.816]		[0.745,0.816]	
NO_X emission intensity of each heat generation technology (tones/PJ)								
Coal boiled	[0.84, 1.08]	[0.84, 1.08]	[0.84, 1.08]	[0.84, 1.08]	[0.84, 1.08]		[0.84, 1.08]	
NG boiled	[[0.17, 0.25]	[0.14, 0.22]	[0.11, 0.19]	[0.1, 0.18]	[0.09, 0.17]		[0.14, 0.22]	
Electricity	[0.214,0.324]	[0.214,0.324]	[0.214,0.324	[0.214,0.324]	[0.214,0.324]		[0.214,0.324]	
PM_{10} emission intensity of each heat generation technology (tones/PJ)								
Coal boiled	[0.23, 0.38]	[0.23, 0.38]	[0.23, 0.38]	[0.23, 0.38]	[0.23, 0.38]		[0.23, 0.38]	
NG boiled	[0.039,0.042]	[0.039,0.042]	[0.039,0.042]	[0.039,0.042]	[0.039,0.042]		[0.039,0.042]	
Electricity	[0.093,0.114]	[0.093,0.114]	[0.093,0.114]	[0.093,0.114]	[0.093,0.114]		[0.093,0.114]	

and other related documents (USDE and USEPA 2000; Iniyan and Sumath 2000; Cormio et al. 2003; Kristoffersen 2007; Fleten and Kristoffersen 2008; Cai et al. 2008, 2009; Carla and Carlos 2011).

Results and discussion

The objective of the ITSDP model is to minimize the costs of the overall system under different gaseous pollutants reduction levels during one year. Solutions provide an effective relationship between the preexisting set of environmental standardize policies and the associated economic interests, for example, improper policies would cause losses and penalties. The results contain a combination of some deterministic, interval and distributional information, it means, therefore, different forms of uncertainties could be reflected out with less error (Li et al., Li et al. 2006a,b). The interval results provide managers multiple decision alternatives and form the basis for further study of trade-offs between electricity and heat

management cost and so-caused gaseous pollutants emission reduction in electricity and heat generation; the binary variable solutions stand for the decisions of facility expansion, so that several alternative schemes are generated; the continuous variable solutions are interrelated to electricity and heat generation and energy resources supply.

Solutions without pollutants emission reduction considered

This case is raised as a reference to show a pattern of energy production and system development without constrains on gaseous pollutants emission reduction (that is 0% pollutants-emission reduction).

Figure 1 shows the results of electricity schemes under this case, from which we can see thermoelectricity would be the largest energy resource among all electricity generation options during the whole year. Because combined heat and power (CHP) is a key energy-saving project as well as technology popularization system

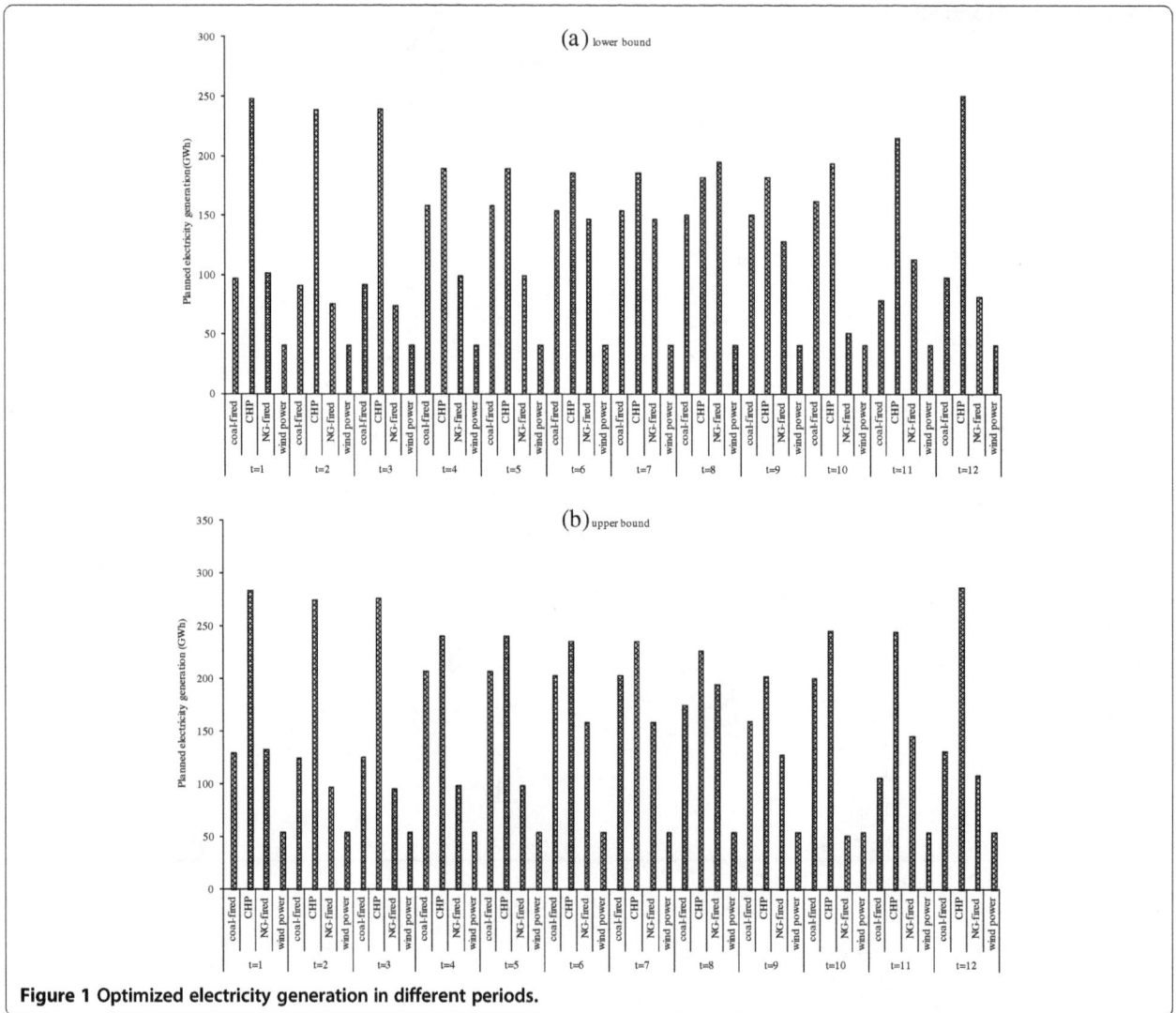

Figure 1 Optimized electricity generation in different periods.

advocated by government. The next most likely methods are, in order, coal-fired power and natural gas-fired power; as the traditional energy resources, they are still an essential part of power supply and share the load. Wind power would provide considerable amounts of electricity as a kind of clean energy; however, the weak wind power system and long distance from load center limit the development of wind farm, which would have a direct impact on wind installed capacity. As seen from Figure 1, electricity production would fluctuate smoothly in one year; overall, the main way of electricity production influenced by some factors (e.g., economy growth, colder/warmer season) would seasonally vary. The pre-regulated coal-fired power generation would be increased from [91.27, 124.30] GWh in February to [157.60, 207.42] GWh in May; for the natural gas-fired power, its plan would be [74.02, 95.36] GWh in March and up to [146.59, 158.32] GWh in July; for the CHP, its pre-regulated targets would be decreased from [247.96,

283.61] GWh in January to [181.97, 226.59] GWh in August. The pre-regulated wind power targets would remain 54 GWh over the year. The solution of planned heat generation is shown in Figure 2. If the planned electricity and heat cannot meet the random demand, the insufficient electricity has to be produced under different demand levels.

Gaseous pollutants (e.g., SO_2, NO_X, PM10) emission associated with energy exploitation mostly from the burning of fuels can be categorized into generation in power supply and generation in heat supply. Figure 3 and Figure 4 present the detailed solutions of these three pollutants emission during a year. Different energy resources would be supplied to these electricity and heat generation technologies to meet the demands; the amount of pollutants emission would be related to energy activities. From Figure 3, the greatest output pollutant would be NO_X in electricity and SO_2 in heat from Figure 4; PM10 would be a small value. SO_2 and NO_X

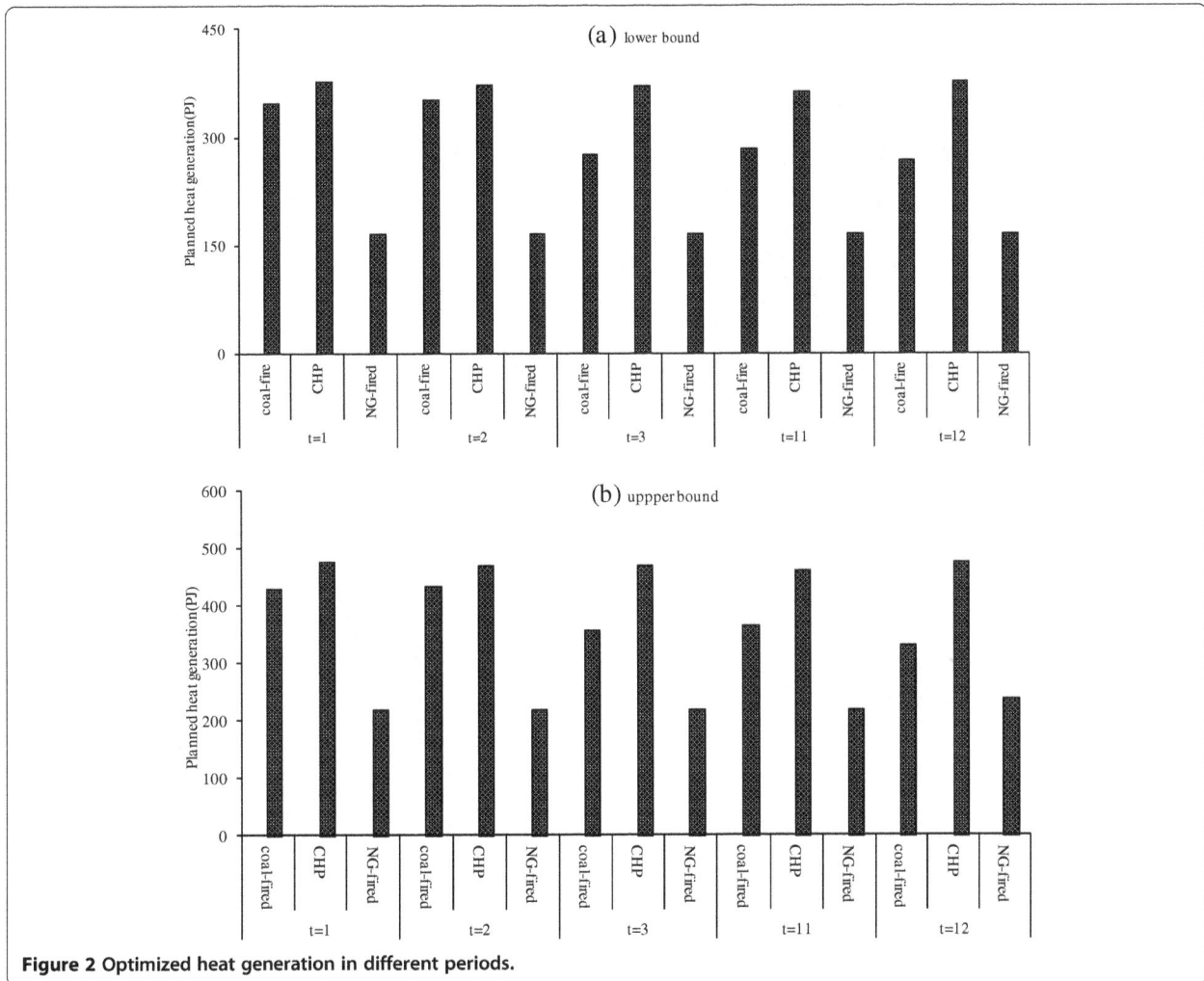

Figure 2 Optimized heat generation in different periods.

are still the main pollutants; in addition, CO_2 would be under control as a major kind of greenhouse gases. The coal-fired power generation and coal boiler heat generation technologies would be the largest pollutants and greenhouse gas emission source.

Solutions under pollutants emission reduction

In this study, two scenarios of pollutants-emission reduction are considered (i.e. 10%, and 20% of total pollutants -emission reduction). The results indicate that increased substantive capacity expansion investment for clean energy (to reduce pollutants emissions) could lead to an increased system cost. The results indicate that the increased capacity expansion investment for clean energy (proven to be effective for pollutants emission reduction) could lead to a higher system cost. Based on the results of the scenario of 0% pollutants emission reduction, coal and natural gas would be supplied to guarantee the electricity and heat demand. Next, Figure 5 shows clean energy resources storage. Compared with the result that no considering pollutants emission

reduction, the amount of wind power would be largely decreased, especially, excess wind power would be stored to meet electricity demand in the future. The utilization of wind power under scenarios of 10% and 20% pollutants emission reduction would increase, compared to the consumption under 0% pollutants emission reduction condition. That is to say, under the scenarios of 10% and 20% pollutants reduction, the fuel-fired power would decrease with pollutants reduction increasing. For example, in April, electricity generated from coal-fired power conversion technology would be [157.60, 207.42] GWh under 0% pollutants emission reduction, [146.30, 159.60] GWh under 10% pollutants emission reduction and [108.85, 145.83] GWh under 20% pollutants emission reduction. Compared the supplies of two energy resources under pollutants emission condition, natural gas supplies would be higher than coal supplies. Therefore, it recommends that natural gas would be more popular than coal in a condition of considering the pollutants emission reduction. This is because the amount of pollutants emission would be confined in a certain level

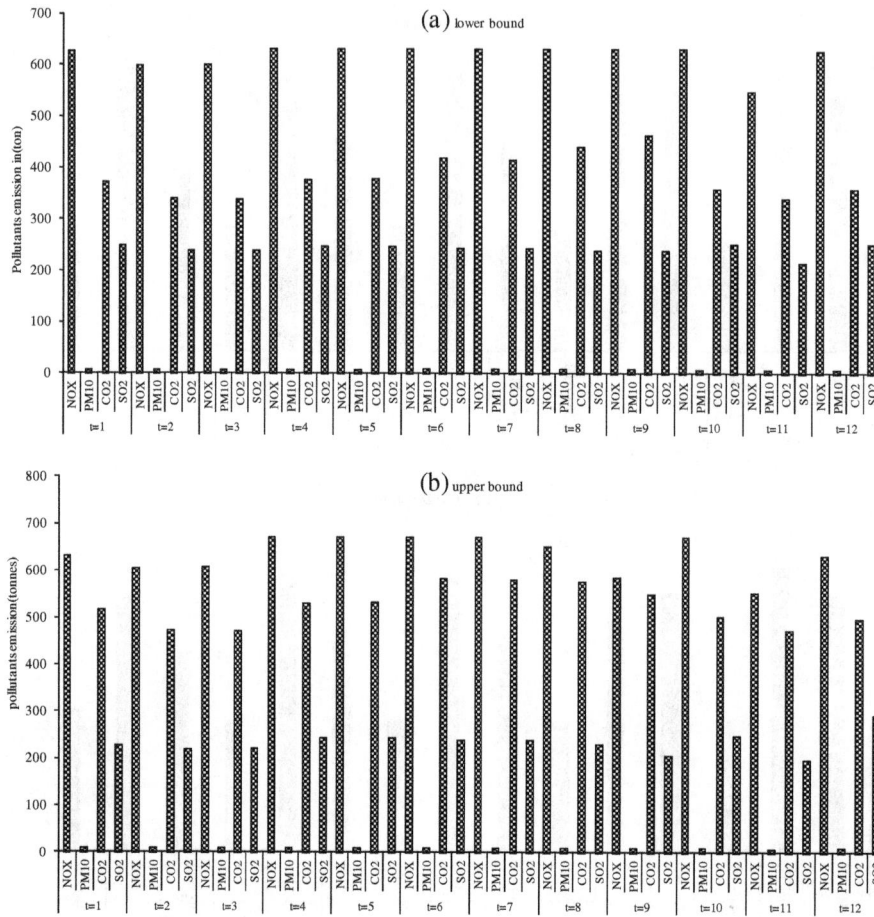

Figure 3 Quantity of pollutants emission in electricity generation.

and coal-fired electricity conversion technology corresponds to a higher pollutants emission rate than natural gas-fired conversion technology. In addition, the increased clean energy (wind power) electricity conversion technology and power conversion technology with higher energy utilization efficiency (CHP) also share the excess electricity load caused by the decreased coal supplies to meet different pollutants emission-cutting policies.

Discussion

Compared the contribution of various electricity generation technologies to the medium power demand, it presents that different electricity conversion technologies have various generation quantities under changed pollutants emission- cutting scenarios. As the above analysis, in the respect of energy type, coal-fired electricity would be the mainly power supply source under 0% pollutants emission reduction. Natural gas-fired electricity conversion technology would play a key role in the power generation activities, coal-fired power would be in a second place and wind power would be the supplement under 10% and 20% pollutants emissions reduction. This is

because coal-fired electricity conversion technology has comparatively low operating and penalty costs and relatively low capital cost for capacity expansion, and the related cost of natural gas-fired electricity conversion technology is a little higher than coal-fired power and the pollutants emission from natural gas-fired power generation process is smaller relatively. The maximum optimized wind power generation would be 0.5 GWh in April, July and October, due to the relatively higher operating cost and capital cost for its storage, which limits the development of wind power. The dominant role of coal-fired electricity would be displaced by some other conversion technologies with an increased demand for pollutants emission reduction. For example, under 20% pollutants emission reduction, the optimized coal-fired power target would decrease to [39.94, 46.20] GWh in January; although fuel-fired power would decrease, the optimized targets of wind power would have a growth and play an important role to meet electricity demand. It indicates that environment-friendly electricity conversion technologies should better serve interests of the pollutants emission reduction and be chosen for power

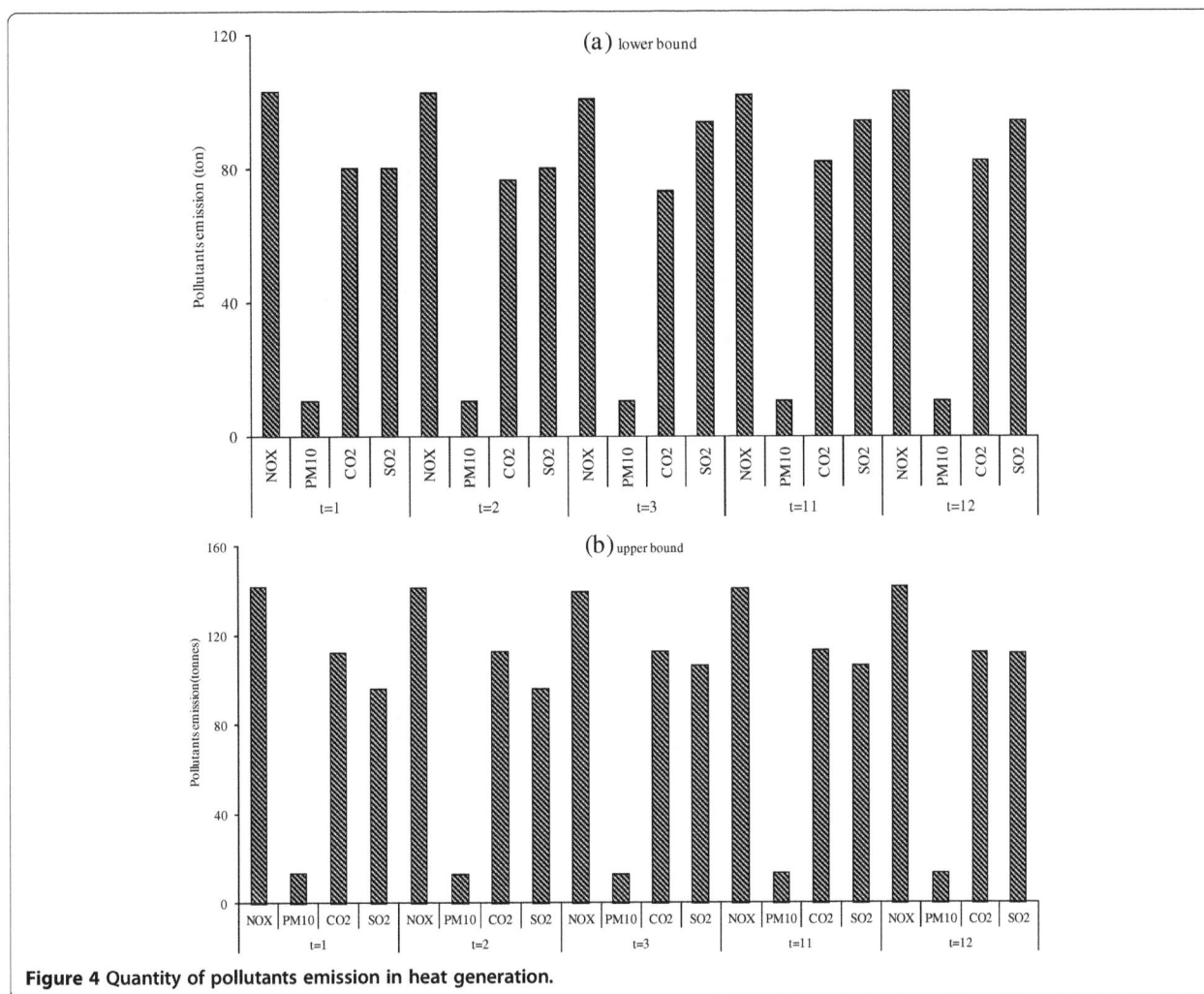

Figure 4 Quantity of pollutants emission in heat generation.

generation to meet the ever-increasing electricity demands and gradually enhancing pollutants emission reduction requirements. These are optimized conclusions in terms of pollutants emission-cutting; from view point of energy saving, CHP would be properly adopted in electricity and heat generation progress.

As shown in Figures 6, 7, 8, the system cost would rise up along with the growing intensity of pollutants emission reduction. Without pollutants emission reduction, the system cost would be $[172.20, 269.96] \times 10^6$, and the system cost would become $[187.64, 297.45] \times 10^6$, $[242.33, 471.41] \times 10^6$ under 10% and 20% pollutants emission reduction respectively. One of the major reason is that the traditional power and heat generation technologies (coal-fired and natural gas-fired) would gradually be replaced by lean energy (wind power)and more effective technologies (CHP) when the restrictions on pollutants emission are considered. Besides, the growing electricity and heat demands also lead to some degree of electricity and heat generating facilities to be expanded, resulting in a high capital cost.

Without ILP, the pollutants emission management and programming problem can also be solved by fixed-mix stochastic programming approach though adopting their mid-point values instead of the interval parameters. Undoubtedly further sensitivity analysis could be undertaken, but the model still cannot efficiently reflect the interactions among these uncertainties because each solution can only provide a single result corresponding to variations of the uncertain inputs. Likewise, if best/worst sub-models are solved, we can only obtain two solutions under extreme scenarios (best condition and worst condition). They are serviceable to judge the possibility that desired goal realizes but not necessarily build a series of stable intervals for decision variables.

Thus, the best/worst case analysis is not efficiently useful to produce decision alternatives. Actually, it is a special type of sensitivity analysis for extreme conditions (Huang et al. 2001).

From the above analysis, we can conclude that the solutions worked out from the ITSDP are value for making decisions of energy resources allocation, capacity

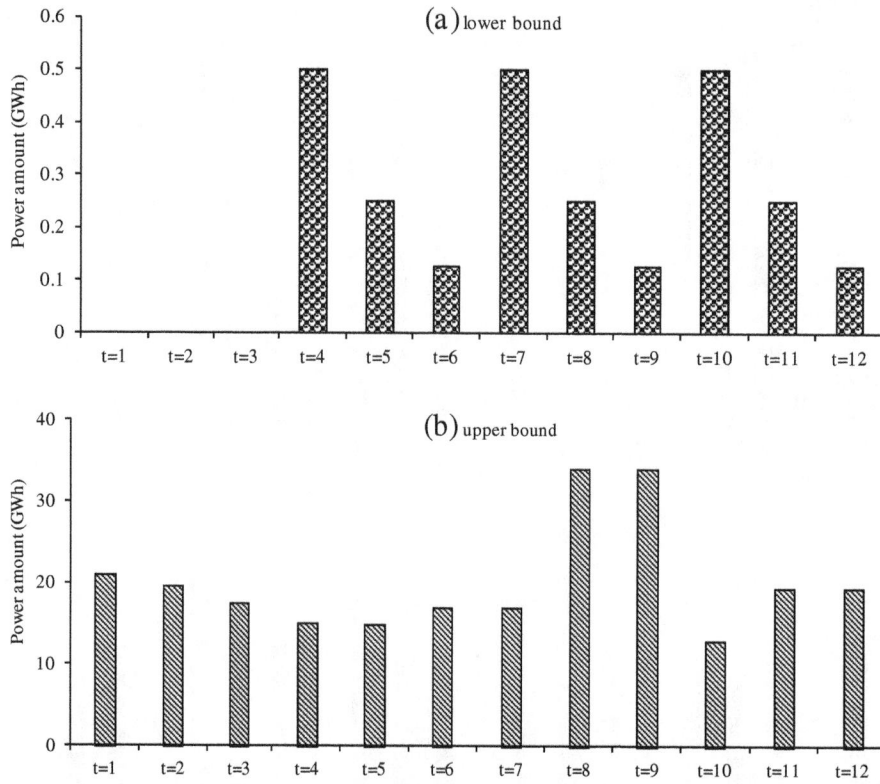

Figure 5 Storage of wind power under 20% pollutants emission reduction.

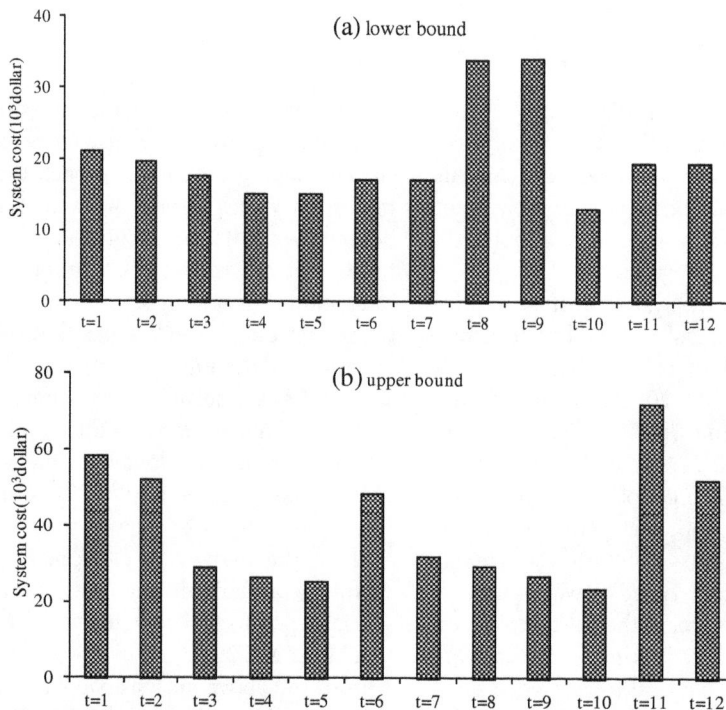

Figure 6 System cost under 20% pollutants emission reduction.

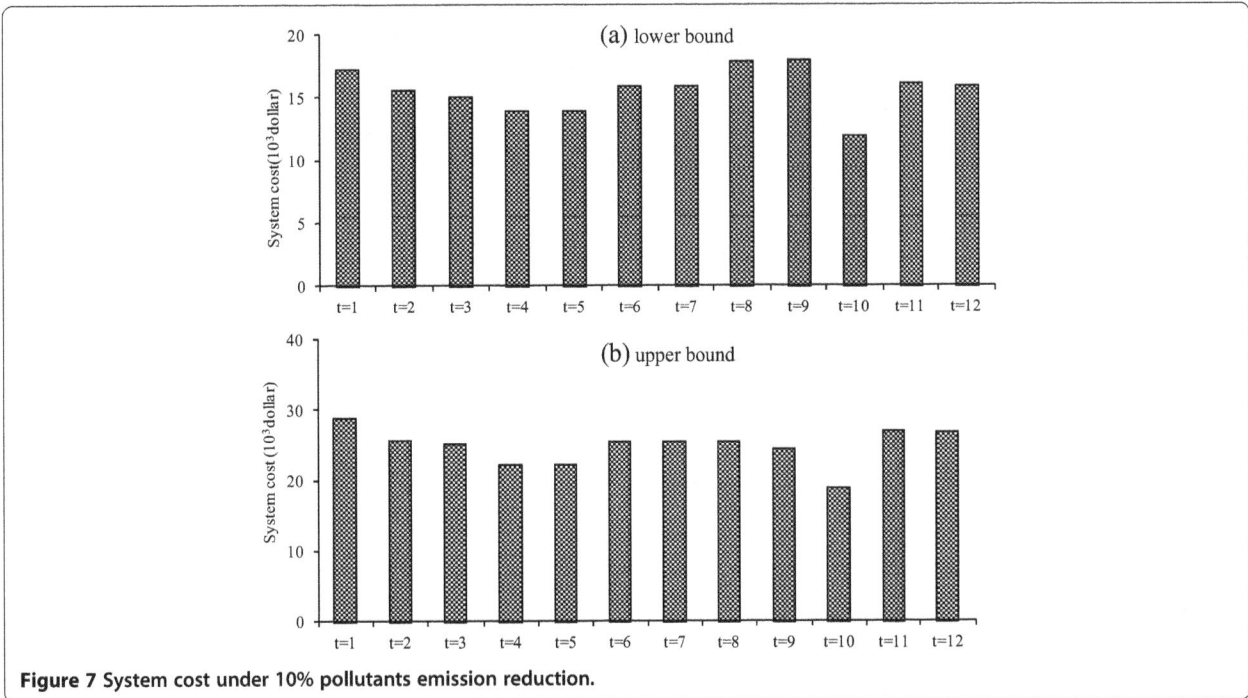

Figure 7 System cost under 10% pollutants emission reduction.

expansion of power and heat generation as well as pollutants emission management. In the progress of decision alternatives generation, the interval solutions are effective to represent various options which reflect environmental-economic trades off. Cost-effective options can be obtained with a least-cost strategy though planning pollutants emission management in regional electricity and heat supply. However, if pollutants emission reduction is considered, the pre-regulated targets of electricity and heat supply from various technologies have to be

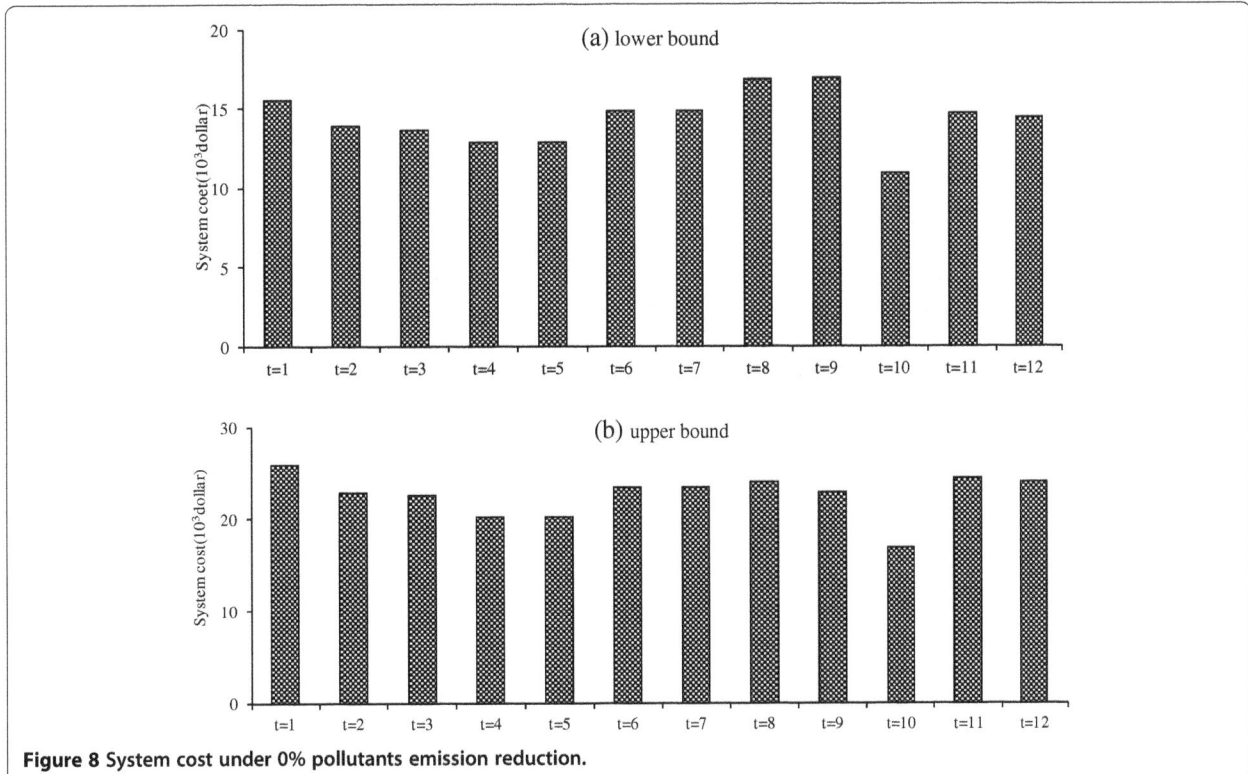

Figure 8 System cost under 0% pollutants emission reduction.

reallocated, the pre-arranged capacity expansion options of power and heat generation technologies would be reselected.

Conclusions

An inexact two-stage dynamic programming (ITSDP) model has been developed for planning pollutants emission management and electricity and heat supply systems under uncertainties. This method based on interval-parameter programming and two-stage stochastic programming so that it allows uncertainties expressed as both interval values and probability distributions to be incorporated within a common optimization framework. Furthermore, ITSDP also addresses various dynamics such as capacity expansion, storage of wind power and pollutants emission reduction scenarios related to different levels of economic implications. Probability distributions of power and heat demand can be integrated into a optimization process under a train of fixed levels by introducing fix-stochastic programming, which has significance advantages in reflecting uncertainties in large-scale problems. Accordingly, the developed model has been applied to a case of electricity and heat supply management planning. The results of this case study suggest that the model is appropriate for reflecting complexities of regional electricity and heat supply management systems and incorporating pollutants emission reduction issue during one year.

The prospect method could help decision-makers identify required management policies under changeable environmental and economic considerations. Nevertheless, there still exists much space for improvement of this model. As a powerful supplement to TSP, ITSDP can reflect the dynamic variations such as sequential structure and simplify large quantities design scenarios in case the problem of "dimension disaster" happen. This study is attempted to integrate IPP and TSP methods into a general framework, and apply the ITSDP for pollutants emission reduction management under uncertainty. In addition, the optimization algorithm is of value on other environmental issues or energy program problems within complex uncertainties. There are also other programming techniques can be integrated with ITSDP for handling more complicated cases, fuzzy programming for instance.

Competing interests

The authors declare that they have no competing interests.

Authors' contributions

The work presented here was carried out in collaboration between all authors. WL and GH defined the research theme. WL developed the model and the solution method based on GH's previous works. WL, XL, GS, and LJ carried out the case study, analyzed the data, interpreted the results and wrote the paper. All authors have contributed to, seen and approved the manuscript.

Acknowledgements

The authors are extremely grateful to the editor and the anonymous reviewers for their insightful comments and suggestions. This research was supported by the Environmental Protection Public Welfare Scientific Research Project, Ministry of Environmental Protection, P.R.China (201309063), the Program for Innovative Research Team in University (IRT1127), and the Research Projector Funding of Electric Power Research Institute of Guangdong Power Grid Corporation, Guangzhou, China (K-GD2012-389).

Author details

[1]MOE Key Laboratory of Regional Energy Systems Optimization, S&C Resources and Environmental Research Academy, North China Electric Power University, Beijing 102206, China. [2]Research Assistant, MOE Key Laboratory of Regional Energy and Environmental Systems Optimization, Resources and Environmental Research Academy, North China Electric Power University, Beijing 102206, China. [3]Research Assistant, Research Institute of Technology Economics Forecasting and Assessment, School of Economics and Management, North China Electric Power University, Beijing 102206, China.

References

Cai YP, Huang GH, Yang ZF, Lin QG, Tan Q (2008) Community-scale renewable energy systems planning under uncertainty-An interval chance constrained programming approach. Renew Sust Energ Rev 13:721–735

Cai YP, Huang GH, Yang ZF, Tan Q (2009) Identification of optimal strategies for energy management systems planning under multiple uncertainties. Appl Energy 86:480–495

Carla O, Carlos HA (2011) A multi-sectoral economy energy environment model: application to Portugal. Energy 36:2856–2866

Chung WS, Tohno S, Shim SY (2009) An estimation of energy and GHG emission intensity caused by energy consumption in Korea: an energy IO approach. Appl Energy 86:1902–1914

Cormio C, Dicorato M, Minoia A, Trovato M (2003) A regional energy planning methodology including renewable energy sources and environmental constraints. Renew Sust Energ Rev 7:99–130

Dong C, Huang GH, Cai YP, Liu Y (2012) An inexact optimization modeling approach for supporting energy systems planning and air pollution mitigation in Beijing city. Energy 37:673–688

Fleten SE, Kristoffersen TK (2008) Short-term hydropower production planning by stochastic programming. Comput Oper Res 35(8):2656–2671

Gustavsson L, Madlener R (2003) CO2 emission cost of large-scale bio-energy technologies incompetitive electricity markets. Energy 28:1405–1425

Heinrich G, Howells M, Basson L, Petrie J (2007) Electricity supply industry modelling for multiple objectives under demand growth uncertainty. Journal of Energy 32:2210–2229

Huang GH, Loucks DP (2000) An inexact two-stage stochastic programming model for water resources management under uncertainty. Civ Eng Environ Syst 17:95–118

Huang GH, Sae-Lim N, Liu L, Chen Z (2001) An interval-parameter fuzzy-stochastic programming approach for municipal solid waste management and planning. Environ Model Assess 6:271–283

Iniyan S, Sumath K (2000) An optimal renewable energy model for various end-uses. Energy 25:563–575

Klaassen G, Riahi K (2007) Internalizing externalities of electricity generation: an analysis with MESSAGE-MACRO. Energy Policy 35:815–827

Kristoffersen TK (2007) Stochastic programming with applications to power systems. Dissertation, University of Aarhus, Denmark

Kwaczek A, Baker-Stariha BD, Macrae KM, Reinsch AE (1996) Modeling air emission control strategies for Saskatchewan. Canadian Energy Research Institute (CERI), Alberta, Canada

Lehtila A, Pirila P (1996) Reducing energy related emissions: using an energy systems optimization model to support policy planning in Finland. Energy Policy 24:805–819

Li YP, Huang GH, Veawab A, Nie XH, Liu L (2006a) Two-stage fuzzy-stochastic robust programming: a hybrid model for regional air quality management. J Air Waste Manage Assoc 56:1070–1082

Li YP, Huang GH, Nie SL (2006b) An interval-parameter multi-stage stochastic programming model for water resources management under uncertainty. Adv Water Resour 29:776–789

Li YP, Huang GH, Nie XH, Nie SL (2008) An inexact fuzzy-robust two-stage programming model for managing sulfur dioxide abatement under uncertainty. Environ Model Assess 13:77–91

Li YP, Huang GH, Chen X, Cheng SY (2009) Interval-parameter robust minimax-regret programming and its application to energy and environmental systems planning. Energy Sources Part B 4(3):278–294

Liu YY (2007) A dynamic two-stage energy systems planning model for Saskatchewan. Dissertation, University of Regina, Canada

Liu L, Huang GH, Fuller GA, Chakma A, Guo HC (2000) A dynamic optimization approach for nonrenewable energy resources management under uncertainty. J Pet Sci Eng 26:301–309

Mehdi H, Ibrahim D, Marc AR (2013) Hybrid solar–fuel cell combined heat and power systems for residential applications: Energy and energy analyses. Journal of Power Sources 221:372–380

Motevasel M, Seifi AR, Niknam T (2011) Multi-objective energy management of CHP (combined heat and power)-based micro-grid. Energy 51:123–136

Sailor DJ (1997) Climate change feedback to the energy sector: developing integrated assessments. World Resource Review 9(3):301–316

USDE and USEPA (2000) Carbon dioxide emissions from the generation of electric power in the United States., http://leaderresources.ca/wp-content/uploads/2014/03/DOE-CO2-Emissions-Generation-Electric-Power-2000.pdf

Zhang C, May MM, Heller TC (2001) Impact on global warming of development and structural changes in the electricity sector of Guangdong Province, China. Energy Policy 29(3):179–203

9

Large-strain consolidation modeling of mine waste tailings

Maki Ito and Shahid Azam[*]

Abstract

Background: Sustainable management of mine waste tailings during operation, closure, and reclamation requires a clear understanding of modeling the large-strain consolidation behaviour of these loose and toxic slurries. A state-of-the-art was presented focusing on process phenomenology and coordinate systems for tailings dewatering thereby devising a simple constitutive equation with a small number of input parameters. A one-dimensional self-weight consolidation model for quiescent conditions was developed using the finite element method. Test data on oil sand fine tailings were used for model training and predictions were made for an upper bound and a lower bound of various tailings types using a 1 m high hypothetical column.

Results: Results indicated that hydraulic conductivity along with specific gravity dictated pore water pressure dissipation and effective stress development with respect to both time and depth. Likewise, volume compressibility and initial solids was found to govern the void ratio reduction and solids content increase with respect to both time and depth.

Conclusions: The developed model requires a small numbers of input parameters and is capable of capturing the behaviour of a wide range of tailings. Depending on field conditions, the model can predict multiple filling conditions and various types of drainage systems in tailings containment facilities by incorporating appropriate boundary conditions.

Keywords: Large-strain consolidation; Mine waste tailings; Numerical modeling

Background

Large volumes of mine tailings (solid minerals suspended in chemical-rich liquids) are generated worldwide as by-products of ore beneficiation. Conventionally, these loose slurries are deposited on ground with perimeter dykes constructed from their relatively coarser fraction. Such facilities are notorious for a high failure rate and the resulting economic, social, health, and environmental issues (Azam and Li, 2010). To reduce tailings footprint, next-generation containment strategies include waste disposal in mined-out pits or in thickening vessels. The storage capacity of these facilities depends on the dewatering behaviour of the deposited material under self weight. The settling process (rate and amount) is influenced by complex physicochemical phenomena at solid-liquid interfaces. Whereas field monitoring and laboratory testing are

* Correspondence: Shahid.Azam@URegina.CA
Environmental Systems Engineering, Faculty of Engineering and Applied Science, University of Regina, 3737 Wascana Parkway, Regina SK, S4S 0A2, Canada

routinely carried out to understand the consolidation behaviour of the placed tailings, these methods are quite expensive, time consuming, and material specific. Sustainable waste management during operation, closure, and reclamation requires a general understanding of tailings consolidation.

The numerical prediction of large-strain consolidation properties of slurries has evolved over the years. Non-linear consolidation equations were independently formulated by Mikasa (1965) and Gibson et al. (1967) and were subsequently modified by Koppla (1970) and Somogyi (1980) to facilitate mathematical solution. Various deposition conditions such as quiescent, staged filling, surcharge loading, and initial solids content were analyzed by Townsend and McVay (1990) and Priestley (2011). Likewise, different forms of the constitutive relationships were employed (Caldwell et al., 1984; Jeeravipoolvarn et al., 2009a) and consolidation models were extended to include sedimentation (Azam et al., 2009; Jeeravipoolvarn et al., 2009b). The main problems with such models are the large number of input parameters and the complexity in solving the constitutive equations. Furthermore, most of the models were developed for material specific consolidation properties and a general purpose large-strain model is currently non-existent (Bartholomeeusen et al., 2002; Priestley et al., 2011).

The main objective of this paper was to study the large-strain consolidation modeling of mine waste tailings. Initially, a stat-of-the-art was established focusing on process phenomenology and coordinate systems for tailings dewatering thereby devising a simple constitutive equation with a small number of input parameters. Next, a one-dimensional self-weight consolidation model was developed for quiescent conditions because of negligible lateral drainage, absence of surcharge loading, and diminishing effect of filling on underlying sediments in the containment facilities. Finite element analysis was chosen for model development because of its robustness in capturing the changes in material coordinates during large strain tailings consolidation. The model was applied to capture the behaviour of a wide range of tailings using the tailings classification scheme of Paul and Azam (2013) that captures physicochemical interactions arising from ore geology and mill processing. Finally, test data on oil sand fine tailings were used for model calibration and predictions were made for an upper bound and a lower bound of various tailings types using a 1 m high hypothetical column.

State of the art

Figure 1 describes the settling phenomenon of a slurry. The self-weight settling versus the elapsed time plot (Figure 1a) for a deposited tailings consists of sedimentation and consolidation. Initially, the slurry settles with a rapid decrease in the interface height from point 'A' through 'B'. This hindered sedimentation refers to the settling of a spatial network of soil particles without measurable effective stress (McRoberts and Nixon, 1976). Simultaneously, a sediment starts to form at the bottom (point 'C') and grows over time following a sediment formation line that eventually meets with the solid-liquid interface at point 'B'. The sediment undergoes consolidation as the solid grains are in contact and transmit effective stresses (Terzaghi et al., 1996). Subsequent deformation between point 'B' and point 'D' is attributed to the expulsion of pore pressure. The boundary between sedimentation and consolidation is not fixed and is governed by the inherent properties of the slurry. Additional deposition results in similar plots for each layer. Overall, all of these interface height versus time plots can be summarized in constitutive relationships for volume compressibility ($e - \sigma'$) and hydraulic conductivity ($k - e$), as shown in Figure 1b and 1c, respectively. These relationships explain the void ratio (e) decrease with an increase in effective stress (σ') along with a corresponding decrease in hydraulic conductivity (k).

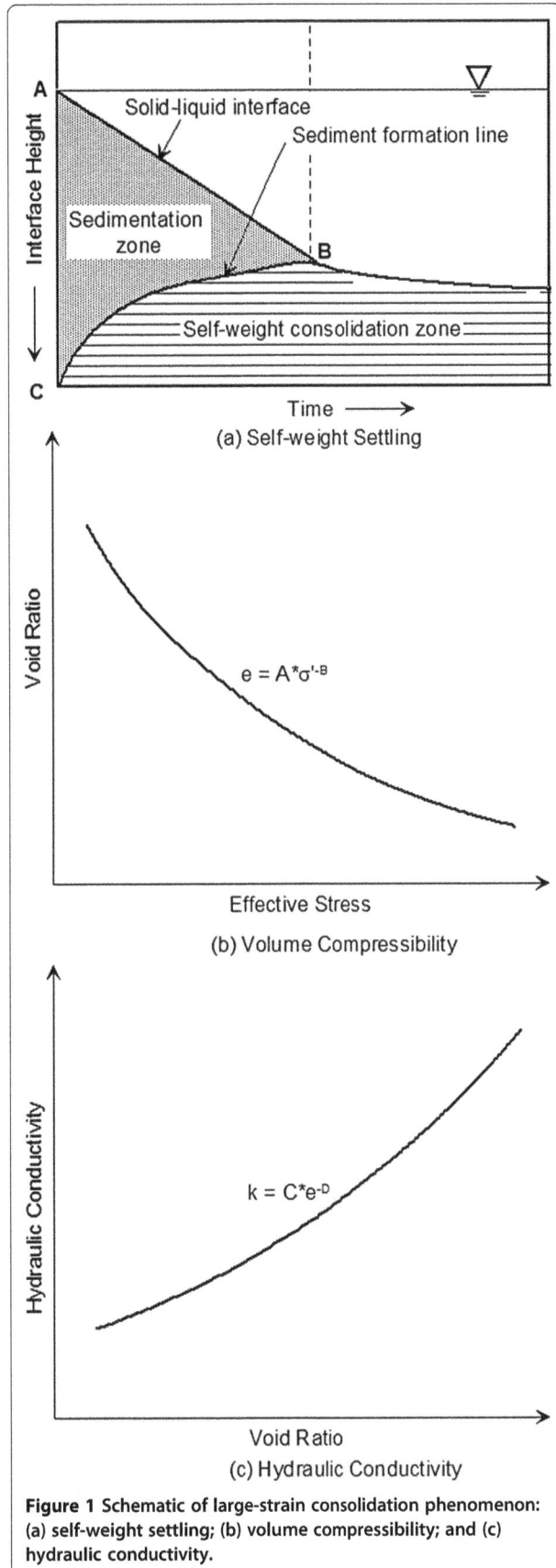

Figure 1 Schematic of large-strain consolidation phenomenon: (a) self-weight settling; (b) volume compressibility; and (c) hydraulic conductivity.

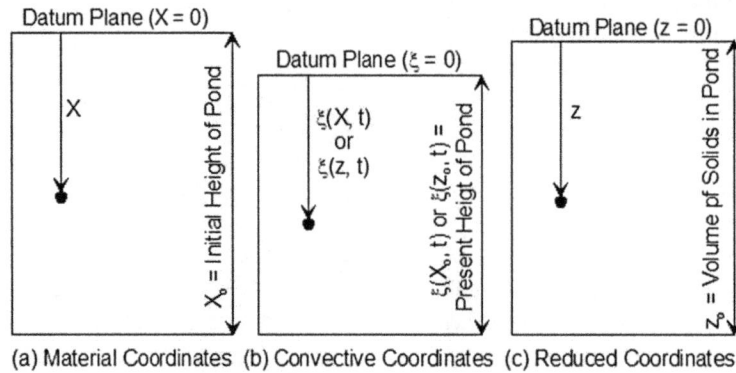

Figure 2 Schematic of coordinate systems for large-strain consolidation modeling: (a) material coordinates; (b) convective coordinates; and (c) reduced coordinates.

Generally, these relationships are best described using power law functions and the fit parameters (A, B, C, D) are used for numerical modeling.

Denoting time by t, depth with respect to datum by x, excess pore water pressure by u, and coefficient of consolidation by C_v, the governing equation for infinitesimal deformation in clays can be written as follows (Terzaghi et al., 1996):

$$\frac{\partial u}{\partial t} = C_v \frac{\partial^2 u}{\partial x^2} \qquad (1)$$

This theory assumes a linear stress-strain relationship, a constant hydraulic conductivity, and infinitesimal strain. The low compressibility of clays allows the use of Eulerian reference frame in which material deformation is related to a fixed plane in space. Because of significant volume changes, the use of Eulerian coordinate system (in which flux rate and soil movement are measured with respect to a fixed reference plane) is not valid for tailings consolidation.

Figure 2 describes the coordinate systems pertinent to tailings consolidation. The material coordinate system

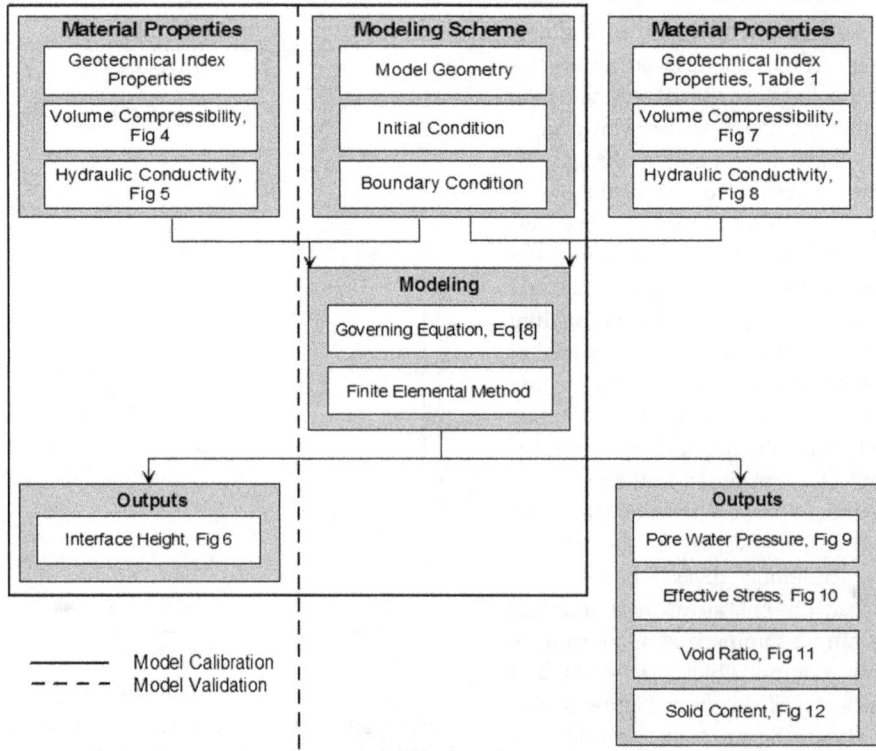

Figure 3 Flow chart of numerical modeling for large-strain consolidation.

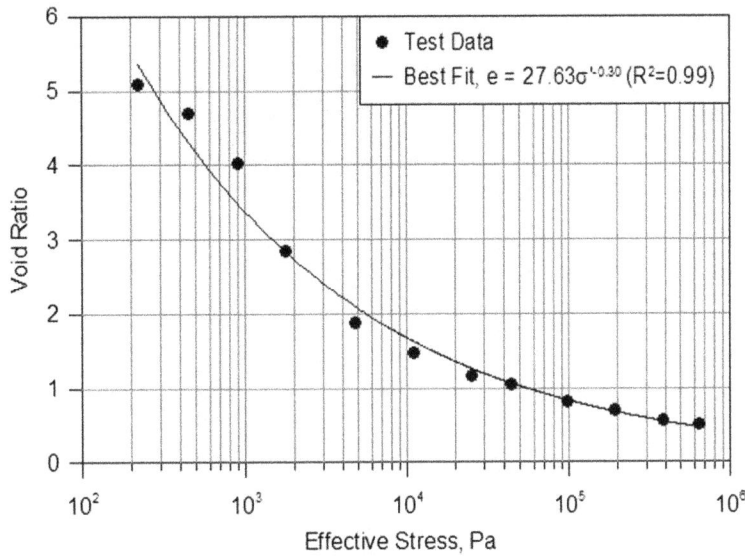

Figure 4 Volume compressibility relationship for oil sand fine tailings.

(Figure 2a) focuses on a particle that occupies position X from a datum ($X = 0$) located at the tailings upper surface at time $t = 0$. The datum ($\xi = 0$) shifts downward during consolidation such that the particle's new position (ξ at time, t) can be represented by the convective coordinate system (Figure 2b). This system focuses on the space occupied by the particle instead of on the particle itself. To solve the governing equation (written in the material coordinate system), the geometry needs to be updated for each time step thereby increasing the computation time. The reduced coordinate system (Figure 2c) focuses on the solid volume (z) occupied between the fixed datum ($z = 0$) and the point of interest at time, t. It simplifies numerical implementation because this system does not require updated geometry at each time step. The datum settlement can be obtained by converting the solution of the reduced coordinate system to the appropriate coordinate system, that is, material coordinate system for $t = 0$ and convective coordinate system for $t > 0$. The coordinate transformation among the three systems uses the following conversions (McNabb, 1960):

$$\frac{\partial X}{\partial \xi} = \frac{1 + e_o}{1 + e} \qquad (2)$$

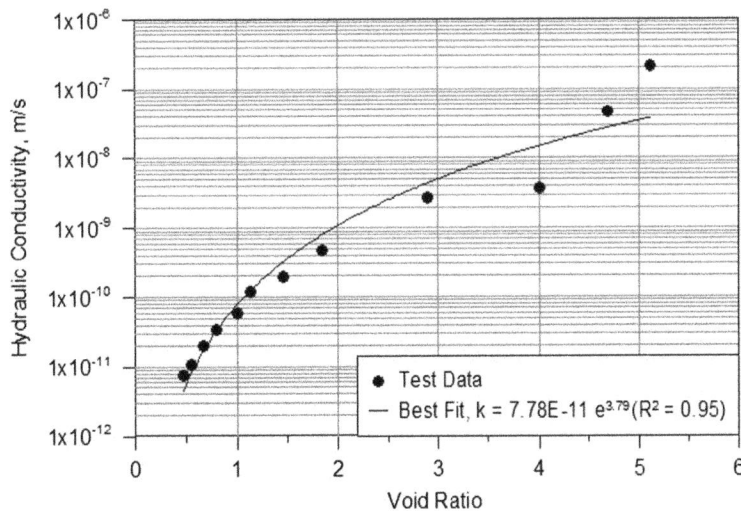

Figure 5 Hydraulic conductivity relationship for oil sand fine tailings.

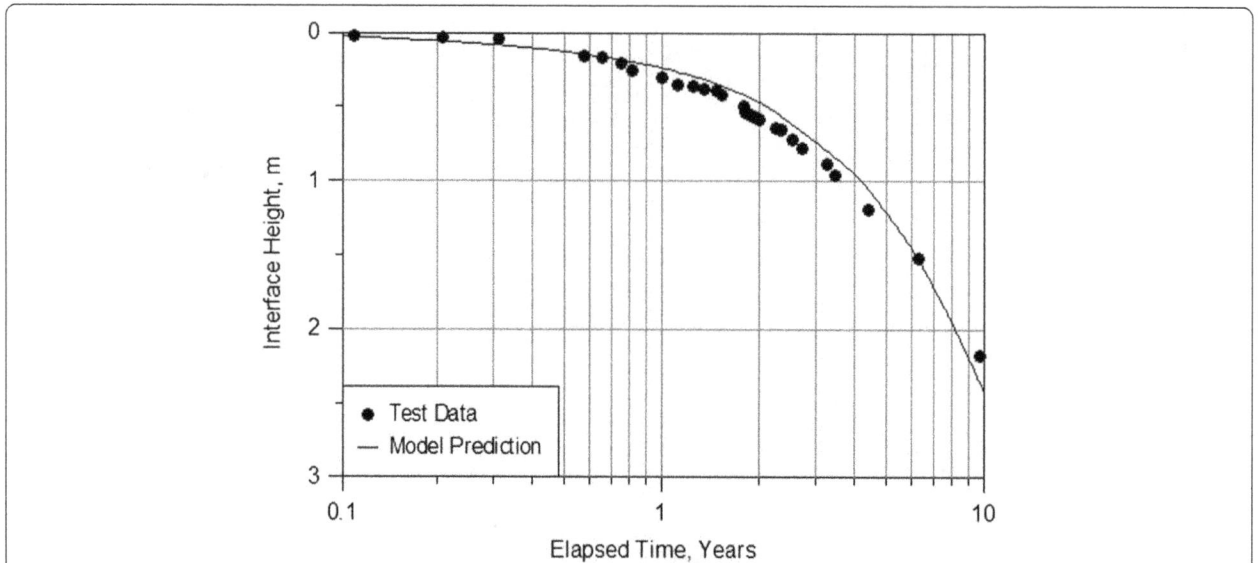

Figure 6 Model training results in terms of interface settlement versus elapsed time.

$$\frac{\partial z}{\partial X} = \frac{1}{1 + e_o} \tag{3}$$

$$\frac{\partial \xi}{\partial z} = 1 + e \tag{4}$$

The reduced coordinate system in the consolidation theory allows for the formulation of a governing equation with inherent large deformations due to the self-weight of materials. Using vertical equilibrium, continuity balance, and a fluid flow relationship, Gibson et al. (1967) developed the following equation for one dimensional large-strain consolidation:

$$\pm \left(\frac{\gamma_s}{\gamma_f} - 1 \right) \frac{d}{de} \left[\frac{k}{1+e} \right] \frac{\partial e}{\partial z} + \frac{\partial}{\partial Z} \left[\frac{k}{1+e} \frac{d\sigma'}{de} \frac{\partial e}{\partial z} \right] + \frac{\partial e}{\partial t} = 0 \tag{5}$$

Koppla (1970) formulated the following equation by rearranging continuity and flow:

$$\frac{\partial}{\partial z} + \left[-\frac{k}{\gamma_w (1+e)} \frac{\partial u}{\partial z} \right] + \frac{de}{d\sigma'} \frac{\partial \sigma'}{\partial t} = 0 \tag{6}$$

Denoting specific gravity of the solids by G_s and the coordinate difference between the surface and the point in question by Δz, Somogyi (1980) updated the above equation using the following time dependent effective stress:

$$\frac{\partial \sigma'}{\partial t} = (G_s - 1)\gamma_f \frac{d(\Delta z)}{dt} - \frac{\partial u}{\partial t} \tag{7}$$

Combining equations 6 and 7 leads to a governing equation in terms of excess pore pressure:

$$\frac{\partial}{\partial z} \left[-\frac{k}{\gamma_f (1+e)} \frac{\partial u}{\partial z} \right] + \frac{de}{d\sigma'} \left[(G_s - 1)\gamma_f \frac{d(\Delta z)}{dt} - \frac{\partial u}{\partial t} \right] = 0 \tag{8}$$

Table 1 Geotechnical index properties of selected fine-grained tailings

Slurry type	G_s	w_L (%)	w_P (%)	I_P (%)	USCS Symbol	Reference
(a) Sedimentary clays						
China clay	2.66	53	32	21	MH	Znidarcic et al., 1986
Georgia Kaolin	2.6	44	25	19	CL	Znidarcic et al., 1986
Phosphate Slime	2.77	100	40	60	CH	Roma, 1976
(b) Residual soils						
Laterite	3.16	83	42	41	MH	Azam et al., 2009
Bauxite	3.05	54	40	14	MH	Cooling, 1985
Uranium	2.81	32	27	5	ML	Matyas et al., 1984
(c) Oil sand tailings						
Ore A–C, RPW	2.55	50	26	24	CH	Miller et al., 2010
Ore A–NC, TRW	2.51	60	31	29	CH	Miller et al., 2010
Ore B–C, RPW	2.48	52	27	25	CH	Miller et al., 2010
Ore B–NC, TRW	2.45	58	28	30	CH	Miller et al., 2010
Ore A–NC, URW	2.5	55	28	27	CH	Miller et al., 2010
Desanded	2.65	57	25	32	CH	Lord and Liu, 1998
Cyclone overflow	2.53	50	21	29	CH	Jeeravipoolvarn et al., 2009a

Extraction Process: *C*, caustic; *NC*, non-caustic.
Pore Water: *RPW*, recycled pond water; *TRW*, treated river water; *URW*, untreated river water.

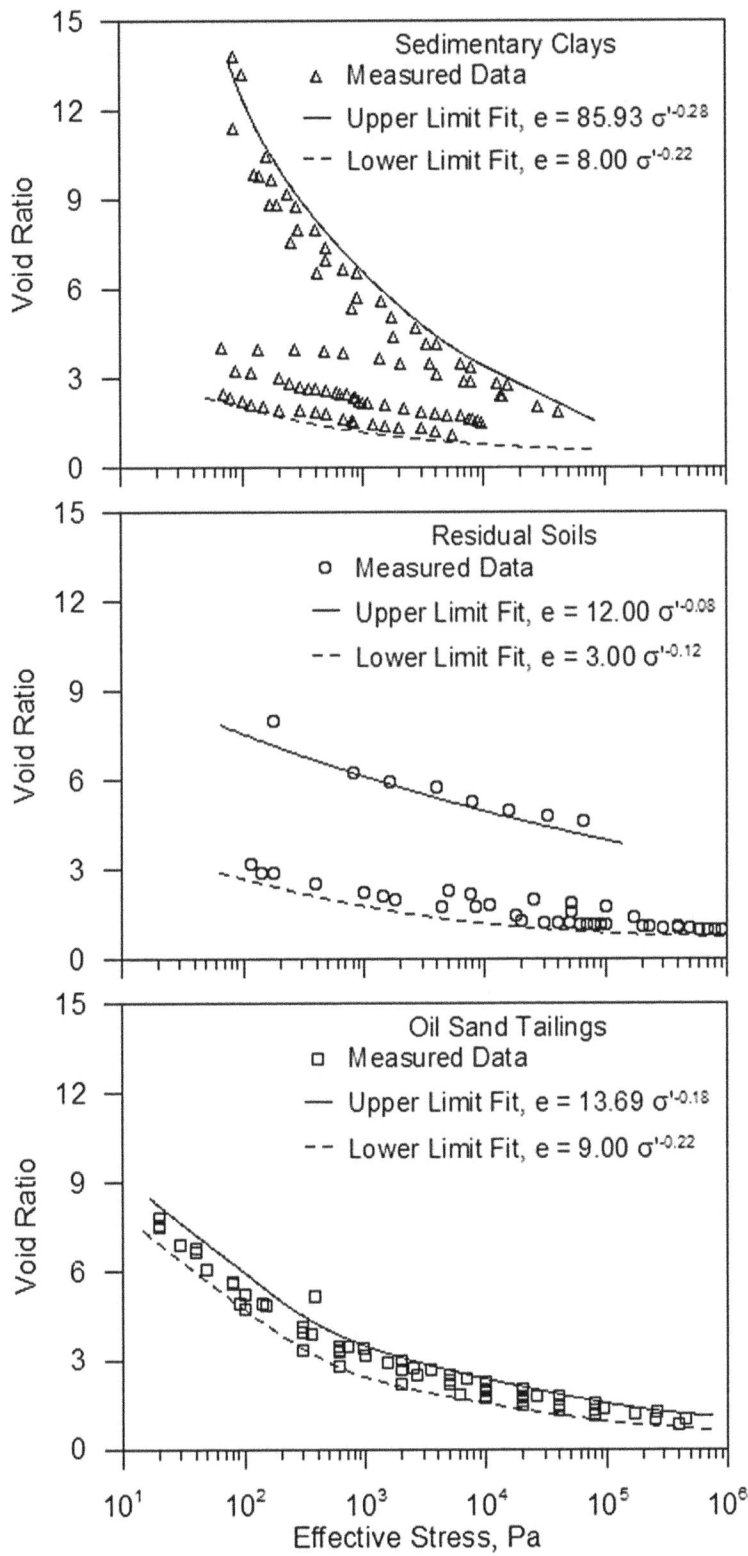

Figure 7 Volume compressibility relationships for various types of tailings.

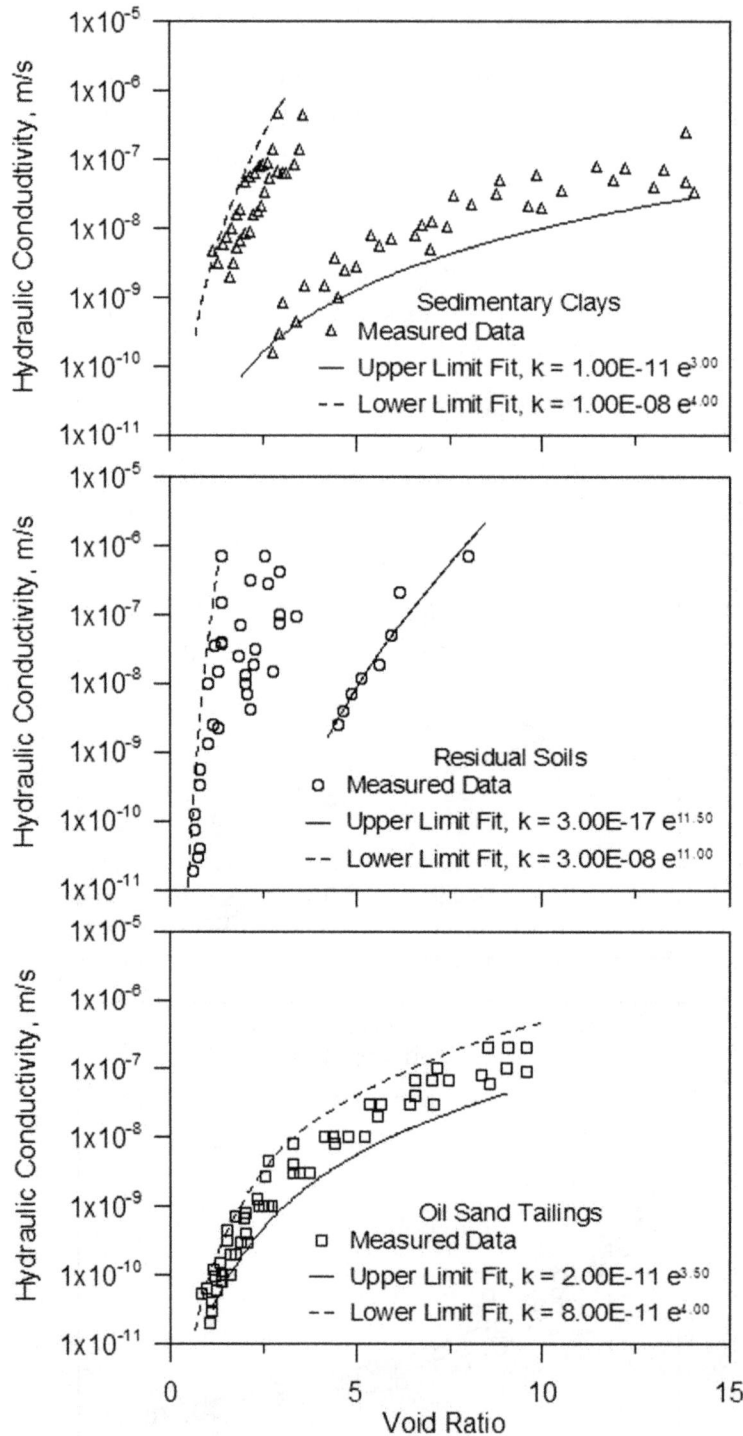

Figure 8 Hydraulic conductivity relationships for various types of tailings.

Model development

Figure 3 describes the large-strain consolidation modeling process. Tailings contained in a column with no horizontal discharge and homogeneous material properties provided the rationale to develop a one-dimensional model. The model required four input parameters related to tailings properties, namely: volume compressibility (A and B), hydraulic conductivity (C and D), initial solids content, and specific gravity. The constitutive relationships were obtained from published large-strain consolidation data.

The model geometry consisted of a vertical tailings column (10 m high for model training and 1 m high for model validation) with no drainage at the bottom and dewatering in the upward direction. Quiescent conditions were simulated using 10 years for model training and 1 year for model validation. The governing equation 8 was solved in terms of pore water pressure by using a general partial differential equation solver FlexPDE. The solver utilized the finite elemental method generating a triangular mesh over a quasi-two dimensional geometry. Finite element analysis was chosen because of its robustness in capturing the changes in material coordinates during large-strain tailings consolidation. The adequacy of the mesh was constantly calculated and the automatic mesh refinement feature of the solver was applied to reduce error to a tolerance of (0.001). Using the script editing facility, the governing equation and material properties were directly put in the model compiler. An

equation analyzer expanded the defined equation and the material properties, performed spatial differentiation, and applied integration by parts thereby reducing second order terms to create symbolic Galerkin equations for use in the weighted residual method. The Galerkin equations were further differentiated to form the Jacobian coupling matrix for improved convergence. Likewise, the solution curvature was also calculated to include time integration for better accuracy. The model outputs were in the form of pore water pressure, effective stress, void ratio, and solid content profiles.

Results and discussion

The model was trained using published data of oil sand fine tailings (G_s = 2.28; initial solids content of 31%) in a 10 m standpipe. Figures 4 and 5 give the volume compressibility and the hydraulic conductivity relationships for the investigated tailings. Power law fits were applied

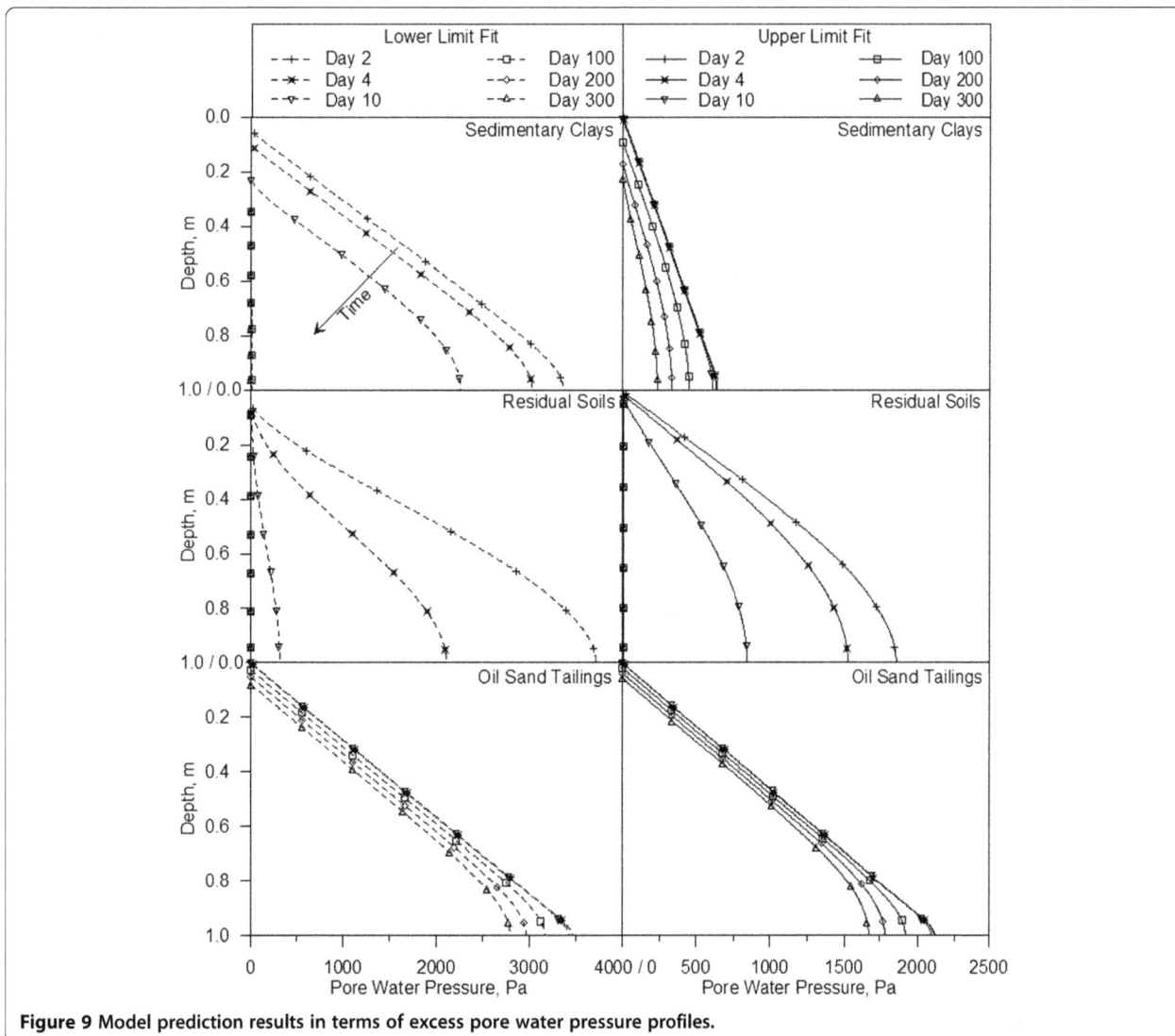

Figure 9 Model prediction results in terms of excess pore water pressure profiles.

to handle the nonlinearity in both relationships. A detailed description of material behaviour was provided by Jeervipoolvarn et al. (2009a).

Figure 6 compares the modeling results with the measured data from Jeeravipoolvarn et al. (2009a) in the form of interface height plotted over 10 years. The modeling results closely approximated the measured data over the entire time. The relatively lower measured values after about 9 months are attributed to an increased possibility for evaporation and leakage from the standpipe and external disturbances due to natural and anthropogenic activity.

The model was validated to determine the effectiveness of capturing the consolidation behaviour of various tailings. Table 1 gives the geotechnical index properties of selected fine-grained slurries. The slurries were classified into sedimentary clays, residual soils and oil sand

tailings based on their geological origin and mining operation. The specific gravity (G_s) of sedimentary clays ranged from 2.6 to 2.77, which is the typical range for such materials (Wesley, 2010). Conversely, residual soils showed higher G_s values due to the presence of ferrous minerals such as goethite and hematite in laterite (Azam, 2005), and gibbsite and hematite in bauxite (Newson et al., 2006); the higher G_s (in comparison with sedimentary clays) of uranium tailings was attributed to the high amount of muscovite (Paul and Azam, 2013). The low G_s (2.55 ± 0.1) for oil sand tailings was due to the presence of bitumen ($G_b = 1.03$) with variations attributable to the variable amount of bitumen in the different samples (Miller et al., 2010). The consistency limits varied over a wide range due to the presence of different types and amounts of clay minerals. Although the consistency limits are influenced greatly by the type of process

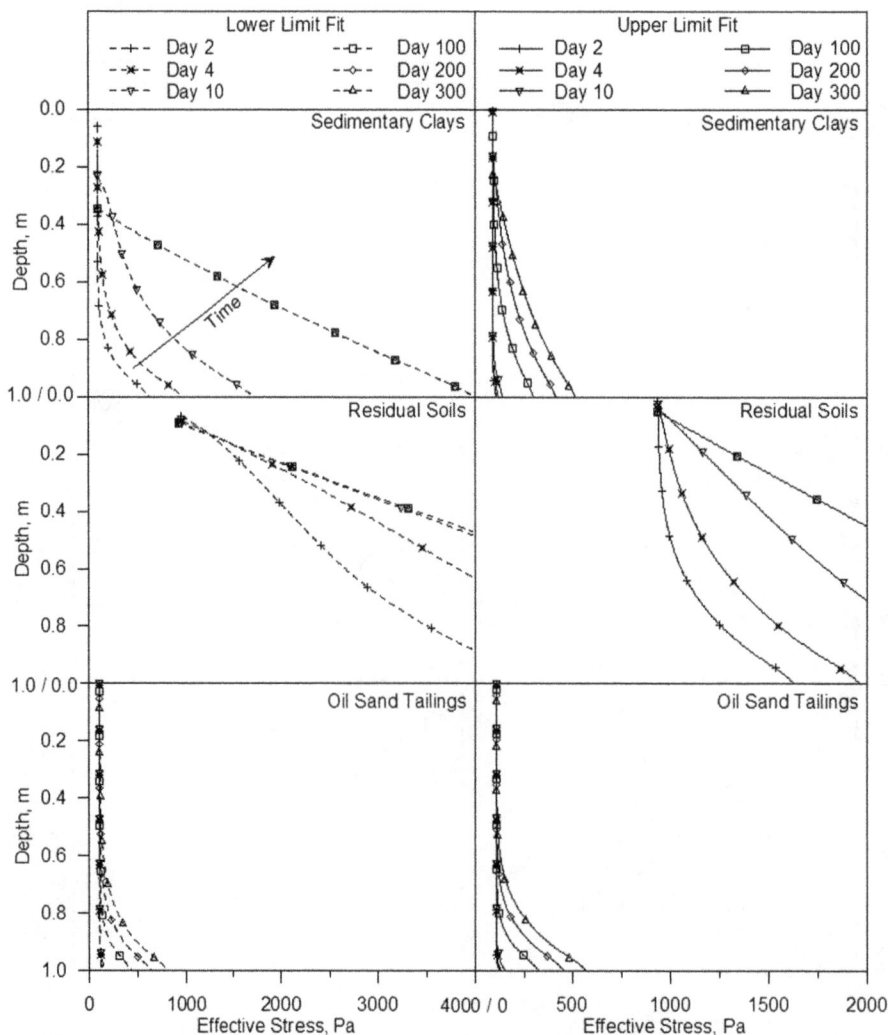

Figure 10 Model prediction results in terms of effective stress profiles.

water, data on the chemical composition of water was not readily available.

Figure 7 plots void ratio versus effective stress data for various types of tailings. Whereas all materials exhibited a decreasing trend, each material group had significant variations in void ratio especially at low effective stresses. Therefore, an upper limit and a lower limit were used to cover the entire range of material properties within each group. Power law equations were found to best describe the volume compressibility behaviour of the investigated tailings.

Figure 8 gives void ratio versus hydraulic conductivity data for various types of tailings. All materials followed an increasing power law function and showed large variations in hydraulic conductivity at high void ratios. Once again, the upper and lower limits (corresponding to those in Figure 7) were used to capture the wide range of hydraulic conductivity for each material group.

Figures 9 and 10 presents the pore water pressure and the effective stress profiles for a 1 m high column. In the first plot, the progress of self-weight consolidation is observed by the gradual increase in the vertical component of the curves that tend to reach the ordinate. Conversely, the same phenomenon is represented by a diminishing nonlinearity in the latter plot. Overall, the lower limit predictions showed better pore water dissipation and effective stress development compared to the upper limit predictions because of low volume compressibility (Figure 7) and high hydraulic conductivity (Figure 8) in the former case. The lower limit of the sedimentary clays and all of the residual soils completed self-weight consolidation within 300 days. On the contrary, the upper limit of the sedimentary clays and all of the oil sands fine tailings were found to be undergoing consolidation up to that time. Overall, results indicated that hydraulic conductivity (Figure 8)

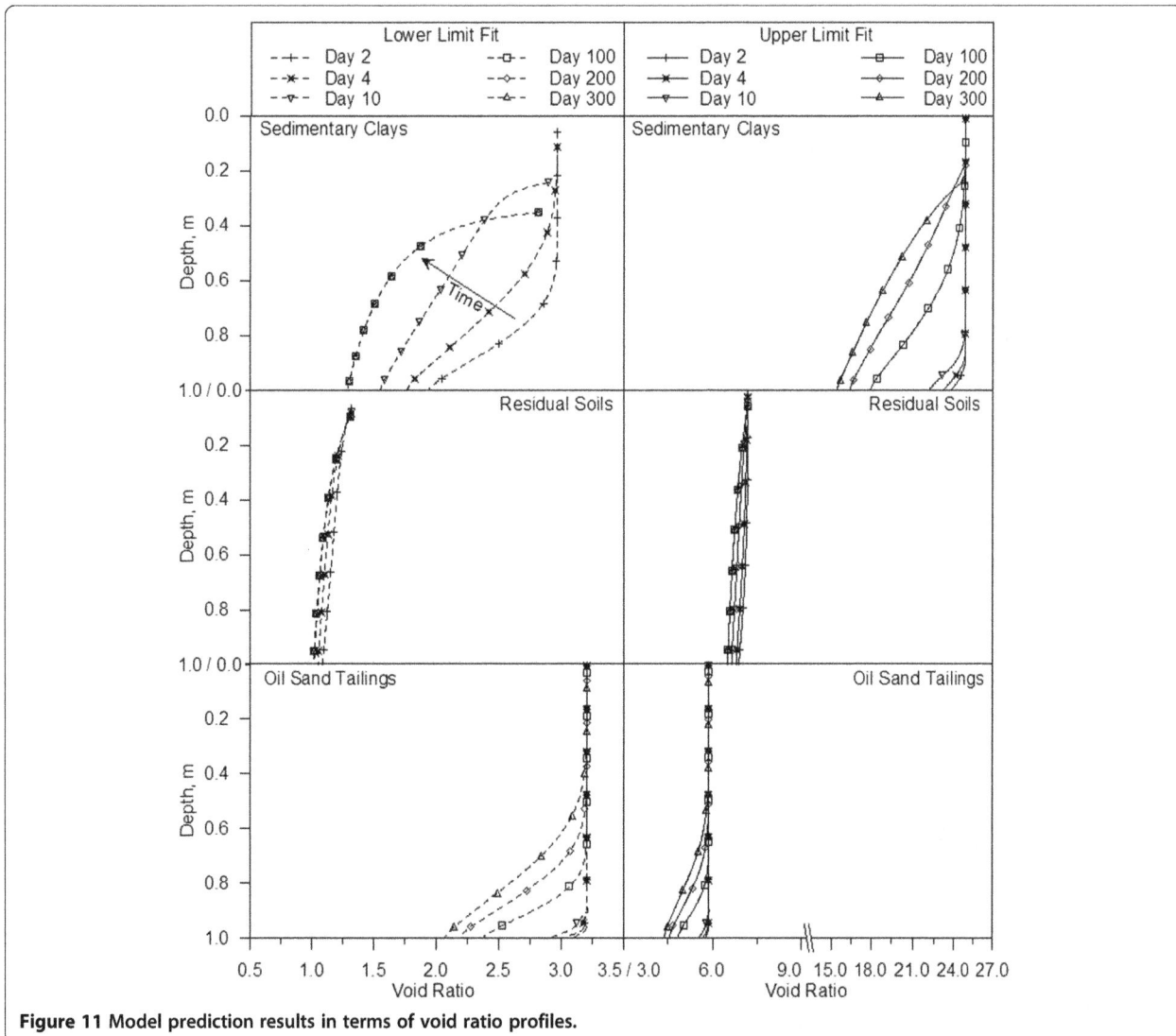

Figure 11 Model prediction results in terms of void ratio profiles.

along with specific gravity (Table 1) dictated the rate of consolidation with respect to both time and depth, as evident from pore water pressure dissipation (Figure 9), and effective stress development (Figure 10).

Figures 11 and 12 shows the void ratio and the solid content profiles for a 1 m high column. In both plots, the progress of self weight consolidation is observed through the curve shape that changes from concave to convex over time. Generally, the predictions with lower limit fits showed better void ratio reduction and solids content increase compared to the upper limit predictions due to the afore-mentioned reasons. Similar to Figures 9 and 10, the lower limit of the sedimentary clays and all of the residual soils finished self weight consolidation within 300 days. Conversely, the upper limit of the sedimentary clays and the entire oil sand fine tailings continued to undergo consolidation during that time. Overall, volume compressibility (Figure 7)

and initial solids was found to govern the amount of consolidation with respect to both time and depth, as given by void ratio reduction (Figure 11) and solids content increase (Figure 12).

Conclusions

A one-dimensional self weight consolidation model for quiescent conditions was developed in a finite element code. The model was calibrated using oil sand fine tailings and validated through a wide range of tailings. The one dimensional model captured the dewatering behaviour because of the homogeneous nature of fine grained tailings. The developed model requires a small numbers of input parameters and capable of capturing the behaviour of a wide range of tailings. Depending on field conditions, the model can predict multiple filling conditions and various types of drainage systems in tailings containment facilities by incorporating appropriate boundary

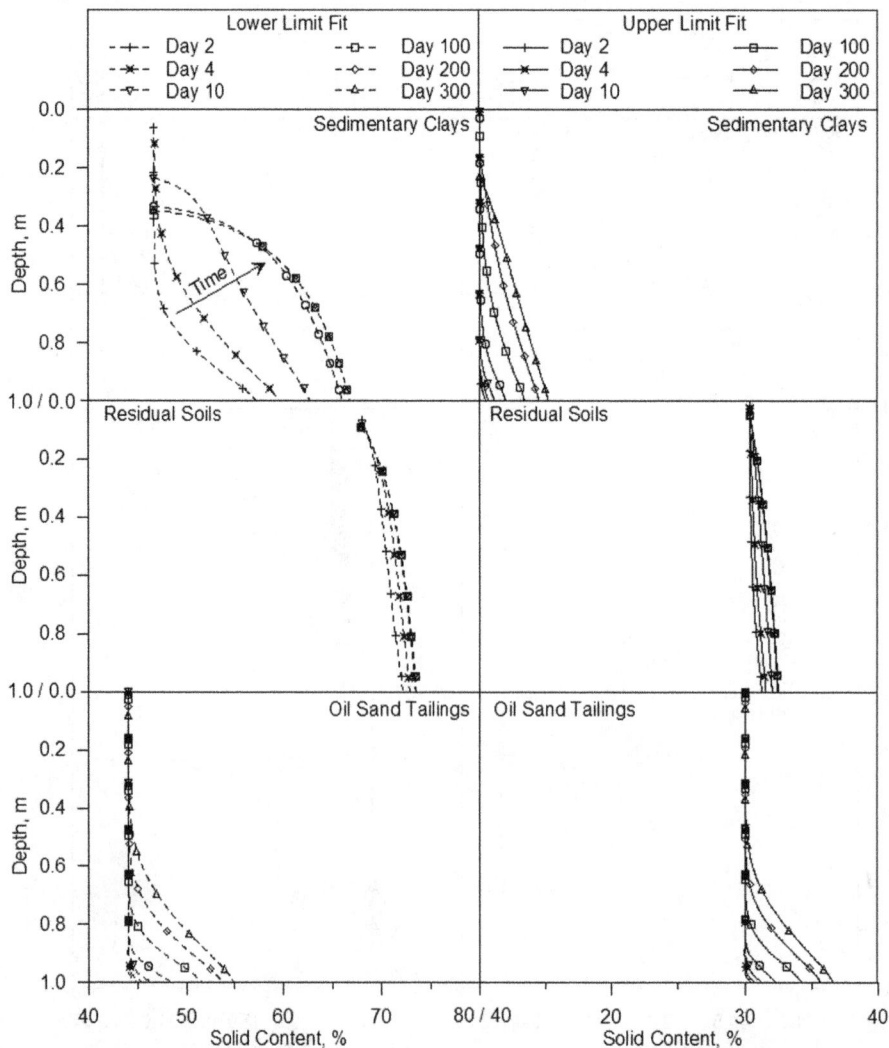

Figure 12 Model prediction results in terms of solids content profiles.

conditions. The general purpose model can be improved to include heterogamous tailings using a multi-dimensional approach. Results indicated that hydraulic conductivity (Figure 8) along with specific gravity (Table 1) dictated the rate of consolidation with respect to both time and depth, as evident from pore water pressure dissipation (Figure 9) and effective stress development (Figure 10), Likewise, volume compressibility (Figure 7) and initial solids was found to govern the amount of consolidation with respect to both time and depth, as given by void ratio reduction (Figure 11) and solids content increase (Figure 12). The model captured the self-weight consolidation behaviour for a wide range of tailings. The load-deformation based model can be extended to include the sedimentation part of slurry settling thereby incorporating the effect of physiochemical interactions.

Competing interests

The authors declare that they have no competing interests.

Authors' contributions

MI carried out numerical modelling and drafted the manuscript. SA provided conceptual guidance and polished the manuscript. Both authors read and approved the final manuscript.

Acknowledgements

The authors acknowledge the financial support provided by the Natural Science and Engineering Research Council of Canada and the computation facilities provided by the University of Regina.

References

Azam S, Chalaturnyk RJ, Scott JD (2005) Geotechnical characterization and sedimentation behavior of laterite slurries, Geotechnical Testing Journal. ASTM 28(6):523–533

Azam S, Jeeravipoolvarn S, Scott JD (2009) Numerical modeling of tailings thickening for improved mine waste management. J Environ Inform 13(2):111–118

Azam S, Li Q (2010) Tailings dam failures: A review of the last one hundred years, Geotechnical News. BiTech Publishers Limited 28(4):50–53

Bartholomeeusen G, Sills GC, Znidarčić D, Van Kesteren W, Merckelbach LM, Pyke R, Carrier WD, Lin H, Penumadu D, Winterwerp H, Masala S, Chan D (2002) Sidere: numerical prediction of large-strain consolidation. Geotechnique 52:639–648

Caldwell JA, Ferguson K, Schiffman RL, van Zyl D (1984) Application of Finite Strain Consolidation Theory for Engineering Design and Environmental Planning of Mine Tailings Impoundments, Sedimentation Consolidation Models - Predictions and Validation, pp 581–606

Cooling DJ (1985) Finite strain consolidation of red mud residue from the alumina refining industry MASc. Thesis. Western Australian Institute of Technology, Perth, Western Australia

Gibson RE, England GL, Hussey MJL (1967) The theory of one-dimensional consolidation of saturated clays. I. Finite non-linear consolidation of thin homogeneous layers. Géotechnique 17(3):261–273

Jeeravipoolvarn S, Chalaturnyk RJ, Scott JD (2009a) Sedimentation–consolidation modeling with an interaction coefficient. Comp Geotech 36:351–361

Jeeravipoolvarn S, Scott JD, Chalaturnyk RJ (2009b) 10 m standpipe tests on oil sands tailings: long-term experimental results and prediction. Can Geotech J 46:875–888

Koppla SD (1970) The Consolidation of Soils in Two-Dimensions and with Moving Boundaries. PhD Thesis. The University of Alberta, Edmonton, Canada

Lord ERF, Liu Y (1998) Depositional and geotechnical characteristics of paste produced from oil sands tailings, Proceedings of 5th International Conference on Tailings and Mine Waste. Balkema, Rotterdam, pp 147–157

Matyas EL, Welch DE, Reades DW (1984) Geotechnical parameters and behavior of uranium tailings. Can Geotech J 21:489–504

McNabb A (1960) A mathematical treatment of one-dimensional soil consolidation. Q Appl Math 17(4):337–347

McRoberts EC, Nixon JF (1976) A theory of soil sedimentation. Can Geotech J 13:294–310

Miller WG, Scott JD, Sego DC (2010) Influence of the extraction process on the characteristics of oil sands fine tailings, Journal of Canadian Institute of Mining. Metallurgy Petroleum 1(2):93–112

Mikasa M (1965) The consolidation of soft clay: A new consolidation theory and its application. Japanese Society of Civil Engineers, Civil Engineering in Japan, pp 21–26

Newson T, Dyer T, Adam C, Sharp S (2006) Effect of structure on the geotechnical properties of bauxite residue. J Geotech Geoenviron Eng, ASCE 132(2):143–151

Paul AC, Azam S (2013) Assessment of slurry consolidation using index properties. Int J Geotech Eng 7(1):55–62

Priestley D (2011) Modeling Multidimensional Large Strain Consolidation of Tailings. MSc Thesis. University of British Columbia, British Columbia, Canada

Priestley D, Fredlund M, van Zyl D (2011) Modeling consolidation of tailings impoundment in one and two dimensions. Proceedings, Tailings and Mine Waste 2011, Vancouver, BC, Canada

Roma JR (1976) Geotechnical properties of Florida phosphatic clay, MS Thesis. Massachusetts Institute of Technology, Cambridge, MA

Terzaghi K, Peck RB, Mesri G (1996) Soil Mechanics in Engineering Practice, 3rd edition. John Wiley and Sons, Inc., Hoboken, New Jersey, USA

Townsend FC, McVay MC (1990) SOA: Large strain consolidation predictions. ASCE J Geotech Eng 116(2):222–243

Somogyi F (1980) Large Strain Consolidation of Fine Grained Slurries. Presented at the Canadian Society for Civil Engineering, Winnipeg, MB

Wesley LD (2010) Fundamentals of Soil Mechanics for Sedimentary and Residual Soils. John Wiley and Sons, Inc., Hoboken, New Jersey, USA

Znidarcic D, Schiffman RL, Pane V, Croce P, Ko HY, Olsen HW (1986) Theory of one-dimensional consolidation of saturated clays: part V, constant rate of deformation testing and analysis. Géotechnique 36(2):227–237

10

Public warning systems for forecasting ambient ozone pollution in Kuwait

Eiman Tamah Al-Shammari

Abstract

Background: In this paper, the performances of different forecasting systems are compared using the daily maximum ozone levels across three locations in Kuwait. The two analytical tools used in this study to forecast daily maximum ozone levels are time series modeling and fuzzy modeling. The structure of the two proposed forecasting models are derived from basic principles, which include a combination of persistence and daily maximum air temperature as input variables.

Results: The two proposed forecasting models /showed significant improvement compared to the pure persistence forecast, which is the model currently used to forecast ambient air pollution in Kuwait. The performance of the two models suggests that daily maximum temperature explains a large proportion of the variation in ozone daily maximum levels.

Conclusions: This study concludes that fuzzy modeling is the most reliable forecasting system, with the lowest number of false positives among the different models.

Keywords: Ground-level ozone, Fuzzy analysis, Ozone forecast, Time series ozone forecast, Time series environmental warning system, Kuwait

Introduction

Industrialization, technical growth, and overpopulation in urban areas of Kuwait have resulted in increased air pollution (Abdul-Wahab et al. 2000), and toxic air pollutants in close proximity to populated areas can have adverse health effects. Ambient ozone is the primary constituent of smog and is a pollutant of concern in industrialized countries, as it can lead to chronic respiratory infection and lung inflammation, aggravate asthma, impair lung defense mechanisms, and reduce the immunity of the human body (Tilton 1989). In the environment, ozone may result in acute foliar injuries, reduce agricultural yield and biomass production, and shift the competitive advantages of plant species in mixed populations (Lefohn et al. 1994). In this respect, surface ozone has become a serious problem in the urban areas of Kuwait, and if it occurs in sufficient concentration, it could threaten both human health and the environment (Jallad and Jallad, 2010).

Ground-level ozone is formed by chemical reactions between nitrogen oxides (NO_x) and volatile organic compounds (VOC) in the presence of heat and sunlight. It is difficult to exactly define the formation and destruction mechanism of ozone. This is because ozone is an extremely reactive pollutant and can be scavenged by its precursors (Dimitriades 1989). As a result, the area of air pollution forecasting through empirical methods has gained importance with the availability of sufficient data. Earlier forecasting models were based on simple empirical data correlations, but the availability of a large amount of information has resulted in development of complex air pollution simulations for forecasting (Telenta et al. 1995).

Management of public warning strategies for ozone levels in densely populated areas requires accurate forecasts of ambient levels. Although ozone prediction models exist or have been proposed for several cities (Robeson and Steyn, 1990; Elsom 1996; Noordijk 1994; Yi and Prybutok 1996), they have not been assessed in realistic conditions. The purpose of this work is to explore the possibility of forecasting daily maximum ozone impacts in urban areas of Kuwait, where most measurements are taken at surface stations. The final goal is to produce a quantitative tool to help authorities monitor ozone

Correspondence: dr.eiman@ku.edu.kw
Kuwait University, Kuwait, Kuwait

pollution, which has become a public health issue in many cities worldwide.

Background

Kuwait is a principality state in the northeastern corner of the Arabian Peninsula with an estimated population of 3.5 million. It is a low lying country, with the highest point being 306 meters above sea-level. The annual rainfall across Kuwait varies from 75 to 150 millimeters, and the country has a desert climate—hot and dry (Federal Research Division 1993).

Kuwait is a major exporter of crude oil (ranked 4th in OPEC's list), with plans to increase production to 3.2 million barrels per day in 2013 (Arab times 2012). Such characteristics make Kuwait a country with air pollution associated with petroleum, petrochemical, and other industrial pollutants (Arab times 2012; Organization of the Petroleum Exporting Countries 2009).

Recently, there has been an increase in air pollution in urban areas of Kuwait. This is the result of rapid industrialization, technical growth, and overpopulation. In the past, it has been observed that toxic air pollutants in close proximity to populated areas can have adverse health effects. Ozone is one such pollutant, and if it occurs in sufficient concentration, it can become a serious problem in urban areas of Kuwait. Jallad and Jallad (2010) observed that ozone pollution levels in the Salmiya residential district exceeded ambient air quality standards during specific times of the year. Therefore, accurate forecasting of surface ozone is required, as it can help with successful implementation of public warning strategies during episodic days in Kuwait.

Most of the air pollution studies carried out in Kuwait during the past two decades were aimed at air pollutant patterns, dispersion, and photochemical mechanisms. Similarly, studies that analyzed ozone levels were focused on comparing ozone levels with international standard limits, assessing health effects of ozone pollution, understanding diurnal behavior of ozone, and studying the seasonal trends in ozone levels (Ettouney et al. 2009a; Ettouney et al. 2009b).

There have been very few studies focused on developing a robust forecasting system which can be used to develop a public warning system. Most of the forecasting systems that have been developed to predict the ambient ozone concentration in Kuwait use meteorological data and precursor concentrations (Ettouney et al. 2009a; Ettouney et al. 2009b; Ettouney et al. 2009c).

Abdul-Wahab et al. (1996) used stepwise multiple regression modeling to predict ozone levels from precursor concentrations and meteorological conditions during daylight hours in the Shuaiba Industrial Area (SIA) of Kuwait. Al-Alawi et al. (2008) applied principal component regression and artificial neural networks to predict ozone concentration in Kuwait's lower atmosphere using the data on seven environmental pollutant concentrations (CH_4, NMHC, CO, CO_2, NO, NO_2, and SO_2) and five meteorological variables (wind speed, wind direction, air temperature, relative humidity, and solar radiation). In their study, Elkamel et al. (2001) used an artificial neural network model to predict ozone concentrations as a function of meteorological conditions and precursor concentrations in the SIA.

Similarly, Abdul-Wahab and Al-Alawi (2002) used neural networks for ozone modeling in the lower atmosphere as a function of meteorological conditions and various air quality parameters in the Khaldiya residential district of Kuwait.

These statistical models are based on semi-empirical statistical relations among available data and measurements. They do not necessarily reveal any relation between cause and effect. They simply attempt to determine the underlying relationship between sets of input data (predictors) and targets (predictands).

However, the complex and sometimes nonlinear relationships of multiple variables can make these statistical models awkward and complicated. Therefore, it is expected that they will underperform when used to model the relationship between ozone and the other variables that are extremely nonlinear. In addition, over the last decade, artificial intelligence (AI) based techniques have been proposed as alternatives to traditional statistical ones for forecasting urban air pollution. Two of the most reliable and feasible AI techniques that have been used for urban air pollution forecasting are neural networks and fuzzy logic. Neural networks are simple mathematical models representing a brain-like system. Although considered as one of the most popular AI methods, neural networks also have inherent drawbacks that impede their practical applications. Scalability, testing, verification, and integration of neural network models into forecasting system are some of the major concerns today. Sometimes, neural network systems become unstable when they are applied to bigger problems. Compared to neural networks, fuzzy logic offers better insights into the forecasting model, and they can form a multivalued logic to deal with reasoning that is approximate rather than precise (Karlaftis and Vlahogianni 2011).

Prediction of ozone levels using a theoretical method (i.e. detailed atmospheric diffusion model) is difficult and empirical analysis is required to develop a forecasting system. A well-evaluated ozone forecast model can raise the possibility of successful ozone control strategy. In addition, forecasting the daily maximum ozone concentrations can help avoid and reduce ozone-related injuries and damages. This research is significant, as it is the first study comparing the ozone forecasting systems across different locations in Kuwait.

Research methodology

Dataset

The study used hourly air pollutants data from January 2007 through September 2011 gathered by the Environmental Monitoring Information System of Kuwait (eMISK, working under the Environment Public Authority of Kuwait). The data was collected from three different locations in Kuwait (see Figure 1): Al-Jahra (Tower A), Al-Mansouriya (Tower B), and Al-Rigga (Tower C). The initial three years of data was used to train the forecasting systems, and data from the remaining years was divided into subgroups to validate the forecasting models. The validation data was divided into subsets based on the availability of the data and seasonal variation.

Geographical location and a history of high ozone episodes were considered in selecting target sites. Tower A is located in Al-Jahra, the capital of the Al-Jahra Governorate of Kuwait and the surrounding agriculturally based Al-Jahra District. Tower B is located in one of the suburbs of the Al-Asimah Governorate, which houses most of Kuwait's financial and business center. Tower C is located in the Al-Ahmadi Governorate, which forms an important part of the Kuwaiti economy, as several of Kuwait's oil refineries are located there.

Model structure

Seinfeld (1986) proposed the atmospheric diffusion equation, from which the theoretical model (Pryor and Steyn 1995; Jorquera, and Acuna 1998) used for this study was developed:

$$\frac{dC}{dt} = -V \cdot \nabla C + \nabla \cdot (K \cdot \nabla C) + Q(x, y, z, t)$$
$$R(\{C_k\}, T, t) + L(x, y, z, t) \tag{1}$$

where $C(x, y, z, t)$ is the concentration, $V = (u, v, w)$ is the wind vector, $K = diag(K_x, K_y, K_z)$ is the eddy diffusivity, Q is the emission rate, R is the net-generation term (balance of chemical production and destruction) and L stands for physical removal processes such as wet and dry deposition. The mathematical expression in equation (1) represents the mass conservation principle and is the most general equation obeyed by any atmospheric pollutant.

In the case of ozone, the mass conservation expression can be simplified, because the emission term is not there and the deposition terms can be ignored. The ozone concentration can be assumed to be uniform in the mixed layer, as long as we take a moving air mass under the conditions of a strong convection.

Figure 1 The geographical location of air quality monitoring stations in Kuwait.

Under such conditions, the column of air satisfies the balance:

$$\frac{dC}{dt} = R(\{C_k\}, T, t) \tag{2}$$

Where $\{C_k\}$ stands for a detailed photochemical mechanism, which includes all the relevant species participating in ozone generation and destruction. These concentrations depend upon the spatial coordinates and time. Assessing the importance of VOC and NO_x in ozone photochemistry is not easy, as there are no previously measured VOC mixing ratios for Kuwait. Thus, to take a simpler approach, a stochastic model is constructed from equation (2). Integrating equation (2) from early morning rush hour up to the time when the ozone concentration is highest leads to the following equation:

$$O_3^{max} - O_3^{min} = \int_{t_o}^{t_f} R(\{C_k\}, T, t)dt \tag{3}$$

Temperature is steadily increasing during this time interval (Figure 2), so using $q = dT/dt$, the above expression can be written in temperature terms as:

$$O_3^{max} - O_3^{min} = \int_{T_{min}}^{T_{max}} R(\{C_k\}, T)\frac{dT}{q} \tag{4}$$

Taking the difference between two consecutive days results in:

$$O_{3,t+1}^{max} - O_{3,t}^{max} = O_{3,y+1}^{min} - O_{3,t}^{min} = \int_{T_{max,t}}^{T_{max,t+1}} R(\{C_k\}, T)\frac{dT}{q} \tag{5}$$

Using the mean value theorem for integrals on the right hand side of equation (5) and dropping the "max" superscript for daily maximum ozone and temperature levels, the theoretical model reduces to:

$$O_{3,t+1} = \alpha O_{3,t} + \beta T_{t+1} + \gamma T_t + \delta \tag{6}$$

Here, parameter α is positive and of order one, β and γ are of about the same magnitude (but differ in sign), and δ has a small value. The planetary boundary layer is well mixed, and the surface measurements are representative of the whole convective mixed layer, when the daily maximum ozone concentration is reached. Furthermore, the temperature recorded at the time when ozone is highest is strongly correlated with the highest temperature recorded that day, so in passing from equations (5) to (6) the interpretation of T_{max} is accordingly modified. Robeson and Steyn (1990) developed an equation similar to equation (6) and used variance analysis to find the best model fit among several slight variations of the equation.

Time series forecasting modeling

Presently three types of stochastic models to forecast air pollutants have been developed using the time-series approach (McCollister and Wilson 1975):

(a) Models based on time series of a single pollutant: AR, ARMA, ARIMA models.
(b) Multivariable time series, where meteorological explanatory variables have been added on: ARX, ARMAX models.
(c) Forecasting models based on a combination of classical and non-parametric methods.

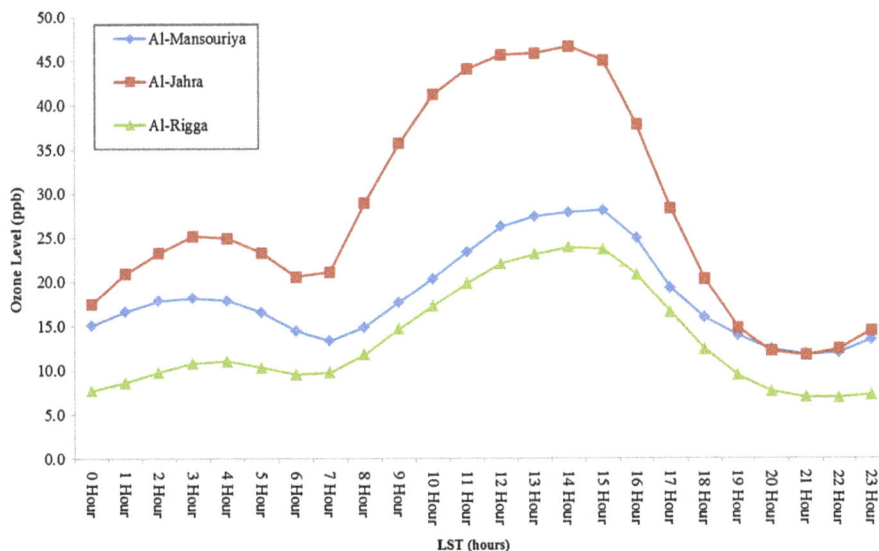

Figure 2 Evolution of typical photochemical pollution at different locations in Kuwait.

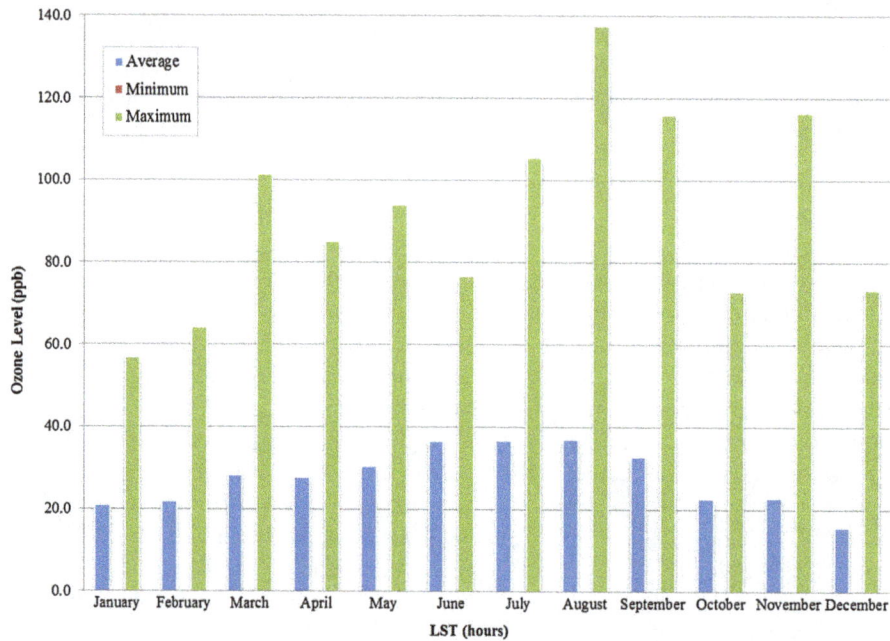

Figure 3 Monthly ozone levels in the Al-Jahra area.

The above citations make it clear that when the aim is on air pollution forecasting, any forecast model ought to be better than the pure persistence forecast (i.e. forecasting for tomorrow what occurred today).

Thus, in order to have an accurate forecast, some persistence and some exogenous, meteorological variables need to be included. Of all the meteorological variables,

air temperature has the strongest correlation with ozone concentrations for two reasons: (1) high air temperatures are an excellent indication of environmental conditions conducive to O_3 production and accumulation (i.e. anti-cyclonic conditions with associated clear skies and light winds), and (2) the rate 'constants' of photochemical reactions are highly temperature dependent (Pryor and

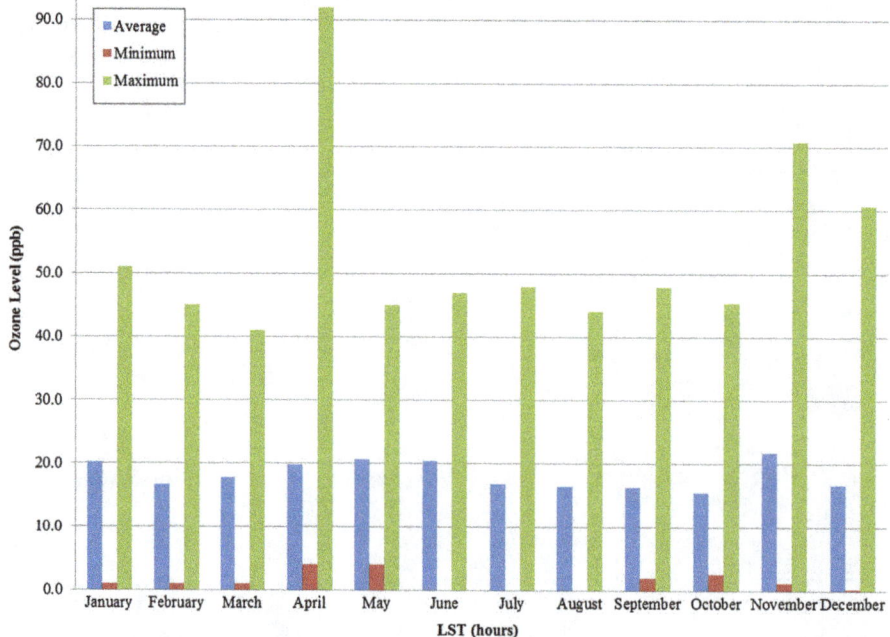

Figure 4 Monthly ozone levels in the Al-Rigga area.

Table 1 Parameter estimates in equation (6) using time series analysis

	α	β	γ	δ
AL-JAHRA	0.624(0.029)	0.166(0.068)	−0.026(0.068)	0.528(0.532)
AL-MANSOURIYA	0.175(0.040)	−0.186(0.116)	0.278(0.115)	0.949(0.959)
AL-RIGGA	0.429 (0.048)	−0.259 (0.144)	0.575(0.142)	0.562(0.572)

Steyn 1995). As a result, air temperature is a reasonable surrogate for the combined effects of wind speed, wind direction, inversion height, and photochemical reaction rate. As discussed earlier, the simple linear equation (6), which considers only surface air temperature as the exogenous variable, will be used. In all cases the actual maximum temperature (T_{t+1}) was used in place of the forecast value (that is, an ex post parameter fit was performed). The errors associated with forecasting T_{t+1} will be assessed ex ante by simulation. Equation (6) has an ARX (1 2 0) structure (Ljung 1987), and its parameters were estimated using the ARX procedure available in MATLAB's System Identification Toolbox (Ljung 1991).

Fuzzy modeling

A fuzzy model is a collection of rules, derived from the original data, which are combined to produce a single output for a given input. A multiple input, single output (MISO) model structure is given by the following set of rules:

$$R^i : \text{if } X \text{ is } A^i \text{ then } y \text{ is } B^i \quad \forall i = 1, \ldots, M \qquad (7)$$

Where $X = (X_1, X_2, \ldots, X_p)$ is a given observation, $A^i = \left(A_1^i, A_2^i, \ldots, A_p^i\right)$ is a premise of the fuzzy model; the points A^i are a basis for the space of the input variables. The consequences of the model are denoted by B^i; they

Table 2 Parameter estimates in model (4) using fuzzy model

	α	β	χ	δ
AL-JAHRA				
Rule 1	0.324	−0.520	−0.128	−0.021
Rule 2	0.657	0.394	−0.516	0.405
Rule 3	1.016	0.040	−0.735	−1.298
AL-MANSOURIYA				
Rule 1	0.293	0.171	0.661	−0.171
Rule 2	0.000	−0.092	0.293	0.134
Rule 3	−0.124	0.988	1.203	−3.290
AL-RIGGA				
Rule 1	0.496	0.500	0.076	0.398
Rule 2	0.621	−5.493	8.288	−5.624
Rule 3	0.367	1.653	0.446	0.287

represent an intensity level associated with the dependent variable y and are the basis of the output space. Equation (7) states that, if a given observation X (for instance, meteorological and air quality measurements) can be associated with a known pattern A^i (for instance, some class of meteorological condition), then a rule specific to A^i will give an estimate of the associated output y (for example, ground-level ozone concentration). By using a fuzzy

Table 3 Statistical comparison of forecasting models

		Persistence	Time Series Modeling	Fuzzy Modeling
AL-JAHRA				
Dataset1 (Episodes = 2)	RMSE	0.999	0.855	0.865
	IA	0.701	0.704	0.720
	Correct Prediction	0	0	1
	False Positive	2	2	0
Dataset2 (Episodes = 9)	RMSE	1.149	0.983	0.966
	IA	0.574	0.554	0.585
	Correct Prediction	1	1	1
	False Positive	8	6	3
Dataset3 (Episodes = 3)	RMSE	1.244	1.040	1.035
	IA	0.479	0.500	0.529
	Correct Prediction	0	0	0
	False Positive	3	3	2
Dataset4 (Episodes = 15)	RMSE	1.152	0.991	0.926
	IA	0.568	0.529	0.655
	Correct Prediction	2	2	3
	False Positive	11	7	6
AL-MANSOURIYA				
Dataset1 (Episodes = 25)	RMSE	0.793	0.885	0.785
	IA	0.822	0.414	0.758
	Correct Prediction	15	1	1
	False Positive	9	5	0
Dataset2 (Episodes = 0)	RMSE	0.941	0.9154	0.861
	IA	0.741	0.3891	0.765
	Correct Prediction	0	0	0
	False Positive	0	0	0
AL-RIGGA				
Dataset1 (Episodes = 12)	RMSE	0.991	0.802	0.801
	IA	0.693	0.701	0.712
	Correct Prediction	5	2	1
	False Positive	7	5	2
Dataset2 (Episodes = 0)	RMSE	1.090	0.862	0.850
	IA	0.643	0.574	0.662
	Correct prediction	0	0	0
	False Positive	0	0	0

clustering classification process, the set of points A^i can be discriminated and then a fuzzy model constructed. A simple choice was proposed by Takagi and Sugeno (1985); they suggested linear rules of the form:

$$R^i : \text{if } X \text{ is } A^i \text{ then } Y^i = C_o^i + \sum_{j=1}^{j=p} c_j^i x_j \qquad (8)$$
$$\forall \cdot i = 1, \ldots, M$$

Then the parameters $\{c^i\}$ can be estimated using ordinary least squares, by minimizing the difference between the observed output and the output of the fuzzy model, given by:

$$Y = \frac{\sum_{i=1}^{M} w_i Y_i}{\sum_{i=1}^{M} w_i} \qquad (9)$$

In order to develop fuzzy models, using fuzzy clustering techniques for parameter identification, the model structure given by equation (6) was used. In fuzzy modeling language, the sets ($O_{3, t}$; T_{t+1}; T_t) and ($O_{3, t+1}$) correspond to the premises and consequences, respectively. The fuzzy C-means algorithm (Rousseeuw et al. 1996; Johanyak, and Kovacs 2011; Sugeno and Yasukawa 1993) was applied to affect the partition of the original data set into M fuzzy clusters; the parameters were estimated using MATLAB's Fuzzy Logic Toolbox.

Model performance
The performance of the two forecasting models was evaluated using two statistical indices: the root-mean-square error (RMSE) and the index of agreement (IA), defined as:

$$RMSE = \sqrt{\frac{1}{N} \sum_{i=1}^{N} (o_i - p_i)^2}$$
$$IA = 1 - \frac{\sum_{i=1}^{N} (o_i - p_i)^2}{\sum_{i=1}^{N} \left(|o_i'| + |p_i'|^2 \right)} \qquad (10)$$

Here o_i and p_i are observed and forecasted ozone maximum values on day i, N is the number of days in the test set, $p_i' = p_i - o_m$ and $o_i' = o_i - o_m$, with o_m, the average observed ozone maximum. The index of agreement is a dimensionless index bound between 0 (showing no agreement at all) and 1 (showing perfect agreement of the time series).

Results and discussions
Evolution of photochemical pollution
The evolution of typical photochemical pollution at different locations in Kuwait is summarized in Figure 2. It was observed that the ozone levels at different locations started rising in the morning (around 7:00 AM), reached maximum levels during the afternoon (11:00 AM – 4:00 PM), and then started falling again in the evening. Similarly, the monthly ozone levels at different locations in Kuwait are summarized in Figure 3 through Figure 4. It was observed that during summers (June to August), the average ozone levels are higher compared to the average ozone levels during the rest of the year. Thus, it can be

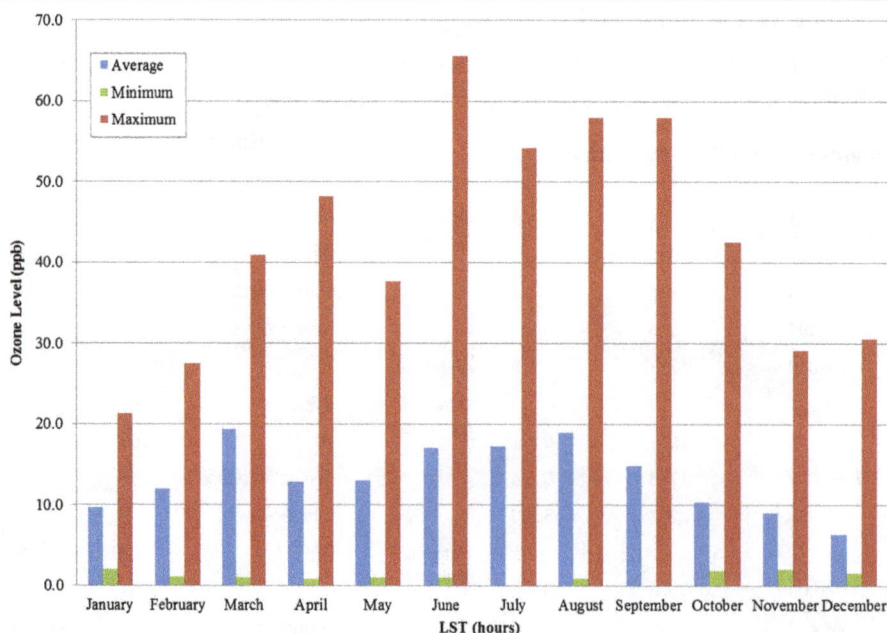

Figure 5 Monthly ozone levels in the Al-Mansouriya area.

concluded that there is a strong association between the ambient temperature and the surface ozone levels.

Empirical findings (time series modeling)

The results in Table 1 shows the parameter estimates for the time series model—presented in equation (6)—developed for the period 2007 through 2009. There is a great amount of fluctuation in the daily maximum temperature during the year; therefore, the data subsets (based on seasonal duration and missing data points) were used individually in the model development process. There are four data subsets used for each of the locations covered under this study. An additional data subset for model validation was also used. It can be seen from the parameter estimates in Table 2 that the fitted values support the model structure derived in equation (6). Additionally, it can be seen that the fitted parameters associated with each location show great amounts of variation during different time periods.

Empirical findings (fuzzy modeling)

The parameters estimated for Towers A, B, and C using fuzzy model are summarized in Table 2. For the fuzzy model, a different number of rules were found for each of the data subsets. Each of these rules corresponds to a linear model valid for a specific cluster obtained from the original data by means of a fuzzy classification. The parameters of these rules are similar to the ones estimated for the time-series models. The predicted output results from a nonlinear combination of these rules.

Comparison of models

The results in Table 3 compare the forecasting for the three towers: Al-Jahra, Al-Mansouriya, and Al-Rigga, using the pure persistence model, the linear time series model, and the fuzzy model. Clearly, the root mean square error shows that the proposed fuzzy forecasting model and time series forecasting model are a significant improvement over the pure persistence forecast ($O_{3,t+1} = O_{3,t}$), for all data subsets. Regarding the observed episodes, the three models achieve a high percentage of correct forecasts, between 66 and 95%. As shown in Figures 3 and 5, the overall quality of the forecast results supports the model structure proposed in equation (6) and the most relevant input variables have been identified. The most significant performance difference among these three forecasts occurs on the number of false positives; fuzzy models are consistently lower, and therefore afford a more reliable forecast. It is worth noticing that the fuzzy forecast performs better in this sense, so it is less sensitive to changes in the parameter values. Nonetheless, all models ought to be calibrated once a year to take into account trends in ozone ground levels and to keep the forecast accurate.

Conclusion

The structure proposed in equation (6) was tested and validated by comparing the outcomes of two different modeling schemes, namely linear time series modeling and fuzzy modeling. These two forecasting systems showed a significant improvement over the pure persistence forecast. The impact of temperature on daily maximum ozone levels was confirmed by deducing the model from basic principles.

Furthermore, the accuracy of the results and the performance of the two models suggest that temperature explains a large proportion of variance in the daily maximum ozone levels. However, the accuracy of the models in forecasting very high ozone concentrations is very low. *Ad hoc* corrections may improve local forecasts, but such models will likely be site specific. Among the different forecasting systems, the fuzzy system showed better performance in terms of lower numbers of false positives for all validation datasets. Thus, fuzzy modeling can be used to develop an effective environmental warning system (EWS) in Kuwait.

Competing interests

The authors declare that they have no competing interests.

References

Abdul-Wahab S, Bouhamra W, Ettouney H, Sowerby B, Crittenden BD (2000) Analysis of ozone pollution in the shuaiba industrial area in Kuwait. Int J Env Stud 57(2):207–224

Jallad KN, Jallad CE (2010) Analysis of ambient ozone and precursor monitoring data in a densely populated residential area of Kuwait. J Saudi Chem Soc 14:363–372

Tilton BE (1989) Health effects of tropospheric ozone. Environ Sci Technol 23:257–263

Lefohn AS, Edwards PJ, Adams MB (1994) The characterization of ozone exposures in rural west Virginia and Virginia. J Air Waste Manag Assoc 44:1276–1283

Dimitriades B (1989) Photochemical oxidant formation: overview of current knowledge and emerging issues, *Atmospheric ozone research and its policy implications*, 35; studies in environmental science. Elsevier Science Publishers, Amsterdam, pp 35–43

Telenta B, Alfksic N, Dacic M (1995) Application of the operational synoptic model for pollution forecasting in accidental situations. Atmos Env 28:2885–2891

Robeson SM, Steyn DG (1990) Evaluation and comparison of statistical forecast models for daily maximum ozone concentrations. Atmos Env 24B:303–312

Elsom D (1996) Smog alert: managing urban Air quality. Earthscan Publications Limited, London

Noordijk H (1994) The national smog warning system in the Netherlands; a combination of measuring and modeling, Air pollution, 2. Pollution Control and Monitoring; WIT Press, Southampton, pp 169–176

Yi J, Prybutok VR (1996) A neural network model forecasting for prediction of daily maximum ozone concentration in an industrialized urban area. Environ Pollut 92:349–357

Federal Research Division (1993) Kuwait: A country study. Kessinger Publishing, Whitefish, Montana

Arab Times (2012) Kuwait plans Oil production increase to 3.2 Million BPD., Retrieved from http://www.gulfbase.com/news/kuwait-plans-oil-production-increase-to-3-2-million-bpd/218917

Organization of the Petroleum Exporting Countries (2009) World Oil outlook. OPEC Secretariat, Vienna, Austria

Ettouney RS, Abdul-Wahab S, Elkilani AS (2009a) Emissions inventory, ISCST, and neural network: modeling of Air pollution in Kuwait. Int J Env Stud 66:181–194

Ettouney RS, Mjalli FS, Ettouney H, Zaki JG, El-Rifai MA, Ettouney H (2009b) Forecasting ozone pollution using artificial neural networks. Mgmt Env Quality 20:668–683

Ettouney RS, Mjalli FS, Zaki JG, El-Rifai MA, Ettouney HM (2009c) Forecasting of ozone pollution using artificial neural networks. Mgmt Environ Quality An Int J 20(6):668–683

Abdul-Wahab S, Bouhamra W, Ettouney H, Sowerby B, Crittenden BD (1996) Predicting ozone levels: a statistical model for predicting ozone levels. Environ Sci Pollut Res 3:195–204

Al-Alawi SM, Abdul-Wahab SA, Bakheit CS (2008) Combining principal component regression and artificial neural networks for more accurate predictions of ground-level ozone. Env Model Soft 23(4):396–403

Elkamel A, Abdul-Wahab S, Bouhamra W, Alper E (2001) Measurement and prediction of ozone levels around a heavily industrialized area: a neural network approach. Adv Environ Res 5(1):47–59

Abdul-Wahab SA, Al-Alawi SM (2002) Assessment and prediction of tropospheric ozone concentration levels using artificial neural networks. Env Model Soft 17(3):219–228

Seinfeld JH (1986) Atmospheric chemistry and physics of Air pollution. John Wiley & Sons, New York

Pryor SC, Steyn DG (1995) Hebdomadal and diurnal cycles in ozone time series from the lower Fraser valley, B. C. Atmospheric Environ 29:1007–1019

Jorquera H, Perez R, Acuna G (1998) Forecasting ozone daily maximum levels at Santiago. Chile Atmos Env 32:3415–3424

McCollister GM, Wilson KR (1975) Linear stochastic models for forecasting daily maxima and hourly concentrations of Air pollutants. Atmos Env 9:417–423

Ljung L (1987) System identification: theory for the user. Prentice-Hall: Englewood Cliffs, New Jersey

Ljung L (1991) System identification toolbox for use with MATLAB. The Math Works Inc, Natick, MA

Takagi T, Sugeno M (1985) Fuzzy identification systems and its applications to modeling and control. IEEE Transact Sys Man Cybernetics 1:116–132

Rousseeuw PJ, Kaufman L, Trauwaert E (1996) Fuzzy clustering using scatter matrices. Comput Stat Data Analysis 23:135–151

Johanyak ZC, Kovacs J (2011) Fuzzy model based prediction of ground-level ozone concentration. Acta Technica Jaurinensis 4:113–125

Sugeno M, Yasukawa T (1993) A fuzzy-logic based approach to qualitative modeling. IEEE Trans Fuzzy Syst 1:7–31

Karlaftis MG, Vlahogianni EI (2011) Statistical methods versus neural networks in transportation research: differences, similarities and some insights. Transp Res 19:387–399

11

The Ontario nuclear power dispute: a strategic analysis

Motahareh Armin[1], Keith W Hipel[1,2]* and Mitali De[3]

Abstract

Background: The Graph Model for Conflict Resolution methodology is used to formally investigate the nuclear power dispute that took place in the Canadian province of Ontario in order to obtain strategic insights into its resolution. This flexible systems methodology is used to study the nuclear conflict at two key points in time, 2008 and 2010.

Results: The results of the 2008 analysis show that the only decision makers involved in the conflict who hold real power are the Federal and Ontario governments, although at the beginning of the investigation other organizations had also been considered as participating decision makers. According to a strategic analysis carried out for the conflict as it existed in 2010, the equilibria or potential resolutions of the 2008 analysis are found to be transitional states leading to the 2010 resolution. Moreover, a negative attitude by the Federal Government can cause an outcome to occur that is not highly preferred by either the Federal Government or the province of Ontario.

Conclusions: By closely following the decision makers' actions, a detailed analysis of the nuclear dispute in Ontario is carried out. Stability, sensitivity, and attitude analyses are performed, and the results are closely correlated with what happened in reality.

Keywords: Ontario nuclear power dispute, Graph model for conflict resolution

Background

The conflict between the Federal Government of Canada and the Ontario Provincial Government has been on-going for the past couple of years. The Ontario Government intends to expand the Darlington nuclear site and plans to procure its reactors from Atomic Energy of Canada Limited (AECL), the company responsible for the building, maintenance, and management of CANDU (CANada Deuterium Uranium) Canadian reactors.

Meanwhile, the Federal Government announced the decision of restructuring and selling or privatizing AECL. Nuclear technology has been very important to Canada since its genesis. Huge investments have been made in this industry by the Federal Government of Canada over the years. If AECL is sold and the Ontario Government does not buy its reactors from this company, it is very possible that no other province will make any purchases from AECL in the future.

CANDU reactors use a technology that is unique among other reactors in the world, and their safety standards are very strict. On March 11, 2011, an earthquake and tsunami occurred in Japan and caused serious damage to the reactors in the affected region. One of the main issues regarding Japanese reactors is that they emit radioactive water, whereas in the CANDUs, the steam and water in the secondary loops are not radioactive. Therefore, in a critical event, when heat release is required, the water in the secondary loop can be let out with no damage to the environment, and new clean water can be used. The other difference is that unlike Japan, all CANDUs are located in areas where tsunamis may not affect them, and environmental factors are taken into careful consideration in the design of CANDU reactors as well as advanced versions thereof (Canadian Nuclear Safety Commission, 2011).

* Correspondence: kwhipel@uwaterloo.ca
[1]Department of Systems Design Engineering, University of Waterloo, Waterloo, Ontario N2L3G1, Canada
[2]Centre for International Governance Innovation, Waterloo, Ontario N2L 6C2, Canada
Full list of author information is available at the end of the article

CANDU technology is a unique treasure to Canada. It would be a disaster for Canada to miss the opportunity of pioneering this exclusive heavy water CANDU technology as well as other versions of its reactors. Experts believe that whatever rearrangements in AECL management take place, the involved parties have to make sure that AECL remains a Canadian-owned company, or else Canada will forfeit this historic opportunity for leadership in nuclear technology, as occurred in the aircraft industry when Canada's Avro Arrow aircraft technology was destroyed (Hipel and Bowman, 2011). Considering the importance of this technology and the conflict over the Canadian nuclear industry, the key purpose of this research is to model the aforementioned conflict by using the Graph Model for Conflict Resolution (GMCR) methodology. This approach realistically models the conflict between two or more players having different options under their control and preferences over what could take place. This academic work is helpful to researchers who would like to have an enhanced strategic understanding of the energy situation in Canada. Moreover, since the analyzed conflict is still an ongoing dispute in Canada, this study could be useful to decision makers and stakeholders involved in the dispute. Following an explanation of the history of the nuclear dispute, the dispute is modelled and analyzed for the situations existing in the years 2008 and 2010. The evolution of the dispute over time, the influence of attitudes on the potential resolution of the 2010 conflict, and other strategic insights are discussed. The application of GMCR to this specific dispute constitutes the first application of this kind of game theory technique to the Ontario nuclear dispute. In the next step, the conclusions are discussed, and finally, the methodology is explained.

Results and discussions
In this section, a brief introduction to the conflict is provided and the conflict in 2008 and 2010 are discussed, and the results are explained.

Introduction to the conflict
In this paper, the Canadian nuclear conflict is modeled with respect to two points in time. The first analysis was conducted in 2008. This issue is an ongoing problem, and the conflict has not been resolved. Since the first analysis, some other related announcements and news items have been published. Therefore, a second analysis was performed in 2010, taking into consideration the more recent information. Although the two analyses (ultimately) involve similar decision makers (DMs) and options, they are different in terms of the relative preferences of the DMs. In each analysis, the DMs and their options are introduced. After the feasible and indistinguishable states are determined, the relative preferences

for each DM are defined. Then, static analysis and other dynamic analyses are performed.

Canada has twenty-two CANDU reactors; twenty of them are located in Ontario, the most populous province, one in Quebec, and one in New Brunswick. Nuclear energy provides about 15 percent of Canada's electricity (AECL, 2008).

In the coming decades, the Liberal Government of Ontario wants nuclear plants to remain the source of half of Ontario's electricity supply. It plans to install two new nuclear reactors, which will provide up to 3,200 megawatts of electricity, and to expand the Darlington nuclear site, in order to address the increasing demand for electricity as well as to reduce greenhouse gas (GHG) emissions. Although Dalton McGuinty, the Ontario Premier, in his election campaign had promised to shut down four coal-fired plants, which are highly polluting, by 2007, in 2006, he decided to postpone this plan to 2014, because there was no proper replacement energy producer (CBC News, 2006). At the beginning of March 2008, Energy Minister Gerry Phillips officially announced that the provincial government was seeking proposals to build a new nuclear plant. He declared that construction should begin in 2012 and electricity should be generated by July 1, 2018 (Benzie and Black, 2008). Organizations that submitted their proposals included AECL, Westinghouse Electric Co. LLC, an American company, and Areva, a French company (Frame, 2008).

As mentioned before, this conflict is analyzed with respect to two different points in time. In the following section, the dispute is explained according to the analyst's information and available publications as of 2008. It is worth noting that the present time in the following section refers to the year 2008.

AECL's reputation
Some incidents have aggravated the position of AECL in this bidding competition and may prevent the Ontario Government from selecting this company as the builder of the new plant reactors.

1) The National Research Universal (NRU) Chalk River reactor is the only nuclear reactor in North America that supplies medical isotopes for molecular imaging, radio therapeutics, and analytical instruments. On November 18, 2007, the Canadian Nuclear Safety Commission (CNSC) ordered the shutdown of the reactor, because it found that AECL had been operating the reactor for 17 months without a back-up emergency power system for cooling pumps, which prevent the reactor core from melting down. In 2006, AECL was ordered by CNSC to upgrade the NRU by installing that system. After two weeks of shutdown, Michael Burns, the chairman of AECL at

the time, resigned and Stephen Harper, the Prime Minister of Canada, accepted his resignation and blamed the Liberal-appointed CNSC for the closure. He pushed an emergency measure through Parliament on December 12, 2007, but the Liberals opposed the measure. The Liberals feared that the NRU was unsafe and required more upgrades. Eventually, the reactor was restarted in late January 2008, when Harper fired Linda Keen, the CNSC's head, who was a Liberal committee member. The outage created a critical and worldwide shortage of the radioactive diagnostic material and is considered a serious negative point in AECL's history (CBC News, 2008; Spears, 2008; Nathwani, 2009).

2) On May 16, 2008, AECL abandoned its plans to complete Multipurpose Applied Physics Lattice Experiment (MAPLE) reactors, which had been started in 1996. These reactors were to serve as a replacement for the NRU at Chalk River. In 2008, the project was millions of dollars over budget and eight years behind schedule. The failure of the MAPLE reactors is a dark point for AECL and has undermined its reputation. As a result of the failure of this project, at the beginning of June 2008, MDS Inc. launched a $1.6 billion lawsuit against AECL (Akin, 2008; Hamilton, 2008b). Moreover, in an expert panel report commissioned by Prime Minister Harper, Goodhand et al. (2009) recommended that a completely new and more flexible isotope reactor be constructed.

3) The safety standards are another problem. After September 11, 2001, the International Atomic Energy Agency (IAEA) established guidelines mandating that reactor builders redesign reactors so that they have the ability to withstand a massive outside shock or explosion. These safety standards must be applied to all new reactor designs. As AECL's reactors do not meet these standards, new regulations could be a major setback for AECL. Jerry Hopwood, vice-president of reactor development at AECL, has accepted the design weaknesses, stating that they would design an advanced CANDU 6 or adapt the old one if needed, to meet the standards [a] (Hamilton, 2007).

The federal Government's view towards AECL

In light of the problems pertaining to AECL, the Federal Government wants to enhance its nuclear credibility. Gary Lunn, Natural Resources Minister at the time, said that it is "imperative" that Ontario purchase new reactors from AECL (Geddes, 2007). Although Harper's government comprehends the importance of AECL as a federal Crown corporation, it does not completely support AECL's efforts to regain its previous reputation.

Harper's government has declared that in order to fulfill the country's growing electricity demands and facilitate the development of Canadian nuclear technology, the government has to decide about AECL's current status. The Conservatives are looking at different business models for AECL, including the sale of ownership stocks to one of several companies that have expressed interest, such as Westinghouse Electric Co. and France's Areva, foreign companies, and Bruce Power Inc., a Canadian company. Natural Resources Minister Lunn said that AECL's status is under review, stating that they are considering all options, from the status quo to a partnership with private investors to a sale to a foreign government. Another incentive for the Federal Government to change the status of AECL is the reliance of AECL's development program on federal tax dollars. Since, as mentioned above, AECL has incurred budget overruns, the sale of its stocks to a private company could alleviate taxpayers' criticisms of the Federal Government. On the other hand, industry observers say the lawsuit pertaining to the MAPLE reactors will make it more difficult for Ottawa to find a private suitor for AECL. This increases the probability that the government will opt to sell AECL to a foreign company or government (Hamilton, 2008b; Puxley, 2007). However, AECL does not want to be privatized. Its spokesman, Dale Coffin, disputed suggestions that AECL needs a strategic private-sector partner to compete in the world.

The Federal Government vs. The Ontario Government

On the other side of the conflict, the Ontario Government is dealing with its own issues. Premier McGuinty stated: "The Ontario Government is unwilling to purchase new reactors from AECL unless it receives assurance that the Federal Government will remain the ultimate backer of AECL". The McGuinty government is concerned about AECL's history and has made it clear that while it would prefer to buy home-grown technology, it is open to purchasing from a foreign company if it means getting the best deal for Ontario's taxpayers. If Ottawa does not support AECL, it will be very hard for it to sell the reactors in Ontario, and if it cannot do so, it will face a difficult time selling them anywhere else in the world (Hamilton, 2007).

In addition, timing and financial issues affect the Liberal Government's decision. McGuinty has promised to shut down all the province's coal-powered plants by 2014. New nuclear plants would be completed by 2018 if everything goes according to schedule. Furthermore, construction of a nuclear plant requires huge investments and compels Ontario's taxpayers to bear a heavy tax burden. Therefore, the Ontario Government wants AECL to be fully financially supported by the Federal Government. The provincial government is in an urgent situation in terms of the need for new power generation. It is unable to wait a very long time for AECL to prove its qualifications, but has to make its decision by March 2009 at the latest (Hamilton, 2008b; McParland, 2008).

Green groups of Canada

The Green Party of Canada, along with other green organizations and environmental groups, and the New Democratic Party (NDP) of Canada are on the other side of the conflict. They have always opposed the use of nuclear energy and believe that the Federal Government does not invest sufficiently in renewable energies. They think Canada has enough clean energy resources and does not need nuclear plants. These groups do not consider nuclear energy a clean energy, because there is still no proper means of nuclear waste management. They express their disagreement through their websites, articles, and speeches. Green groups are also concerned about the costs and consider nuclear energy generation to be expensive. Hence, in this dispute, green groups would agree with privatizing or selling AECL, as there would then be no need for the Federal Government to spend much money on AECL's funding, and taxpayers will not suffer. It would furthermore be easier for green groups to oppose AECL as a private nuclear organization, as it would not have governmental support (New Democratic Party, 2008; Sierraclub, 2008; Harris, 2008).

Analysis in 2008

The state of the conflict in 2008

AECL is in trouble; it has not sold a single reactor in ten years (by 2008). A new president was appointed in December 2007 to change the situation and return it to its once leading position in the nuclear industry. In the February federal budget, it received an appropriation of $300 million to support research and develop new technology. However, AECL still cannot convince the Ontario Government to buy its advanced reactors. A few key factors can possibly change AECL's situation. In June 2008, AECL announced that it had signed an agreement with the Nuclear Power Institute of China to collaborate on the "design, research, development, and demonstration" of "low uranium consumption CANDU technologies". Moreover, AECL is working with South Korea on a process called "direct use of spent pressurized water reactor fuel in CANDUs" (DUPIC). DUPIC is unique and can give Canada the opportunity to solve many problems in a nuclear energy market increasingly dominated by light-water reactors. DUPIC also gives existing and new CANDU 6 reactors a chance to minimize the environmental risks. As nuclear experts point out, the existence of the DUPIC project alone gives the Federal Government a new option to give AECL another chance. It could be a point of strength for the company.

If AECL accomplishes good results with these projects and keeps achieving satisfactory contracts, it might change the Federal and Provincial Governments' views (Hamilton, 2008a).

Different parties and groups in Canada are concerned about AECL's future. Ontario Energy Minister Duncan believes that AECL would be worth far less if Ottawa privatized it. If AECL were sold to a foreign company, thousands of skilled workers would lose their jobs at a time when the province has already lost thousands of industrial jobs.

Decision makers, options, states and relative preferences

According to the background of the conflict, the DMs of the dispute are listed below:

- The Federal Government
- The Ontario Government
- AECL
- Green Groups

Hereinafter, FG, OG, and GG denote Federal Government of Canada, Ontario Government, and Green groups, respectively.

The four aforementioned DMs and their options, and the current state of the conflict are shown in Table 1. In order to better represent and discriminate states, each state is defined as follows:

$$s_i = (x_1 x_2 x_3, x_4 x_5, x_6, x_7), \quad x_j \in \{Y, N\}, \quad j = 1, 2, \ldots, 7,$$

where $x_1 x_2 x_3$ are the options that belong to FG, $x_4 x_5$ belong to OG, x_6 belongs to AECL, and x_7 belongs to GG. In addition, $x_j = Y$ indicates that the j-th option is chosen and $x_j = N$ indicates that it is not. It is a large conflict, so GMCR II software developed by the Conflict Analysis Group in the Department of Systems Design Engineering at the University of Waterloo (Fang et al., 2003a,b; Hipel et al., 1997) was used to perform various types of analyses.

The infeasible states are also determined. These states must be eliminated from the analysis. A number of infeasible states are observed in this conflict, and after removing those states, 48 states remain.

- The options of the Federal Government and the Ontario Government are mutually exclusive. Therefore, considering the options mentioned in Table 1, the states listed below should be removed:

  - Federal Government: (Y Y -, - -, -, -), (Y - Y, - -, -, -), and (- Y Y, - -, -, -)
  - Ontario Government: (- - -, Y Y, -, -)

Table 1 DMs and options in 2008 analysis

DMs	Options	Status Quo
Federal Government (FG)	1. Sell less than 50% of AECL's stocks, and keep control of it	N
	2. Sell or privatize AECL	N
	3. Fully support AECL	N
Ontario Government (OG)	4. Buy reactors from AECL	N
	5. Buy reactors from a foreign company	N
AECL	6. Convince both governments that it is capable of fulfilling its mandates	N
Green Groups (GG)	7. Continue their protests against nuclear power	Y

Table 2 The option prioritization table for the Federal Government (FG)

Preference Statements	Explanation
-3	The most important thing for FG is not to support AECL, and not to invest more money into it.
2	Next, FG prefers to privatize AECL.
-1 & -2 & -3 & 4	FG mostly prefers that OG buys its reactors from AECL.
-1 & -2 & -3 & -4 & -5	It prefers that both governments take no action.
2 & 4	Next, it prefers to privatize AECL and prefers OG to buy its reactors from AECL.
1 & 4	Next, it would like to sell less than half of AECL's stocks and OG to buy its reactors from AECL.
3 & 4 & -5	After that, it prefers the case that it supports AECL, and OG buys its reactors from AECL.
3 & -4 & -5	It next prefers the situation in which it supports AECL, and OG waits.
-1 & -2 & -3 if 5	If OG makes a foreign purchase, FG prefers to do nothing.
6	Redesigning the reactors by AECL is one of its least priorities.
-7	AECL working on the DUPIC project is also of less importance for FG.

Based on the background study of the conflict, the preferences of the three DMs are determined using the option prioritizing method (Explained in Methods). Some explanations are provided and used to form the relative preferences.

- The Federal Government (FG) is contemplating the future of AECL: To sell less than 50% of AECL, sell it all or privatize it, or support it. However, its negotiations are not clear to other parties and to the public. The Federal Government prefers AECL reactors be sold to Ontario, so that AECL can gain credit to sell more reactors to other countries.

After the states presenting this situation, the Federal Government next prefers the states in which neither the Federal nor the Ontario Government take any action. The Federal Government would prefer to sell AECL if it is faced with complaints from taxpayers or the Ontario Government. The least favoured states for this DM are the ones that represent the support of AECL by the Federal Government. Table 2 demonstrates the specific way that state prioritization is managed in GMCR II for FG. In this table, preference statements are listed from most important at the top to least important at the bottom. The numbers in the left column of Table 2 refer to the option numbers given in Table 1, where a negative sign means that the option is not taken. Notice that the most important preference for the Federal Government is not to fully support AECL by not taking option 3 (denoted by −3). Assuming transitive or ordinal preferences, an algorithm can take the prioritized preference statements of the Federal Government in Table 1 and rank the states from most to least preferred where ties are allowed.

- The Ontario Government (OG) most prefers to select home-grown technology if AECL is successful

in the redesign process. In addition, it would prefer that the Federal Government supports AECL. AECL being supported by FG is much more important to OG than redesigning CANDU reactors. It expects the Federal Government to support AECL in completing its projects on schedule. Next, it prefers the future of AECL to be resolved and to buy reactors from this company when AECL is trusted. As the Ontario Government does not want to waste time, it would rather purchase reactors from a foreign company if the future of AECL is undetermined.

- AECL is trying to complete its projects in order not to be sold. The most desirable states for AECL are the ones in which AECL is not sold. Among these states, it is more preferable for AECL to be supported, and it is also very important for AECL to sell its reactors to the Ontario Government.

- Green Groups (GG) are against nuclear energy. They declare their opposition via speeches and websites.

Static analysis

The equilibrium states (along with their corresponding stability types) determined by running the stability analysis in GMCR II are listed horizontally as:

- (N Y N, N N, N, Y): state 27, GMR, SMR,

- (N Y N, N N, Y, Y): state 39, GMR, SMR, and
- (N Y N, Y N, Y, Y): state 43, Nash, GMR, SMR, SEQ.

These states indicate what actually took place in reality. State 43 is the most stable equilibrium state, as its stability type is Nash, which also means that state 43 is equilibrium according to SEQ, GMR, and SMR. Because 43 is Nash equilibrium, no DM can unilaterally move to a more preferred state from state 43. States 27 and 39 are also equilibria in this conflict, but only according to SMR and GMR stability, which means sanctioning DMs can move to less preferred states when blocking the given DM's unilateral improvement, which is a movement by a DM to a more preferred state. What happened in reality (as of 2010) is that the Ontario Government chose AECL as the vendor of the reactors, and then the Federal Government decided to attempt to privatize AECL. After this decision was made, the Ontario Government postponed the purchase of the two reactors, because the future of AECL was very uncertain. Therefore, the state that took place was state 43, but states 27 and 39, which can be presented as (N Y N, Y N, N -, Y), happened after state 43, where a dash means Y or N. Also, since states 27 and 39 are only GMR and SMR equilibria, the stability of these states is not as strong as an SEQ equilibrium in which the sanctions by other DMs against unilateral improvements by a focal DM can only be levied using unilateral improvements. Here, one may conclude that neither state 27 nor state 39 are the final equilibria of the conflict, so the dispute will not finish at this point.

Status Quo analysis

To apply status quo analysis, the current state of the conflict has to be determined. Next, the analyst investigates the way the conflict has evolved from its initial state in 2008. In the current state of the conflict, FG and OG are not taking any action; AECL is working on the DUPIC project and on the design of Advanced CANDU Reactors (ACRs), and consequently, trying to satisfy its customer, OG, and its owner, FG (option 7); and GG is protesting. This set of options represents the status quo state 25 (N N N, N N, N, Y). In this case, the game develops from state 25 (N N N, N N, N, Y) to states 43, 27, and 39. The evolution of the conflict is shown in Table 3. In each level, one DM can move the conflict from the existing state to another state. The arrows, along with an assigned DM, show which DM is moving the conflict.

Although it is shown that the 2008 conflict will finish at states 27 and 39 (which is what happened in reality), later the authors show that in fact the conflict will again move to state 43, and states 27 and 39 are the transition

Table 3 Evolution of the conflict

Federal Government					
1. Sell less than 50% of AECL	N	N	N	N	N
2. Privatize AECL	N	N	N FG →	Y	Y
3. Support AECL	N	N	N	N	N
Ontario Government					
4. Buy from AECL	N	N OG →	Y	Y OG →	N
5. Buy from a foreign company	N	N	N	N	N
AECL					
6. Satisfy FG and OG	N AECL →	Y	Y	Y	-
Green Groups					
7. Protest	Y	Y	Y	Y	Y
	25	37	29	43	27&39

states in the 2010 conflict. It should be noted that the number assigned to each state is not the same in the 2008 and 2010 conflicts.

Sensitivity analysis

In order to gain more insights, some sensitivity analyses have been run. Sensitivity analyses can help the analyst to have a better understanding of the conflict and the static analysis. For example, in this conflict, it seems that although the Green Groups, GG, have an option to protest against the investment in and the use of nuclear energy, they are not as powerful as the other three DMs. The main reason for this is that all the other DMs, in contrast to GG, are in favour of nuclear energy. It seems logical that the fourth DM, being the only anti-nuclear DM, does not have a considerable effect on the result of the conflict. To see how much this anticipation is correct, GG is omitted from the game, and another static analysis is executed, and the equilibrium states are shown below:

- (N Y N, N N, N): state 3, GMR, SMR,
- (N Y N, N N, Y): state 15, GMR, SMR, and
- (N Y N, Y N, Y): state 19, Nash, GMR, SMR, SEQ.

These equilibria verify the aforementioned expectation, as the indicated results are essentially the same as the results in the previous analysis. Option prioritization tables show that the decision of GG regarding whether to protest or not is among the least important issues for all DMs except for GG. Therefore, the results show that GG is not an important DM in this conflict. In other words, GG cannot significantly influence the other DMs' decisions regarding the future of AECL.

Another idea is to eliminate AECL as well. Although it might seem that AECL's attempt to obtain the trust of FG and OG would affect the decisions of those two DMs, it is possible that their decisions are independent of AECL's achievements. If this possibility is in fact true, AECL would not actually be a DM in the conflict, and only the two governments would be important DMs. Thus, if AECL were to be omitted from the analysis, and AECL is not a DM of any real consequence in the conflict, the two equilibria one would expect to obtain would be: (N Y N, Y N), which would be equivalent to state 43, and (N Y N, N N), which is the same as states 27 and 39. The related analysis results verify the fact that AECL is not an influential DM. The resulting equilibrium states for this analysis are shown below:

- (N Y N, N N): state 3, GMR, SMR,
- (N Y N, Y N): state 7, Nash, GMR, SMR, SEQ.

In order to more deeply investigate the effect of AECL as a DM on the results of the modeling, the preferences of this DM are changed. If AECL is not significantly affecting the game, changing its preferences should not change the equilibria regarding the future of this company. This change is applied to the game with three DMs, in which only GG is omitted. Different arrangements of AECL's relative preferences are made at this level. The results show that the equilibria do not change, although their types do with some changes in the relative preferences. Therefore, one may conclude that AECL does not, in fact, have a considerable effect on the other two DMs.

Finally, since only two DMs are found to be influential in this conflict, the status quo table is reproduced to show the evolution of this smaller conflict (Table 4).

Analysis in 2010

Previously, the Ontario nuclear dispute was analyzed for 2008. This dispute, however, is an ongoing conflict in the province. Therefore, in the following sections, an attempt is made to perform a new GMCR analysis of the

same conflict, but with an updated background. In addition, regarding energy issues in the province, the Government of Ontario has recently published an updated energy report (Ontario Government, 2010).

Updated background of the conflict

To summarize the history of the conflict in 2010, and to update the background of the dispute in 2008, the timeline of the nuclear project is described as follows (Ontario Government, 2009). In March 2008, the Ontario Government announced a two-phase competitive procurement process to choose a preferred nuclear reactor vendor. After holding a series of confidential meetings with the vendors in June 2008, Infrastructure Ontario announced that all three vendors that had submitted Phase 1 Proposal Submissions received 'satisfactory' ratings and would be invited to proceed to Phase 2 of the Request For Proposal (RFP). Subsequently, Infrastructure Ontario released Phase 2 of the Nuclear Procurement Project RFP. At this stage, and in November 2008, the first analysis was performed. From February to May 2009, all three respondents submitted Phase 2 Proposals, and the Government of Canada announced that it was proceeding with a restructuring of AECL. In June 2009, the Ontario Government announced that the Nuclear Procurement Project RFP was being suspended due to concerns about pricing and uncertainty regarding AECL's future.

The Centre for International Governance Innovation (CIGI) published an important report in November 2009. This report studies the nuclear industry in Ontario and briefly investigates the expansion of the Darlington nuclear site, the most important nuclear project in Canada. The report stated that by 2008, Ontario had planned to invest $40 billion to replace and refurbish its nuclear generating capacity, and subsequently, in February 2009, bids to build a new facility at Darlington were accepted by the province (Cadham, 2009).

The CANDU design, proposed by AECL, is proudly Canadian in that Team CANDU represents the provider of Ontario's existing installed nuclear facilities. The AECL CANDU design was the only design to fully meet the requirements of Infrastructure Ontario among the three vendors, consisting of AECL, Areva Group and Westinghouse Electric Co. Hence, for this and other reasons, the province selected AECL's technology as the winner.

In July 2009, George Smitherman, the Energy Minister of Ontario, said that the government wanted to negotiate with Ottawa to reduce the bidding price. Smitherman declared that AECL's bid was "billions" above what Ontario had anticipated. Ontario Power Generation (OPG) had estimated the cost of the installation of the two ACRs at $3,000 per kilowatt, compared to $10,800,

Table 4 Evolution of the 2-DM 2008 conflict

Federal Government							
1. Sell less than 50% of AECL	N		N		N		N
2. Privatize AECL	N		N	FG →	Y		Y
3. Support AECL	N		N		N		N
Ontario Government							
4. Buy from AECL	N	OG →	Y		Y	OG →	N
5. Buy from a foreign company	N		N		N		N
	1		5		7		3

the price offered by AECL. One of the major reasons for the offer being this high is that the design of the ACR is not yet complete, but Ontario is disinclined to pay for the cost of the research and development (R&D) process. On the other hand, before the offer was submitted by AECL, the Harper government had told the company that its bidding price must cover all the costs of R&D, and that AECL should not count on future sales to put off the cost overruns. In the case in which the Federal Government decides to keep AECL as a federal Crown corporation and not sell it, the government needs to ensure that the Ontario nuclear project is commercialized in an attempt to preserve AECL's value and to avoid federal taxpayers subsidizing Ontario ratepayers (McCarthy and Howlett, 2009).

The AECL's restructuring is currently under scrutiny. In 2007, the Federal Government hired the National Bank to provide independent financial advice and to help find the best way to carry out the mandates of AECL. The National Bank put forward some solid recommendations. In its recommendations, it was suggested that AECL has two concurrent mandates: commercial goals involving the selling and servicing of reactors, and R&D with regards to projects and technology. The bank advised that at least 51% of AECL be sold and encouraged the Government of Ottawa to improve AECL's standing in the international market. Natural Resources Canada published a report in May 2009 to present the ideas they received from the National Bank and other consultants. AECL's failure in handling its projects shows that in the past, the two opposing mandates have not worked well together, and AECL has fallen short on many of its objectives. Some of AECL's unsuccessful projects were mentioned above. The former Minister of Natural Resources, Lisa Raitt, advised: "The best chance to take advantage of this nuclear renaissance is to divide the two of them and seek global participation." Raitt suggested that the designing and building of reactors is very expensive, and Canadian taxpayers cannot shoulder this burden on their own, so AECL needs a strategic alliance in order to compete in the world (CTV, 2009). As stated in the project procurement, the Government of Canada announced the restructuring of AECL in May 2009. It also hired N.M. Rothschild & Sons to provide financial advice and available options and received their financial analysis on the restructuring plan of AECL in October 2009, but the report is confidential due to commercial confidentiality considerations.

As stated by Raitt, the company's research-and-development division, Chalk River laboratories, will continue to be government-owned, but with private-sector management (CBC News, 2009). The reactor business and its attractive maintenance and refurbishment activities would then be offered for sale on either a majority or minority ownership. Some parties accuse the government of wanting to sell AECL in order to balance its budget deficit. According to the former Minister of Natural Resources, however, this decision is about bolstering the industry. However, this reconstruction is not desired by the Ontario Government as stated by Smitherman: "The government of Canada needs to do the work that they are doing now to clarify the future ownership of AECL, and when they have clarified that, to sharpen their pencils substantially so that the people of the province of Ontario can renew their nuclear fleet with two new units from that company" (McCarthy and Howlett, 2009).

Having discussed the points of view of the two governments, it can be concluded that Ontario will not move until Ottawa clarifies AECL's ownership status. The other key issue is AECL's bidding price. Thus, the uncertain future of AECL and the high price were two important factors that led Ontario to postpone the project. According to the background information, it is nonetheless clear that Ontario mostly prefers to buy reactors from AECL.

The government of Canada, on the other hand, can decide between several options: privatizing and restructuring AECL, selling it to a Canadian or a foreign corporation, or keeping it public and consequently helping AECL to decrease the price. It has been a long time since the announcement of selling AECL, but it has not yet happened. Therefore, there is still the possibility that the government will not privatize the federal Crown corporation. Industry insiders say that the companies that are interested in partnership with AECL are an international company, US-based Westinghouse Electric Co., Canadian engineering giant SNC-Lavalin Group Inc., and Bruce Power, a Canadian-owned consortium that operates a nuclear station in Ontario. The bidding process closed on June 30, 2010 (The Globe and Mail, 2010). Sources close to negotiations say that only SNC-Lavalin Group Inc. and Bruce Power have submitted their bids to partner with AECL (McCarthy, 2010). What happened in reality was that in June 2011, the Ottawa government announced the sale of AECL to SNC-Lavalin Group Inc. (McCarthy, 2011).

Since there is opposition against selling or privatizing AECL, this decision is not an easy one for the government to undertake (The Star, 2009). AECL possesses an internationally competitive reactor design and employs thousands of Canadian workers, and AECL's supporters argue that it is not beneficial for the governments to let this company be sold. Canadian nuclear analyst David Jackson says that the problem with dividing AECL into two parts and privatizing one of them is that "no potential purchaser would want to buy an ACR with no assured R&D backup and thus, in effect the restructuring is the end of ACR" (Cadham, 2009). On the other hand, the Federal Government's supporters believe that selling AECL is a sound decision as this company has been a burden on taxpayers for a long time and it has not been a cost-efficient investment. In addition, they think selling to a foreign company, rather than a Canadian company, would be a much better decision. A domestic sale will not change the pressure on taxpayers. Furthermore, selling AECL to a domestic buyer will not help the company to regain its reputation and become a competitive player in the international market (McCarthy, 2010).

There are also other provinces, such as New Brunswick and Saskatchewan, that have indicated an interest in buying new reactors. In July 2010, however, New Brunswick announced that it would not choose AECL as the provider of its reactors, and is instead turning to Areva Group, which is a company that was interested in buying AECL, but dropped out from the bidding. This decision, along with what happened in the Ontario contract, are considered major setbacks for AECL, and complicate the decision of the Federal Government regarding selling off AECL (The Globe and Mail, 2010). AECL was counting on the Darlington project to galvanize its huge resources to launch its new Advanced CANDU Reactor (ACR). Therefore, in the end, if Ontario, as the largest customer, decides not to buy any reactor at all, it would be unlikely that other provinces would consider AECL's unproven, first-of-its-kind ACR technology as a serious option.

The problem gets more complicated when one considers the massive amount of money that Ontario has spent on its nuclear industry. It operates more CANDU reactors than all of the other Canadian provinces or countries combined. Moreover, AECL has about 5000 employees, and privatizing it will lead to a large number of job losses, which is not desired by any of the DMs or the political parties in the country (McCarthy and Howlett, 2009).

Decision makers, options, states and relative preferences
In the 2010 analysis, there are only two DMs, the Federal and Ontario Governments. According to the updated background, neither AECL nor the Green Groups are important DMs. Consequently, their decisions do not affect the final decision that the Federal and Ontario Governments will make (Cadham, 2009). Since it is the responsibility of the Federal Government to financially support AECL, if this government does not provide sufficient funding for the company, the possibility that AECL can compete with its foreign rivals and win the contract becomes very low. The other DM that can seriously affect the outcome of the conflict and the future of AECL is the Ontario Government. Although AECL's suggested bidding price for the expansion of Ontario's nuclear power stations is very high, if the Ontario Government accepts AECL's

offer, it is possible that AECL could remain as a public company. Regarding the DUPIC project, which AECL is working on, the published news and interviews of the officials of the two governments do not indicate that they pay much attention to the performance of AECL on this project. The Federal and Ontario Governments are more concerned about AECL's progress in building and selling reactors. Therefore, in the 2010 analysis, there are only two main DMs, the Federal Government and the Ontario Government. Regarding the options for the two governments, Table 5 shows the options for each DM. As can be seen, the options for each DM change with respect to the previous analysis in 2008, shown in the upper part of Table 1.

In the most recent analysis (2010), the first two options of the Federal Government change from what they were in 2008, since in 2010, it is determined that if the Federal Government decides to restructure AECL, it will privatize it, and not sell only less than half of its stocks. The options, however, are selling to a domestic or an international organization. The reason for this is that, in practice, the Federal Government is studying the privatization of one part, R&D, and selling the other part, CANDU. The Ontario Government has the same options as in 2008. However, after examining the vendors' bids, the province recognized that if the final decision is to buy new reactors, the vendor would definitely be AECL, so its option to purchase reactors from a foreign company would be essentially eliminated. Taking into consideration the latest announcements and the updated background, as well as the options, the relative preferences of the DMs will also change.

Regarding the infeasible states, the options for the two DMs are mutually exclusive. The Federal Government cannot privatize and support AECL at the same time. Similarly, the Ontario Government cannot choose its two options simultaneously. Therefore, the following infeasible states are removed from the game, and 12 feasible states remain, which are shown in Table 6.

- FG: (Y Y -, - -), (Y - Y, - -), (- Y Y, - -)
- OG: (- - -, Y Y)

Rather than using option prioritization to determine the relative preferences of the second DM, as was done

Table 5 DMs and options in 2010 Analysis

DMs	Options
Federal Government	1. Sell AECL (to a foreign company)
	2. Sell AECL (to a Canadian company)
	3. Support AECL
Ontario Government	4. Buy from a foreign company
	5. Buy from AECL

in the previous analysis, direct ranking is employed whereby the states are ranked from most to least preferred by the analyst. This method works well for small conflicts, such as the small 12-state conflict considered here. In Tables 7 and 8, the ranking of states for the Federal and Ontario Governments are directly defined.

Static analysis

After defining the DMs, their options and preferences, static analysis is performed to investigate the final possible outcomes. States 1 and 11 are found to be the equilibria of the conflict. The equilibria in this case are shown below. It is worth noting that in order to find the equilibria, Sequential and Nash stability are considered.

- (N N N, N N): state 1, SEQ,
- (Y N N, N Y): state 11, Nash, GMR, SMR, SEQ.

It should be noted that by DMs being conservative, more states are stable, and the uncertainty of the game rises. In other words, if the DMs are very conservative, other concepts of stability should be considered. The reason for this is that if the DMs refrain from moving to different states on account of the possibility of losing benefits, the conflict will stop in more states and there will be more equilibria, compared with the situation in which the DMs accept the risks and move from one state to another. However, in this specific conflict the DMs are not conservative. As an example, more than a year ago, the Federal Government announced the restructuring AECL, but it has not yet sold it. In addition, Ontario Government announced AECL to be the best company, but postponed its decision about buying reactors from it. These examples show that the DMs do accept some risks and do move from state to state. Therefore, GMR and SMR stability concepts are ignored, and SEQ and Nash stability are being considered.

As of July 2011, what happened in reality was that the commercial reactor business of AECL was sold to SNC-Lavalin, a Montreal-based company. Specifically, on June 27, 2011, Ottawa announced the sale of the reactor

Table 6 List of feasible states

Federal Government												
1. Sell AECL (to a foreign company)	N	Y	N	N	N	Y	N	N	N	Y	N	N
2. Sell AECL (to a Canadian company)	N	N	Y	N	N	N	Y	N	N	N	Y	N
3. Support AECL	N	N	N	Y	N	N	N	Y	N	N	N	Y
Ontario Government												
4. Buy from a foreign company	N	N	N	N	Y	Y	Y	Y	Y	N	N	N
5. Buy from AECL	N	N	N	N	N	N	N	Y	Y	Y	Y	Y
State Number	1	2	3	4	5	6	7	8	9	10	11	12

Table 7 Direct ranking box in GMCR II for the Federal Government

Federal Government												
1. Sell AECL (to a foreign company)	N	Y	N	Y	N	N	Y	N	N	N	N	N
2. Sell AECL (to a Canadian company)	Y	N	N	N	Y	N	N	Y	N	N	N	N
3. Support AECL	N	N	N	N	N	N	N	N	N	Y	Y	Y
Ontario Government												
4. Buy from a foreign company	N	N	N	N	N	N	Y	Y	Y	N	N	Y
5. Buy from AECL	Y	Y	Y	N	N	N	N	N	N	Y	N	N
State Number	11	10	9	2	3	1	6	7	5	12	4	8

Table 8 Direct ranking box in GMCR II for the Ontario Government

Federal Government												
1. Sell AECL (to a foreign company)	N	N	N	Y	N	N	Y	N	Y	N	N	N
2. Sell AECL (to a Canadian company)	N	N	Y	N	N	Y	N	Y	N	N	N	N
3. Support AECL	Y	N	N	N	N	N	N	N	N	Y	Y	N
Ontario Government												
4. Buy from a foreign company	N	N	N	N	Y	Y	Y	Y	N	N	Y	N
5. Buy from AECL	Y	N	Y	Y	N	N	N	N	N	N	N	Y
State Number	12	1	11	10	5	7	6	3	2	4	8	9

business of AECL to this Canadian company for $15 million. This change in ownership will cause a loss of jobs in Ontario: 1,200 out of about 2,000 employees at AECL's commercial division will move to the new CANDU Energy division of SNC-Lavalin in Quebec (Babbage, 2011; McCarthy, 2011). Although Ontario's decision to buy two new Enhanced CANDU 6 reactors has not yet been officially announced, the analysis results of this paper are in fact very close to what happened in the real world. In particular, state 11, the equilibrium of the conflict, represents the situation in which AECL is sold to a Canadian company (corresponding with SNC-Lavalin's ownership of CANDU Energy), and Ontario buys two new reactors from AECL.

Since this conflict is not very big, the static analysis can also be performed by hand without the software. The reachable list is shown in Table 9. In Table 10, the states are ranked from most to least preferred for each DM. The unilateral improvements are listed under each state for each DM. The type of stability for a given state and DM is written above the state in the ranking of states for the DM. In Table 10, the letters r, s, and u stand for rational stability, sequential stability (SEQ), and unstable, respectively. When a state is stable for all of the DMs, it is an equilibrium indicated by E. Notice, for example, that state 11 has no unilateral improvement written below it for both FG and OG. Therefore, the state is rational (r) for each DM and constitutes an overall equilibrium (E). State 1 is sequentially stable (s) for FG since both of its unilateral improvements can be blocked by OG. In particular, if FG moves from state 1 to 3, OG can then move from state 3 to 7, which is less preferred to state 1 by FG. Likewise, if FG moves from state 1 to 2, then OG can move from 2 to 6, which is less preferred to state 1 by FG. Since both unilateral improvements by FG from state 1 can be blocked by OG, state 1 is sequentially stable (s) for FG. Moreover, because state 1 is rational (r) for OG, it forms an overall equilibrium (E).

According to the definitions provided in the last section (Methods) regarding stability concepts, Nash and SEQ equilibria are found and shown in Table 10. State 1, where no DM takes any action, is one of the equilibria. This is a reasonable solution, since for a long time, the governments did not announce any specific decision. Although the Federal Government had announced the restructuring, as of May 2011, FG had not introduced the purchaser. The Ontario Government, on the other hand, has postponed its plan.

In the remaining part of this section and the next one, the states that could occur in 2008 and 2010 are designated by A and B, respectively. In order to compare the equilibria of the 2008 and 2010 analyses, one should compare the results of the 2010 conflict:

- (N N N, N N): state B1, SEQ,
- (N Y N, N Y): state B11, Nash, GMR, SMR, SEQ.

with the results of 2008:

- (N Y N, N N): state A3, GMR, SMR,
- (N Y N, Y N): state A7, Nash, GMR, SMR, SEQ.

Table 9 Reachable states

State	FG	OG
1	2, 3, 4	5, 9
2	1, 3, 4	6, 10
3	1, 2, 4	7, 11
4	1, 2, 3	8, 12
5	6, 7, 8	1, 9
6	5, 7, 8	2, 10
7	5, 6, 8	3, 11
8	5, 6, 7	4, 12
9	10, 11, 12	1, 5
10	9, 11, 12	2, 6
11	9, 10, 12	3, 7
12	9, 10, 11	4, 8

Table 10 Stability analysis

Overall Equilibria	E	×	×	×	×	E	×	×	×	×	×	×
Individual Stability	r	u	u	r	s	s	s	r	u	u	u	u
FG State Ranking	11	10	9	2	3	1	6	7	5	12	4	8
		11	10		2	3		6	7	9	1	5
			11			2			6	11	2	6
										10	3	7
Individual Stability	r	r	r	r	s	u	u	u	u	u	u	u
OG State Ranking	12	1	11	10	5	7	6	3	2	4	8	9
					1	11	10	7	6	12	4	1
								11	10		12	5

State B11 is a very strong equilibrium since it is Nash stable. This state is somewhat similar to state A7 in the 2008 analysis. State A7 is also Nash stable and is a stronger equilibrium than state A3. Therefore, the relationship between the results of the two conflicts can be listed below:

- States A1 and B1 constitute similar states which are the initial states of the 2008 and 2010 conflicts, respectively.
- State B2 (which is similar to A3) is a transition equilibrium in the 2010 conflict and the equilibrium of the 2008 conflict.
- State A7 or B11 are the strong equilibria in both conflicts.

Now the question that needs to be addressed is why is state A7 not the finishing point of the 2008 analysis? The answer is that state A7 (or B11) are actually the equilibria, but state A3 is a very considerable transition state. This nuclear conflict is very complicated since the DMs have changed their decisions on several occasions. That is why the conflict moved from state A7 to A3 in the 2008 conflict (Table 4).

Status Quo analysis

To investigate how the conflict evolves and moves from its 2008 state, state A1 or B1, to state B11, status quo analysis is performed. In a status quo analysis, one determines if a state of interest can be reached following a path from a starting or status quo state. The evolution of the conflict is shown in Table 11.

State B2, which is equivalent to state A3, is the transition state. State A3 is the ending point of the 2008 conflict, but can be considered as a starting point of the 2010 conflict. The status quo table of the 2-DM 2008 conflict (Table 4) can be compared to Table 11. Although the initial state is shown to be state 1 in Table 11, this state is actually the very first point of the conflict in 2008. Therefore, state A1 is the initial state of the 2008 conflict, and state A3 (B2) is its ending. Finally, state

B11 is the equilibrium in the 2010 game. The main reason for the difference between the two stages of the conflict (2008 and 2010) is the available information. Hence, one can understand how the changes in the situation over time can affect the final result.

Attitude analysis

Attitude analysis (Inohara et al., 2007; Bernath Walker et al. 2009) is another analysis that is performed on this conflict. In the original form of static analysis, each DM is considered to have a positive attitude towards him/herself, and to be neutral towards other DMs. This means that each DM does not carry out an action that harms him/herself, but may move to a state that harms or benefits other DMs. The way DMs' attitudes can be shown is indicated in Table 12. The symbol eij $\in \{-, 0, +\}$, i, j = 1, . . . , n, where n is the number of players, denotes the attitude of player i towards player j. $e_{ij} = -$, $e_{ij} = 0$, $e_{ij} = +$ indicate negative, neutral, and positive attitudes towards the opponent, respectively. Table 12 also shows the attitudes of players in the original form of static analysis.

It is worth noting that the analysis that is shown in Table 10 is performed by taking into consideration the attitude set indicated in Table 12. However, another set of attitudes may be assumed in this game. Two different governments constitute the DMs in this conflict. The Federal Government is run by the Conservative Party of Canada, and the Ontario Government is run by the Liberal Party. Therefore, aside from the situation of the Ontario contract and the future of AECL, the game involves the political conflicts between these two politically opposed DMs. According to the background of the conflict, after the Ontario Government announced that it favoured AECL's proposal, the Federal Government announced the restructuring of the company. The implementation of a new nuclear station in Ontario is a vital requirement for the province. However, although the Federal Government is keen on being a pioneer in

Table 11 Evolution of 2010 conflict

Federal Government							
1. Sell AECL to a foreign company	N	FG →	Y		Y	FG →	N
2. Sell AECL to a Canadian company	N		N		N	FG →	Y
3. Support AECL	N		N		N		N
Ontario Government							
4. Buy from a foreign company	N		N		N		N
5. Buy from AECL	N		N	OG →	Y		Y
	1		2		10		11

Table 12 Tabular representation of attitudes in a regular 2-DM game

	Player 1	Player 2
Player 1	$e_{11} = +$	$e_{21} = 0$
Player 2	$e_{21} = 0$	$e_{22} = +$

nuclear technology, it seems that it does not have a plan to help Ontario in this matter, and this may be the reason for the political conflicts. Therefore, a new attitude arrangement (Table 13) is considered, and a new static analysis is performed (Table 14). According to the attitudes in Table 12, from each state, FG can move to a state among the reachable states that is less preferred by OG and more preferred by itself. OG can move to the states that are more preferred by itself, ignoring FG's preference. The corresponding unilateral moves to each state are indicated below the ranking of states for each DM in Table 14.

With the new set of attitudes, three states are found as the equilibria of the conflict, states 4, 10, and 11. State 10 represents the situation in which FG sells AECL to a foreign company, and OG buys reactors from AECL. State 4 is a state in which the Federal Government supports AECL, and the Ontario Government decides not to buy any new reactor. State 11 represents the situation in which FG sells AECL to a domestic company, and OG buys reactors from AECL. The results show that if, in reality, the Federal Government has the intention of harming the Ontario Government, and at the same time, benefitting itself, it is possible that the final outcome of the game is state 4, which is among the least preferred states for both DMs.

Conclusions

The dispute over the expansion of the Darlington nuclear site in Ontario is formally studied in this paper for two different points in time: the fall of 2008 and spring of 2010. According to the background information about this conflict in 2008, four DMs were considered to be involved in the game: the Federal and Ontario Governments, AECL, and green groups. However, after several sensitivity analyses, the authors arrived at the conclusion that the key DMs are the Federal and Ontario Governments. The status quo analysis is also provided and the evolution of the conflict from March to fall 2008 is explained. The conflict did not terminate at this point and is still ongoing. The results and information of the

Table 13 Attitudes in the new analysis

	FG	OG
FG	$e_{11} = +$	$e_{12} = 0$
OG	$e_{21} = -$	$e_{22} = +$

Table 14 Stability Analysis ($e_{11} = e_{22} = +$, $e_{12} = -$, $e_{21} = 0$)

Overall Equilibria	E	E	×	×	×	×	×	×	×	×	E	×
Individual Stability	r	r	r	r	s	u	r	u	u	u	r	r
FG State Ranking	11	10	9	2	3	1	6	7	5	12	4	8
				2	3		6	7	9			
				2			6		11			
									10			
Individual Stability	r	r	r	r	s	u	u	u	u	r	u	u
OG State Ranking	12	1	11	10	5	7	6	3	2	4	8	9
					1	11	10	7	6	12	4	1
							11	10			12	5

analysis in 2008, and an updated background in 2010, are used to perform another analysis in 2010. Only the two governments are counted as the main DMs at this time, and their options have changed from the situation in 2008. The equilibria found in this analysis for 2010 suggest that the conflict will ultimately move to a state in which the Federal Government sells AECL to a domestic company, and the Ontario Government buys its reactors from AECL or its new CANDU owner. The other outcome is the initial state, in which no DM takes any action. This state actually was stable in reality for a period of time. So far, the results are closely correlated with what happened in reality, as the Federal Government sold AECL to SNC-Lavalin, a Canadian company (Babbage, 2011; McCarthy, 2011). The decision of the Ontario Government is still unknown, but the analyses suggest that Ontario will buy its reactors from AECL. In addition, in the attitude analysis, it is discovered that if the Federal Government has a negative attitude towards the Ontario Government, it is possible that the final outcome is a state that is among the least preferred states for both DMs.

Methods

The Graph Model for Conflict Resolution (GMCR) is the approach that is used to analyze this conflict. The GMCR methodology was developed by Fang et al. (1993) based upon earlier work by Fraser and Hipel (1984) and Howard (1971). This decision technology can be used to analyze disputes among different parties with different options or decision choices, and different preferences or value systems. It is for this reason that this methodology has been chosen to be used in this research. GMCR constitutes a flexible approach to the representation, analysis, and understanding of a strategic conflict. It also facilitates modifications to the way in which the conflict is represented, thereby supporting sensitivity and what-if analyses. Finally, descriptions and comparisons of a range of different approaches for formally studying conflict are provided in two books edited by Hipel (2009a,b).

GMCR is able to provide a DM with suggestions for reaching possible resolutions either independently or in cooperation with others. In some cases, win/win resolutions can be reached even though this may initially appear to be difficult to achieve. Figure 1 illustrates the way in which a conflict study is carried out in practice. GMCR, along with its associated decision support system GMCR II (Fang et al., 2003a,b; Hipel et al., 1997), is used to model the conflict and analyze the current situation.

The most important part of applying the GMCR model to a real-world conflict is the background investigation. Searching the news, talking to experts, and reading the related published articles help the analyst acquire a proper understanding of a conflict and develop a realistic model. Therefore, accurate and comprehensive information plays a fundamental role. In fact, the whole "Modeling" section, as shown in Figure 1, as well as the "Interpretation and Sensitivity" stage in the "Analysis" part, directly depends on the analyst's findings. In other words, the analyst determines the DMs, their options, infeasible states, and relative preferences. The analyst's

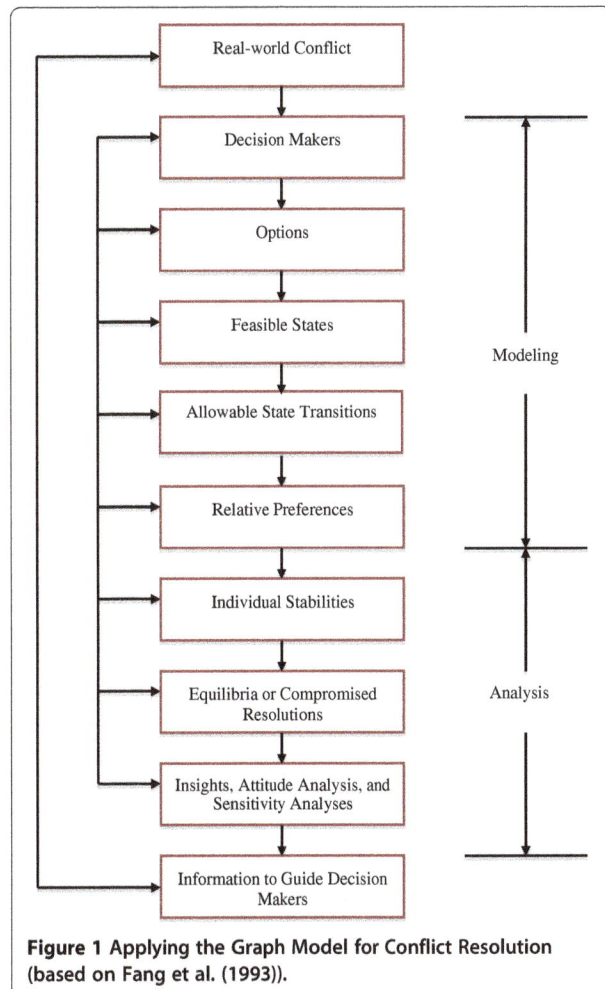

Figure 1 Applying the Graph Model for Conflict Resolution (based on Fang et al. (1993)).

decision about executing different sensitivity analyses, and the interpretations and suggestions to actual DMs for an ongoing conflict depend on the available data. In some cases, the analyst may work directly with a DM or the DM's representative involved in a dispute, while in other situations, he or she may be working with an interested stakeholder who has no direct decision-making power in influencing the dispute under study but is affected by its outcome.

Decision makers, options, feasible states, and allowable state transitions

To use the graph model methodology, one must first model the dispute in terms of the DMs. A DM is a person or a group who plays a role in a conflict and has one or more decisions to make, or alternatives to choose. Besides the DMs and options, states should be defined. A state is any combination of chosen options. A sample state, under the name of Status Quo, is shown in Table 1. The letter "Y" means that the DM selects an option under its control, while "N" denotes the refusal of a DM to choose the option. A dash ("-") indicates that for the DM to choose or not to choose an option is inconsequential. Some states cannot occur in reality. These states are considered "infeasible" and have to be removed from the conflict.

After determining the feasible states, the allowable transitions should be determined. From each state, a specific DM can only move to a certain set of states. This set, which is referred to as a DM's Reachable List, consists of the states in which other DMs' choices, including Y, N, or -, are not different from the initial state. When a DM on his/her own moves from one state to another, this move is called a unilateral move. If the state to which the DM moves to is more preferred by him or her, the move is called a "unilateral improvement".

Relative preferences and static analysis

After generating a complete set of feasible states, the analyst must determine the relative preferences, in which, for each DM, states are ranked from most to least preferred, where ties are allowed. Three methods can be used in GMCR II to define the relative preferences of each DM: Direct Ranking, Option Weighting, and Option Prioritizing.

Subsequently, according to a rich range of solution concepts describing how people or organizations may behave under conflict, a stability analysis of the conflict is carried out to calculate the stable states for each DM. A state that is stable for all DMs in the dispute is called an equilibrium, which suggests a possible resolution to the conflict.

A range of stability definitions have been defined for determining stable states in a dispute in which it is not advantageous for a DM to move away from a state under study. The four useful stability definitions given in Table 15 consist of Nash (Nash, 1950, 1951), sequential (Fraser and Hipel, 1984), general metarationality (GMR) (Howard, 1971), and symmetric metarationality (SMR) (Howard, 1971). Qualitative explanations of how each stability definition works along with its overall characteristics are provided in Table 15, while Fang et al. (1993) furnish mathematical definitions within the paradigm of GMCR. When a state is stable for all of the DMs according to a specific stability definition, the state constitutes an equilibrium or possible compromise resolution.

Interpretation and sensitivity analysis

After carrying out a stability analysis, one may perform appropriate sensitivity analyses. In this case, the DMs, the options, or the relative preferences can be changed to obtain a better understanding of the issue and ascertain how the equilibrium results are affected.

There are also other types of analyses that can be carried out within the GMCR framework to interpret the results, including attitudes (Inohara et al., 2007; Bernath Walker et al. 2009), coalitions (Kilgour et al., 2001; Inohara and Hipel, 2008a,b), strength of preference (Hamouda et al., 2004, 2006; Xu et al., 2009a), misperceptions (called hypergames) (Wang et al., 1988), emotions (Obeidi et al., 2005, 2006, 2009a, 2009b), preference uncertainty (Li et al., 2004; Hipel et al., 2011),

Table 15 Solution concepts and human behaviour (Fang et al., 1993; Hipel et al., 1997)

Solution Concept	Stability Description	Foresight	Knowledge of Preference	Disimprovement	Strategic Risk
Nash	DM cannot move unilaterally to a more preferred state	Low	Own	Never	Ignores risk
SEQ	All DM's unilateral improvements are sanctioned by subsequent unilateral improvements by others	Medium	All	Never	Takes some risks
GMR	All DM's unilateral Improvements are sanctioned by subsequent unilateral moves by others	Medium	Own	By opponent	Avoids risks
SMR	All DM's unilateral improvements are sanctioned, even after response by the DM	Medium	Own	By opponent	Avoids risks

conflict dynamics (Li et al., 2005), and matrix stability calculations (Xu et al., 2009b). By applying different analyses, the outcomes can be more deeply interpreted and additional strategic insights may be achieved.

Endnotes

[a]After NRU restarted in January 2008, another shutdown happened in 2009. On May 14, 2009, NRU was shut down due to a loss of electrical power in Ontario. On May 15, when the experts were investigating the reactor, they observed a small leak of heavy water within the facility. Therefore, the NRU was kept out of service for repair (AECL, 2009). On August 17, 2010, the NRU was returned to operation (NRU Canada, 2010). The performance of NRU, as an important supplier of medical isotopes in the world, is critical, and the repeated shutdowns of this reactor diminished the reputation of AECL, the company responsible for it.

Abbreviations

AECL: Atomic Energy of Canada Limited; GMCR: Graph Model for Conflict Resolution; DM: Decision Maker; CANDU: CANada Deuterium Uranium; ACR: Advanced CANDU Reactors; OPG: Ontario Power Generation; R&D: Research and Development; CIGI: Centre for International Governance Innovation; RFP: Request For Proposal; GG: Green Groups; OG: Ontario Government; FG: Federal Government; DUPIC: Direct Use of spent Pressurized water reactor fuel In CANDUs; GMR: General Metarationality; SMR: Symmetric Metarationality; SEQ: Sequential Stability; NDP: New Democratic Party; MAPLE: Multipurpose Applied Physics Lattice Experiment.

Competing interests

The authors declare that they have no competing interests.

Authors' contributions

MA modeled the conflict and performed the analysis. KWH and MD revised the manuscript critically for important intellectual content and improved the paper by their important comments and guidance. All authors read and approved the final manuscript.

Author details

[1]Department of Systems Design Engineering, University of Waterloo, Waterloo, Ontario N2L3G1, Canada. [2]Centre for International Governance Innovation, Waterloo, Ontario N2L 6C2, Canada. [3]School of Business and Economics, Wilfrid Laurier University, Waterloo, Ontario N2L3C5, Canada.

References

AECL (2008) CANDU Reactors. http://www.aecl.ca/Reactors/CANDU6.htm. Accessed September 25, 2008

AECL (2009) Restarting Safely, Reassuring Canadians. http://www.nrucanada.ca/en/home/projectrestart/NRUOutage.aspx. Accessed December 8, 2010

Akin D (2008) Over Budget and Behind Schedule, Reactor Replacement Axed. http://www.canada.com/topics/news/story.html?id=ba8921b1-5d42-4122-983e-3e0a3008ccea. Accessed October 22, 2008

Babbage M (2011) AECL sold to SNC-Lavalin Group for $15M. http://www.citytv.com/toronto/citynews/life/money/article/139957–aecl-sold-to-snc-lavalin-group-for-15m. Accessed July 17, 2011

Benzie R, Black D (2008) Darlington to Get 2 New Reactors. http://www.thestar.com/News/Ontario/article/444474. Accessed October 10, 2008

Bernath Walker S, Hipel KW, Inohara T (2009) Strategic Decision Making for Improved Environmental Security: Coalitions and Attitudes in the Graph Model for Conflict Resolution. J Syst Sci Syst Eng 18(4):461–476

Cadham J (2009) The Canadian Nuclear Industry: Status and Prospects. http://www.cigionline.org/publications/2009/11/canadian-nuclear-industry-status-and-prospects. Accessed Sep 16, 2011

Canadian Nuclear Safety Commission (2011) Is our energy superpower vision slipping away?. http://nuclearsafety.gc.ca/eng/mediacentre/updates/march-11-2011-japan-earthquake-comparison-candu-and-bwr-reactors.cfm. Accessed April 4, 2011

CBC News (2006) McGuinty Takes Blame for Broken Promise on Coal plant Closures. http://www.cbc.ca/canada/toronto/story/2006/11/16/coal-responsibility.html. Accessed September 15, 2008

CBC News (2008) Nuclear Safety Watchdog Head Fired for "Lack of Leadership": Minister. http://www.cbc.ca/canada/story/2008/01/16/keen-firing.html. Accessed December 8, 2008

CBC News (2009) Canada Eyes Sale of Stake in AECL Reactor Business. http://www.cbc.ca/money/story/2009/05/28/aecl-future-sale.html. Accessed December 8, 2010

CTV (2009) Feds Announce Major Shakeup of Nuclear Agency. http://www.ctv.ca/servlet/ArticleNews/story/CTVNews/20090528/AECL\shakeup\090528/20090528?hub=Canada. Accessed December 8, 2010

Fang L, Hipel KW, Kilgour DM (1993) Interactive Decision Making. Wiley, New York

Fang L, Hipel KW, Kilgour DM, Peng X (2003a) A Decision Support System for Interactive Decision Making, Part 1: Model Formulation. IEEE T Syst Man Cy C 33(1):42–55

Fang L, Hipel KW, Kilgour DM, Peng X (2003b) A Decision Support System for Interactive Decision Making, Part 2: Analysis and Output Interpretation. IEEE T Syst Man Cy C 33(1):56–66

Frame A (2008) A Very Political Power Play. http://www.thestar.com/comment/Columnist/article/349255. Accessed September 20, 2008

Fraser NM, Hipel KW (1984) Conflict Analysis. Models and Resolutions, Elsevier, New York

Geddes J (2007) Harper Embraces the Nuclear Future. http://www.macleans.ca/canada/features/article.jsp?content=20070507\105095\105095. Accessed October 25, 2008

Goodhand P, Drouin R, Mason T, Turcotte E (2009) Report of the Expert Review Panel on Medical Isotope Production. Minister of Natural Resources, Ottawa, Canada, http://www.marketwire.com/press-release/Natural-Resources-Canada-Expert-Panel-Delivers-Report-on-Isotopes-1085758.htm. Accessed December 20, 2010

Hamilton T (2007) Could Reactors Withstand Blast? Report That Regulator Will Impose New Safety Standards May Pose a Big Hurdle for AECL Nuclear Sale in Ontario. http://pqasb.pqarchiver.com/thestar/access/1196866631.html?FMT=ABS&FMTS=ABS:FT&type=current&date=Jan+19,+2007&author=Tyler+Hamilton&pub=Toronto+Star&edition=&startpage=F.1&desc=Could+reactors+withstand+blast. Accessed October 25, 2008

Hamilton T (2008a) A Chance for Nuclear Industry to Clean up Its Act. http://www.thestar.com/article/293658. Accessed September 23, 2008

Hamilton T (2008b) AECL Failed to Disclose Loss of Radioactive Part: Report. http://www.thestar.com/News/Ontario/article/467742. Accessed October 20, 2008

Hamouda L, Kilgour DM, Hipel KW (2004) Strength of Preference in the Graph Model for Conflict Resolution. Group Decis Negot 13:449–462

Hamouda L, Kilgour DM, Hipel KW (2006) Strength of Preference in Graph Models for Multiple Decision-maker Conflicts. Appl Math Comput 179(1):314–327

Harris J (2008) Nuclear power: Expensive, Irresponsible and Unnecessary. http://www.greenparty.ca/en/node/4218. Accessed September 25, 2008

Hipel KW (ed) (2009a) Conflict Resolution, Volume 1. Eolss Publishers, Oxford, United Kingdom, ISBN-978-1-84826-120-4 (Adobe e-Book), ISBN-978-1-84826-570-7 Library Edition (Hard Cover)

Hipel KW (ed) (2009b) Conflict Resolution, Volume 2. Eolss Publishers, Oxford, United Kingdom, ISBN-978-1-84826-121-1 (Adobe e-Book), ISBN-978-1-84826-571-4 Library Edition (Hard Cover)

Hipel KW, Bowman C (2011) Is our energy superpower vision slipping away? Toronto Star, March 16, 2011. http://www.thestar.com/opinion/editorialopinion/article/955245–is-our-energy-superpower-vision-slipping-away. Accessed April 4, 2011

Hipel KW, Kilgour DM, Bashar MA (2011) Fuzzy Preferences in Multiple Participant Decision Making. Scientia Iranica, Transactions D: Computer Science & Engineering and Electrical Engineering, special publication dedicated to the lifelong achievements of Professor Lotfi A Zadeh 18(3):627–638

Hipel KW, Kilgour DM, Fang L, Peng X (1997) The Decision Support System GMCR in Environmental Conflict Management. Appl Math Comput 83(2 and 3):117–152

Howard N (1971) Paradoxes of Rationality: Theory of Metagame and Political Behaviour. MIT Press, Cambridge, MA

Inohara T, Hipel KW (2008a) Coalition Analysis in the Graph Model for Conflict Resolution. Syst Eng 11(4):343–359

Inohara T, Hipel KW (2008b) Interrelationships among Noncooperative and Coalition Stability Concepts. J Syst Sci Syst Eng 17(1):1–29

Inohara T, Hipel KW, Walker S (2007) Conflict Analysis Approaches or Investigating Attitudes and Misperceptions in the War of 1812. J Syst Sci Syst Eng 16(2):181–201

Kilgour DM, Hipel KW, Fang L, Peng X (2001) Coalition Analysis in Group Decision Support. Group Decis Negot 10(2):159–175

Li KW, Hipel KW, Kilgour DM, Fang L (2004) Preference Uncertainty in the Graph Model for Conflict Resolution. IEEE T Syst Man Cy A 34(4):507–520

Li KW, Kilgour DM, Hipel KW (2005) Preference Uncertainty in the Graph Model for Conflict Resolution. J Oper Res Soc 5:699–707

McCarthy S (2010) Ottawa's plan to sell AECL threatens future of Canada's nuclear industry. http://www.theglobeandmail.com/report-on-business/industry-news/energy-and-resources/ottawas-plan-to-sell-aecl-threatens-future-of-canadas-nuclear-industry/article1792762/. Accessed December 7, 2010

McCarthy S, Howlett K (2009) AECL's Future in Doubt as Ontario Suspends Nuclear Power Plans. http://www.theglobeandmail.com/news/national/ontario-suspends-nuclear-power-plans/article1200469/. Accessed December 8, 2010

McCarthy S (2011) Ottawa to sell AECL to SNC-Lavalin. http://www.theglobeandmail.com/report-on-business/ottawa-to-sell-aecl-to-snc-lavalin/article2078110/. Accessed July 18, 2011

McParland K (2008) Dalton McGuinty's Opportunity to Slash Verbal Gas Emissions. http://network.nationalpost.com/np/blogs/posted/archive/2008/03/26/dalton-mcguinty-s-opportunity-to-slash-verbal-gas-emissions.aspx (September 25, 2008)

Nash JF (1950) Equilibrium Points in N-player Games. Proc Natl Acad Sci 36(1):48–49

Nash JF (1951) Non-cooperative Games. Ann Math 54(2):286–295

Nathwani J (2009) Canada's Medical Isotope Crisis: A Way Forward. J Policy Engagement 1(4):12–13, Available at www.ocepp.ca

New Democratic Party (2008) NDP Platform. http://www.ndp.ca/platform/environment/newenergyeconomy. Accessed October 17, 2008

NRU Canada (2010) Delivering on Our Commitment to Restart. http://www.nrucanada.ca/en/home/default.aspx. Accessed December 1, 2010

Obeidi A, Hipel KW, Kilgour DM (2005) The Role of Emotions in Envisioning Outcomes in Conflict Analysis. Group Decis Negot 14(6):481–500

Obeidi A, Hipel KW, Kilgour DM (2006) Turbulence in Miramichi Bay: The Burnt Church Conflict over Native Fishing Rights. J Am Water Resour Assoc 42(12):1629–1645

Obeidi A, Kilgour DM, Hipel KW (2009a) Perceptual Graph Model Systems. Group Decis Negot 18(3):261–277

Obeidi A, Kilgour DM, Hipel KW (2009b) Perceptual Stability Analysis of a Graph Model System. IEEE T Syst Man Cy A 39(5):993–1006

Ontario Government (2009) Nuclear Procurement Project Update. http://www.news.ontario.ca/mei/en/2009/06/nuclear-procurement-project-update.html. Accessed July 19, 2009

Ontario Government (2010) Ontario's Long-term Energy Plan. http://www.mei.gov.on.ca/en/energy/. Accessed December 7, 2010

Puxley C (2007) Sale of Canada's Nuclear Giant Shouldn't be Negotiated in Private: Lib- erals. http://energyquest4nanticoke.ca/nukegiant.htm. Accessed November 28, 2008

Sierraclub (2008) Towards a Nuclear-Free Canada. http://www.sierraclub.ca/national/programs/atmosphere-energy/nuclear-free/index.shtml Accessed September 23, 2008

Spears T (2008) Reactor Connected to Second Safety Pump. http://www.canada.com/topics/news/national/story.html?id=0e63c2ea-fd59-4a80-b047-b3880f30c729. Accessed November 5, 2008

The Globe and Mail (2010) New Brunswick throws a wrench into AECL sale plans. http://www.nuclearcounterfeit.com/?p=3445. Accessed December 7, 2010

The Star (2009) Privatization of AECL Radioactive Issue for Ottawa. http://www.thestar.com/Opinion/article/612823. Accessed December 20, 2010

Wang M, Hipel KW, Fraser NM (1988) Modeling Misperceptions in Games. Behav Sci 33(3):207–223

Xu H, Hipel KW, Kilgour DM (2009a) Multiple Levels of Preference in Interactive Strategic Decisions. Discret Appl Math 157(15):3300–3313

Xu H, Hipel KW, Kilgour DM (2009b) Matrix Representation of Solution Concepts in Multiple Decision Maker Graph Models. IEEE T Syst Man Cy A 39(1):96–108

12

GIS-based multi-criteria analysis for land use suitability assessment in City of Regina

Jiapei Chen

Abstract

Background: Land use suitability assessment is a key factor in any urban and suburban planning and decision-making processes. The assessment is evaluated by a series of criteria involving socio-economic needs. To deal with the conflicting, disproportionate and multiple criteria, land usage will be characterized with respect to the preferences and importance. Meanwhile, many spatial decision problems can be typically analyzed and interpreted visually by applying GIS for mapping and analysis. Accordingly, this study is to introduce the Multi-criteria decision analysis into the land use suitability analysis along with the existing perspective evolving the role of GIS.

Results: For case study, it is the first time to conduct the identification of the current land use situations in City of Regina by using GIS, combined with multi-criteria analysis ideology for existing condition analysis with three criteria referring to social and economic factors. Finally, the study identified five suitability levels, which has revealed the trends, challenges and prospects of land use analysis for urban extension in City of Regina.

Conclusion: Comparing the simulated land use suitable classes to the existing land use pattern, difference and optionality have been presented. The results are providing valuable information for the urban extension for policy and decision makers. To improve the accuracy and the reliability of the real-world case study, criteria selection and weight assignments ranks the first place. The integrated use of data analysis and Multi-criteria Decision Analysis approach, in a GIS context, resulted in a visible assessment of current land use in Regina.

Keywords: Land use suitability; Multi-criteria decision analysis; GIS; Urban extension; Regina

Background

Due to the increasing population and economic growth, human activities have continuous impacts on land use. Those impacts might lead to a series complexities toward environment and land resources development (Huang and Xia 2001). Issues related to population and land use competition has emphasized the need for more effective land use planning and policies. In Canada, in particular, the population has been projected to be 47.7 million by 2036 and 63.8 million by 2061, in comparison to 33.7 million in 2009. This pressure of the population will increase the requirement of land use enormously. Land use suitability assessment is to afford a reasonable and sustainable manner for land resource development. Meanwhile, the growing demand for urbanization, land resources used for a variety of purpose will interact and compete with each

other. Rational and sustainable decision support based on the land use suitability assessment is an issue of great concern to governments and land users. Thus, an effective way to assess suitability of the current land use and provide policy support to the future land use is desired.

Previously, various methods of spatial analysis for land use are commonly used in the suitability assessment studies. The problem of land use suitability assessment have often been tackled using multi-criteria decision analysis (MCDA) since 1980s (Antoine et al. 1997, Collins et al. 2001, Kiker et al. 2005, Sharifi et al. 2006, Kunwar et al. 2010). Cheng et. al., has reported an integrated MCDA linear programming approach to support selection of an optimal landfill site (Cheng et al. 2003). To create visualized suitability map for users and decision makers, the integration of MCDA and GIS has been widely promoted for solving spatial problems in urban assessment and planning (Phua and Minowa 2005). (Malczewski

Correspondence: jiapeichen88@gmail.com
Faculty of Engineering and Applied Science, University of Regina, S4S 0A2
Regina, SK, Canada

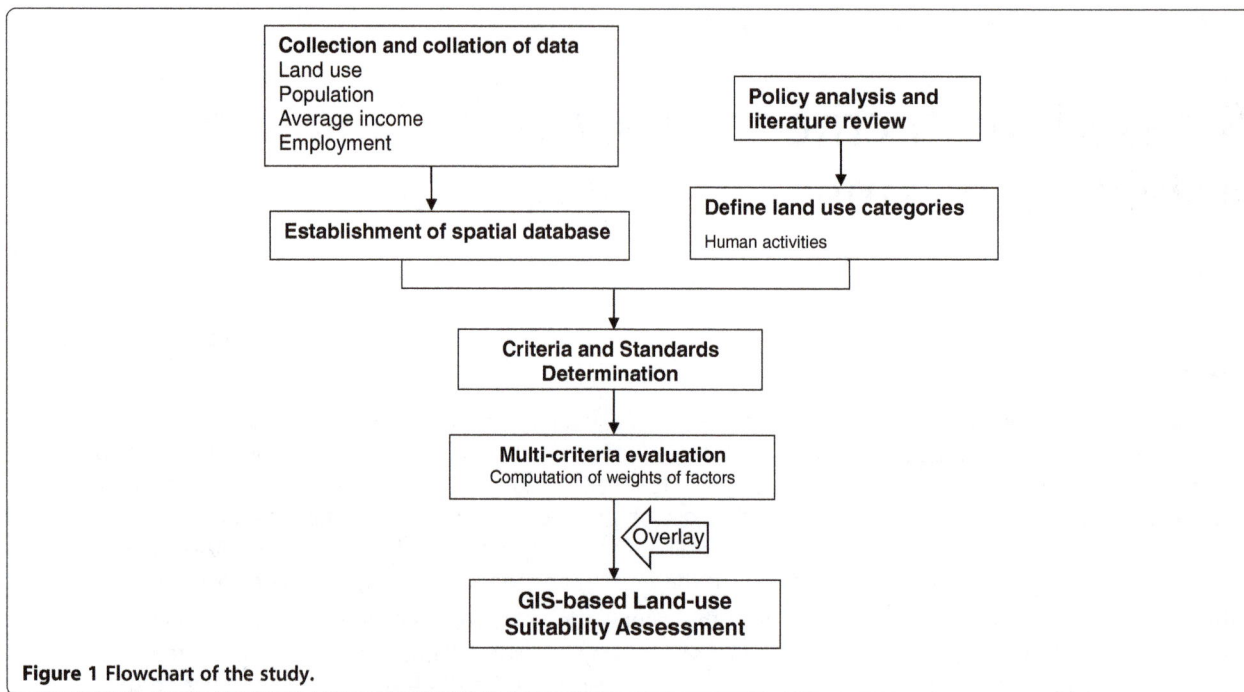

Figure 1 Flowchart of the study.

2006) conducted a survey of literature of the GIS-based Multi-criteria analysis from 1990 to 2004. There has been a substantial acceleration in the number of the GIS-MCDA articles published in this field. Joerin et al. (2001) put forward an outranking multi-criteria analysis method and output a set of land use suitability maps incorporating complex criteria integrating several stakeholders' points of view. Meanwhile, some case studies were performed using the method of integrating GIS and multi-criteria analysis. Hanyang lake area located in Wuhan city, China, was studied with this comprehensive method to analyze the suitability of future land use

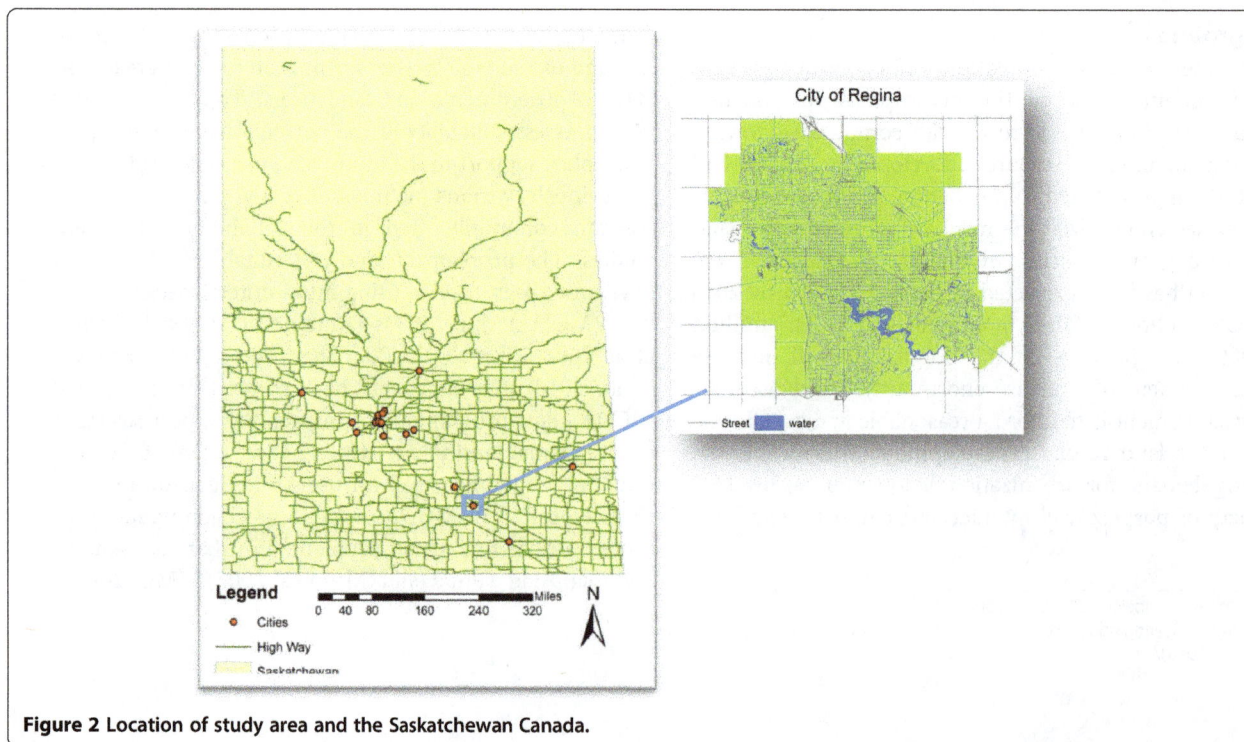

Figure 2 Location of study area and the Saskatchewan Canada.

Land Use in City of Regina

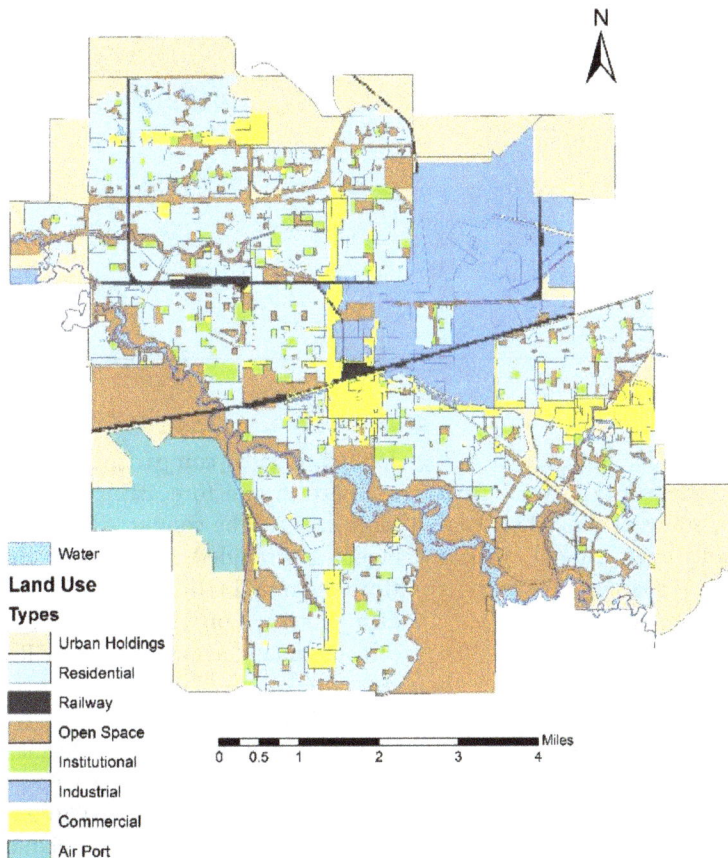

N

Water

Land Use

Types

Urban Holdings
Residential
Railway
Open Space
Institutional
Industrial
Commercial
Air Port

Miles
0   0.5   1   2   3   4

Figure 3 Current land use pattern in Regina.

Current land use types in Regina

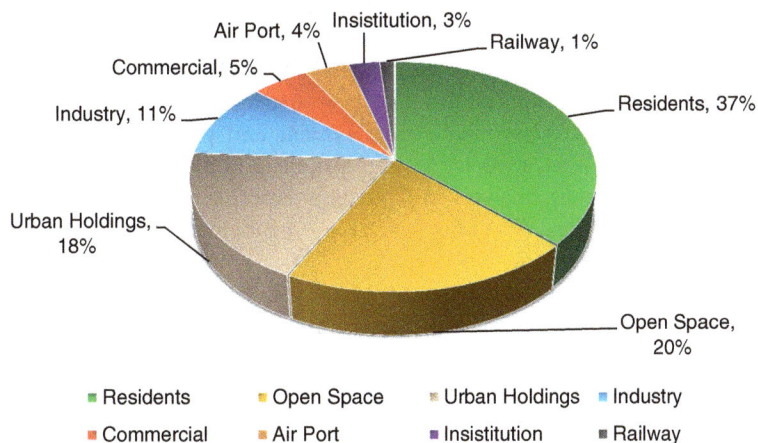

Air Port, 4%   Insistitution, 3%

Commercial, 5%   Railway, 1%

Industry, 11%   Residents, 37%

Urban Holdings,
18%

Open Space,
20%

■ Residents   ■ Open Space   ■ Urban Holdings   ■ Industry
■ Commercial   ■ Air Port   ■ Insistitution   ■ Railway

Figure 4 The current land use types and proportion in Regina.

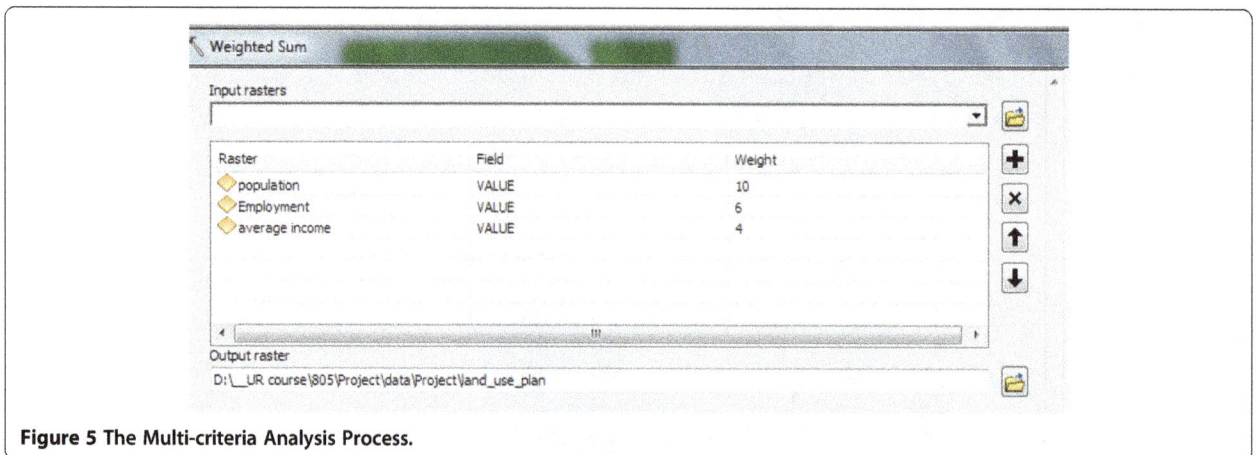

Figure 5 The Multi-criteria Analysis Process.

according to specified requirements, preferences and predictions in Yong Liu et al.'s research (Liu et al. 2007). Dai et al. (2001) conducted a study of the urban land use planning based on GIS and multi-criteria analysis method, which was applied in Lanzhou city, and its vicinity in north-western China.

However, the research has seldom focused on the Prairie Provinces in Canada, especially for the city with less proportion of population. For the further development and urbanization, the land use suitability assessment will play an important role. Thus, the primary aim of this study is to provide recommendation for integrated and sustainable plan for land use development referring to accessibility, economic, integration and environment by means of weighting the selected criteria. Since the required criteria are heterogeneous and measured on various scales in this method. In addition, the data is limited, this research will only analyze the current situation and future exploitation forecasting based on simplified factors, such as population and income.

The assessment is often assessed by a number of criteria. To deal with the inconsistent, incommensurate and multiple criteria, a comprehensive method is needed to analysis the land use scheme under current situation. Then it can be applied to the land use forecasting and planning. The individual criteria will be characterized with respect to the significance and preferences. Meanwhile, spatial decision problems can also be typically analyzed and demonstrated on the base of visualization. Accordingly, the study is to introduce the Multi-criteria decision analysis into the land use suitability analysis along with the existing perspective evolving the role of Geographical Information Systems (GIS).

In this study, the existing land use situation was analyzed and evaluated. Then, based on the analysis the importance of each factor or criterion, the multi-criteria analysis system for land use suitability will be applied. Finally, comparison between the existing land use pattern and forecasting land use possibility of the same area will be conducted. The flowchart is shown in Figure 1. Provided data sets in order to generate and store them in a GIS framework, while, as far as the last data sets are concerned, MCDA was involved in the calculation and acquisition in a GIS environment.

The study area

The study area is the City of Regina, located in southern Saskatchewan, Canada (as shown in Figure 2). The coordinate of the site is 50°27′17″N, 104°36′24″W. The area of the city is 118.87 km^2. Total population of the study area was 179,246 according to census record in 2006. As the capital city of Saskatchewan, city of Regina ranks the second largest in the province, and is a cultural and commercial metropolis in southern Saskatchewan.

The current land use map is presented in Figure 3. There are eight main categories in this study area: Airport, Commerce, Industry, Resident, Institution, Open space, Railway, and Urban holdings. Figure 4 shows the proportion of different land use types. As see form the graph, most of the land in Regina is used as residential area, and then is the open space.

It is reported that strong economic growth has led to employment and population growth in the Population, Employment and Economic Analysis of city of Regina. Thus population, employment and income are chosen as social and economic effect factors. The data source and

Table 1 The weight of each criterion

Criteria	Wj
Population	10
Employment	6
Average_income	4

information are obtained from TerraServer of TERRA Lab, website of Bureau of Saskatchewan and Canada Statistics. On the basis of "A Guide to the Municipal Planning Process in Saskatchewan", a comprehensive policy framework to guide physical, environmental, economic, social, and cultural development in municipality is provided. As social development is one of the most important factors in land use planning, it is considered separately. The data that will be used in the multi-criteria analysis would be related to population, average income of particular census, and employment.

Methods

Multi-Criteria Decision Analysis (MCDA), developed in the environmental of Operation Research, aids analysts and decision-makers in situations in which there is a need for identification of priorities according to multiple criteria. This usually happens in situations where conflictive interests coexist (Gomes and Lins 2002). MCDA can incorporate both geographical data and stakeholders' preferences into quantified values for assessment and further decisions (Malczewski 2004). The GIS analysis, if integrated with a procedure of data analysis and structuring, can be usefully developed when data are available but the decision context cannot indicate how these data have to be used to produce information and support decisions. The GIS support the solution of complex spatial problems, providing the decision-maker with a flexible environment in the process of the decision research and in the solution of the problem. The visualization of the context, structure of the problem and its alternative solutions is one for the most powerful components of a decision support system (Gomes and Lins 2002). Thus the integration GIS-MCDA has the objective of the supporting decision-makers, providing them with ways to evaluate several alternatives, based on multiple, conflictive criteria.

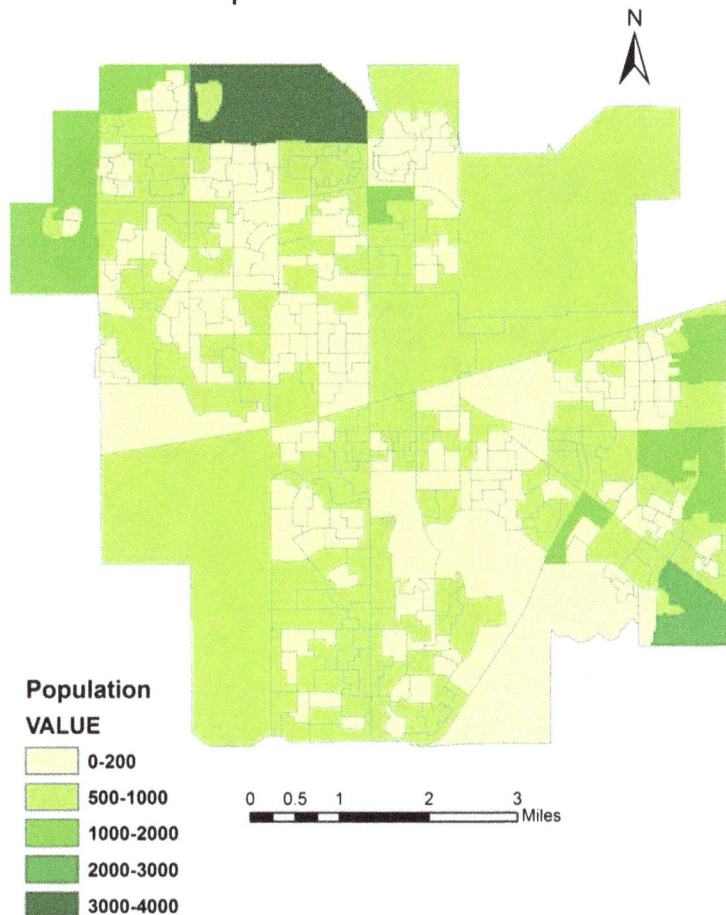

Figure 6 The population distribution of Regina.

GIS

GIS is a set of tools for inputs, storage and tetrieval, manipulation and analysis, as well as outputs of spatial data (Malczewski, 1999). ArcGIS is acknowledged to be a powerful tool in solving the spatial problems. ArcGIS by ESRI GIS and mapping software was applied for spatial data analysis and mapping in this study. All the related data were collected from Terrasever and Terra lab at University of Regina. Land-use maps and administrative information were input into GIS digitally to establish a new geo-database, then overlapped with each other.

Meanwhile, policy analysis based on community plans and literature reviews were completed, serving as foundation for land use type categories. According to available data, land use for human activity were divided into five suitability levels. In this process, multi-criteria analysis method was used for classifying and weighing criteria. Quantitative analysis is necessary for multi-criteria analysis, including scoring, ranking and weighting.

Finally, an output map of the land use suitability with five classes was displayed and a comparison was conducted between the new land use pattern and the preexisting land use status. The Halme et al. approach introduces the decision-maker's preference in the efficiency analysis, by explicitly locating his most preferred solution vector on the efficient frontier.

The same authors highlight that when systematically exploring the neighborhoods of the Most Preferred Solution (MPS), one does not know explicitly the decision-maker's value function, but its form becomes known when the end of the search for MPS is reached.

Weight product

Weight product (WP) method has been introduced centuries ago and been advocated in the past few years. WP is a relative simple multiple attribute utility methods. Since WP is easily understood by decision makers and is easy to be conducted, this method have been widely applied in

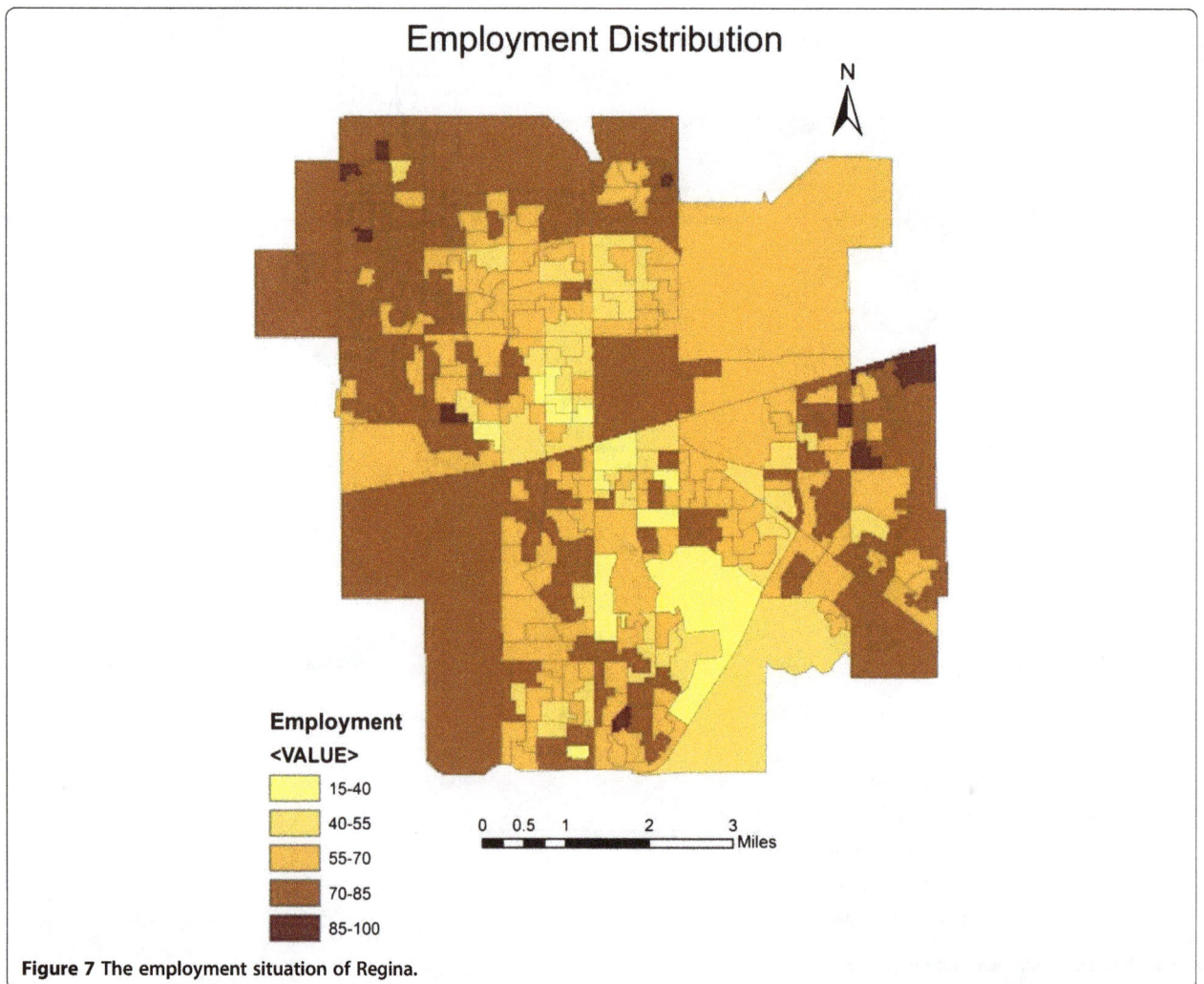

Figure 7 The employment situation of Regina.

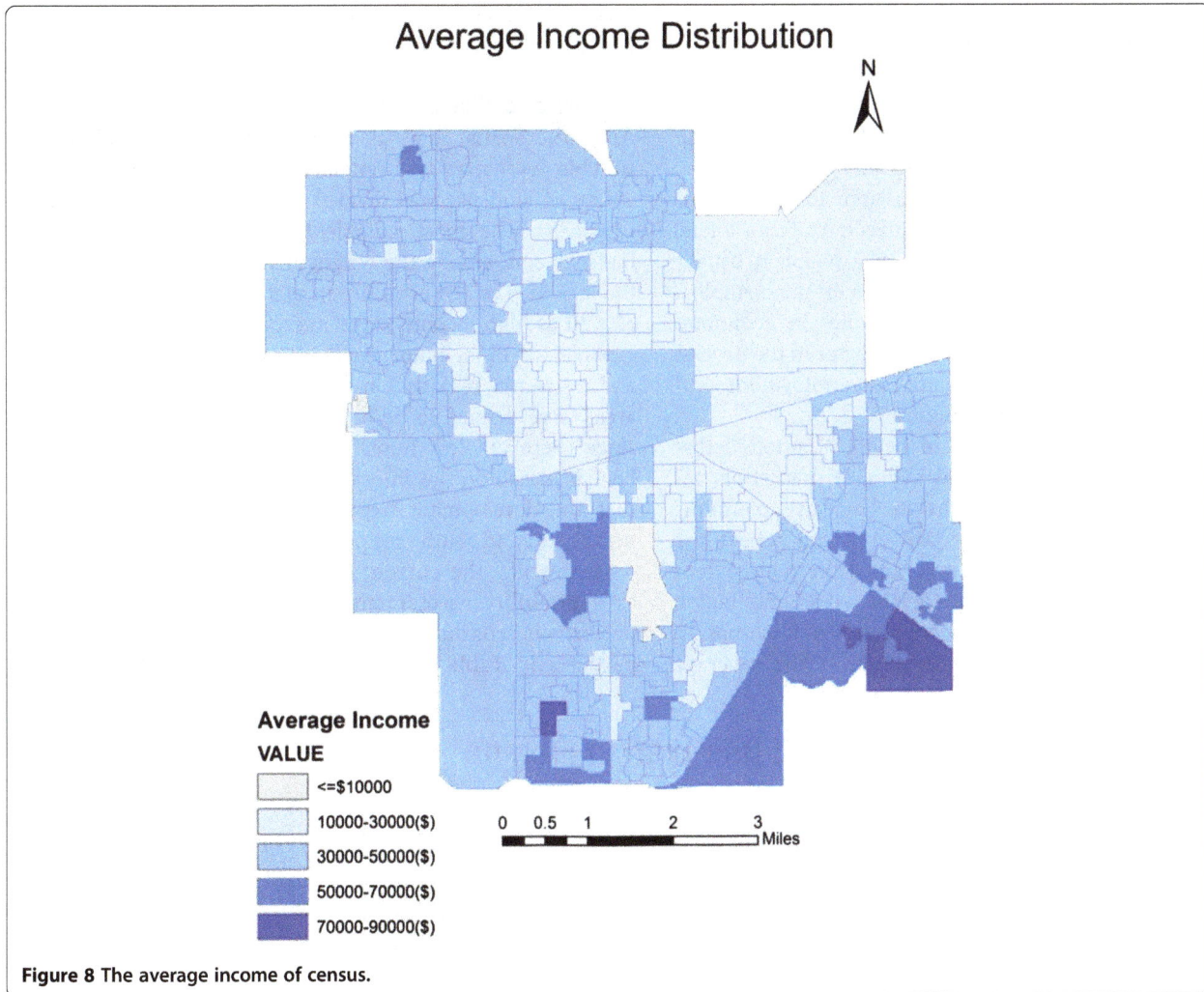

Figure 8 The average income of census.

many fields. In this study, based on the WP method, three factors including population, employment and average income, hold significance in municipal land use planning. In the multi-criteria analysis process as showed in Figure 5, they were assign different weight to obtain a total score of every region based on the following formula (Gomes and Lins 2002):

$$S_f = \sum_{j=1}^{3} W_j * A_i \qquad (1)$$

Where W_j –- is weight of each criterion j = 1,2,....m,

And A_i—is the normalize the value of each grid cell, i = 1,....., n.

The general MCAD approach for this case may be seen in (1), where X_i, Y_i,..., represent the value of the criterion X, Y..., for the alternation i; λ are the decision variables that represent the decision-maker's preferences

for the alternative i, i = 1,...,n. For this case study, λ vector representing the decision maker preferences.

$$max \sum_{i=1}^{n} X_i * \lambda_i$$

$$max \sum_{i=1}^{n} Y_i * \lambda_i$$
$$\vdots$$
$$s.\ t. \sum_{i=1}^{n} \lambda_i = 1$$

$$\lambda \geq 0, \quad i = 1, ..., n.$$

Defining the criteria

The expansion of land has overwhelmingly been a response to fast-rising population decades ago, so population is considered the most essential drive force of land exploitation. The weight of each criterion has been shown in Table 1. Meanwhile, as the increasing of urban population, the urbanization of Regina fringe is an inevitable trend. Study of population can not only help analysis the existing land use pattern, but also assist land use trend forecasting. Based on the above-mentioned consideration, that population is specified the maximum weight among the criteria. The population distribution of Regina is shown in Figure 6.

The employment is expanded to labor force, which is in turn expanded to population equivalents. From a land use point of view, although a city or a region is usually studied as a whole, it is also necessary to examine employment changes brought about by changes in economic. Therefore, employment is defined as a factor in the process of land use or land development. The employment situation of Regina is showed in Figure 7.

Furthermore, as like Hok Lin Leung mentioned that income investigate is essential in house marketing, income is equivalent important in land use planning. The income census of a certain area can reflect the consumption level and trend in a certain extent (Leung 2003). Thus, as one of the economic factor, income is also taken into account when it comes to the land use suitability assessment. The average

income of a certain census tract in Regina is showed in Figure 8.

Results and discussion

After the spatial analysis through ArcGIS, a land use suitable level map was obtained. In this map, the suitability of land use was divided into 5 classes (shown in Figure 9). The most suitable level was defined as the highest score among the weighted values, marking red in the legend. The red areas are in high rate of employment, as well as high income with large population, so they are capable for more human activities area like commercial and residential regions. On the contrary, the least suitable areas were defined to be more suitable for open sources, since there are less human beings and low income or benefit. Based on this interpretation, it could be predicted that the development tendency of Regina would be at north-west and south-east part of the city.

Meanwhile, the current land use map is jointed into the suitability assessment output (Figure 10). Comparison is conducted between these two land use patterns in Table 2. As shown in Table 2, there is a

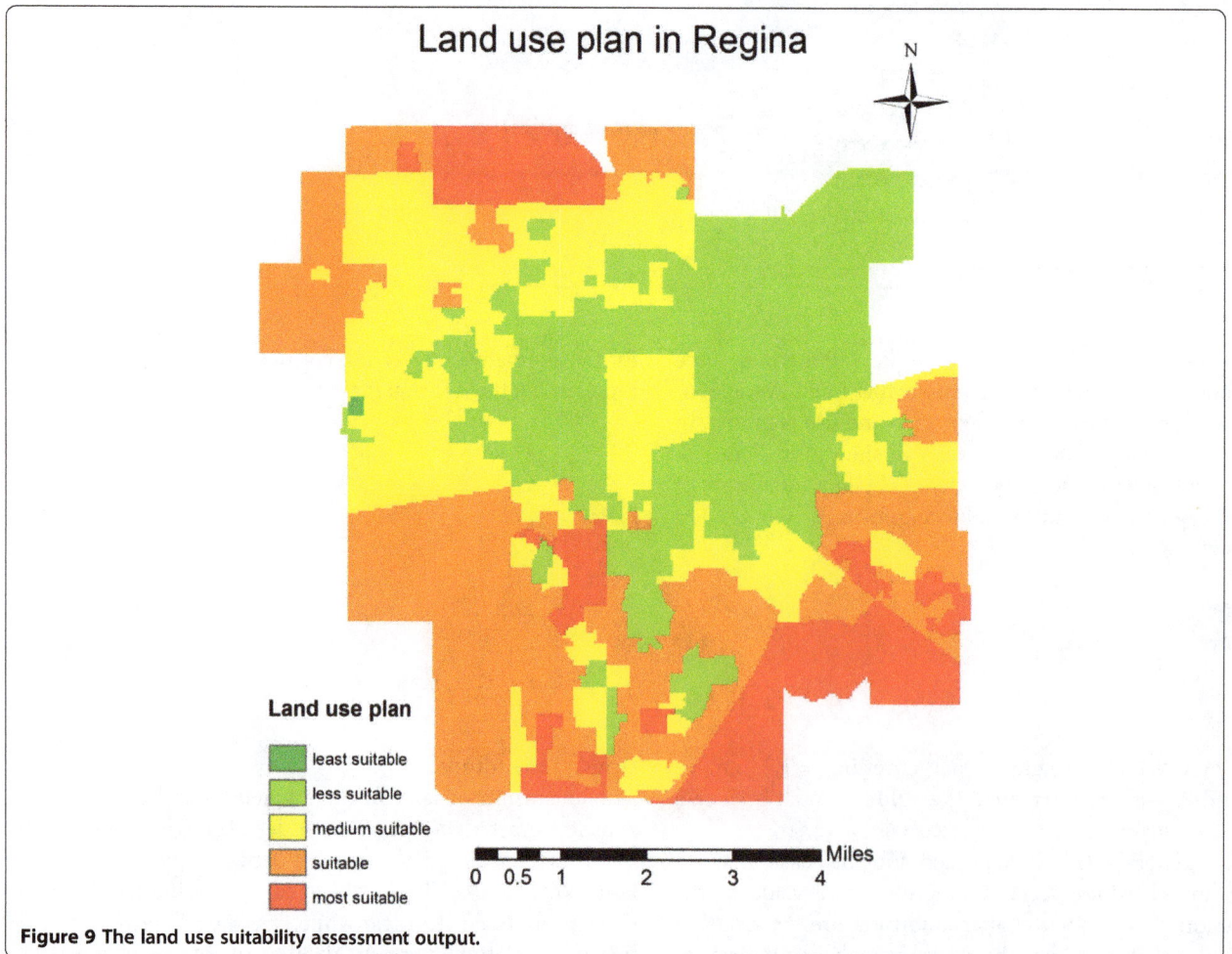

Figure 9 The land use suitability assessment output.

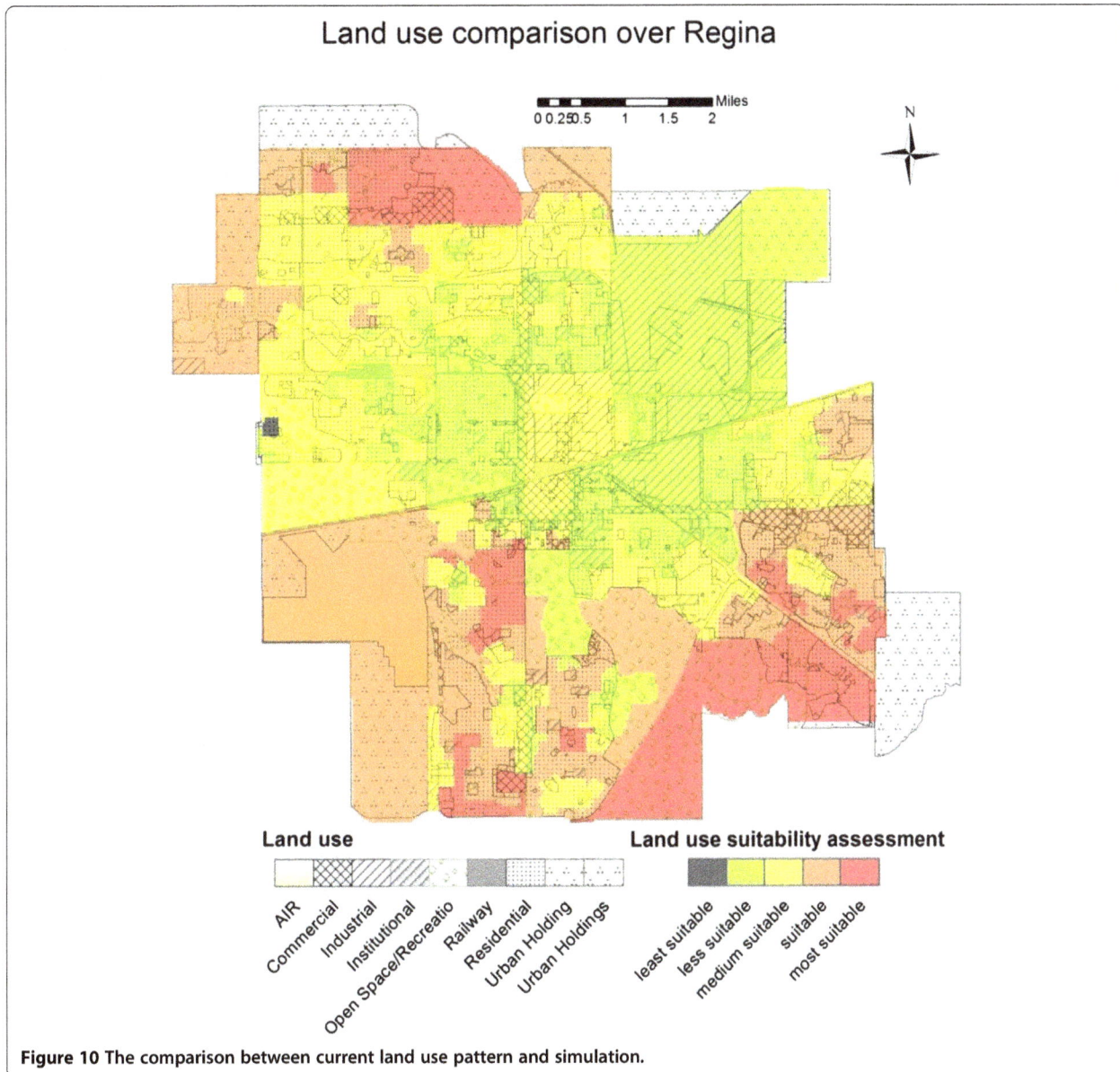

Figure 10 The comparison between current land use pattern and simulation.

Table 2 The comparison between simulation and actual land use

Suitable level	Simulation land use	Actual land use
Least suitable	Urban holdings	A black spot
Less suitable	Air	Industry
	Open space/Recreation	Residential
Medium suitable	Industry	Residential
	Railway	Commercial
		Open space/Recreation
Suitable	Commercial	Air
	Institutional	Urban holdings
Most suitable	Residential	Commercial
		Urban holdings

significant distinction between the simulated land use and the actual land use pattern. The results shows that there are significant difference between the simulated land use availability and the existing land use conditions. Although we can say it is not an ideal plan, we can still find and learn something faulty from the process of analysis and make it available to improve in the next time.

In the study, the priority is the criteria selection and determination. In real-world case study, a large number of criteria should be determined, referring to environment, ecology, social and economic. Whilst in this research, only economic factor were considered. This can be considered as a single stage test of the complicated land use plan procedure. Moreover, the weight of criteria should be determined by both experts and stakeholders. In this research, the weight was just determined by literal evaluation from municipal land use project plan.

Secondly, the simulation of suitable level is simplified, whilst in the real world, the development pattern is complicated. In the analysis, development scenarios in land use are not easy to pin down. There are no universal rules as to what they should or should not include, and there are no tools available to generate them automatically. F.C. Dai et al. set two scenarios in their study of urban fringe land use planning. One is mainly based on the actual development simulation, and the other scenario is mainly depended on the government planning. Thus, setting scenarios might increase the accuracy of the simulation as well as the reliability.

Moreover, as it is under a small scale analysis, the simplified definition is considered to be acceptable. However, the definition of suitability level results in unprofessional. Take the black spot in the least suitable level for example. It is hard to explain its appearance. It might be assumed to be the fault in the multi-criteria analysis procedure. And beyond the public opinion survey and social investigation, the suitability assessment seems to be groundless. Thus, the result might be better if the operational research is implemented in the future study.

Conclusion

Assessing the land use development problem has become an important task under the increasing aware of land resource conservation. Land use suitability assessment is a practical tool to make decision on land use development. However, less works could be found that use quantized index to evaluate the land resource usage at Prairie Provinces in Canada. This study applied GIS combined with multi-criteria decision analysis ideology in the City of Regina to assess land use condition with three important criteria referring to social and economic. Finally, a map with five suitability levels can be obtained.

When comparing the simulated land use suitable classes with the existing land use pattern, we can find that: the significant differences between these two land use patterns reflect the limitation of the approach and data availability. Two possibilities are put forward to increase the reliability and accuracy of land use suitability assessment, where the research is potentially valuable. Urban land use categories are complicated process. Recommendations made for future studies is to improve the efficacy and objectivity of local land use evaluation to support the land use suitability assessment and avoid the subjectivity. As for the accuracy of the real-world case study, the criteria selection and the weight assignment should be widely and deeply discussed and researched.

The results represent the potential of GIS-based evaluation for urban planning purpose. However, it needs to be emphasized that the reliability of the assessment results depends on a multitude of factors ranging from the quality of the database to the potential errors in the GIS. Meanwhile, the modeling results are highly significant to the weights applied. The determination of a multitude of factors and weights for the various factors is one of the most important challenges in the future.

Competing interests
The author declares that she has no competing interests with anyone.

Author's contributions
JC carried out the data collection, methods, results analysis and finish the manuscript.

Acknowledgement
This research has been conducted within University of Regina. The Terrasever and Terra lab provided the equipment and materials for this research. This research is also supported by Dr. Joe Piwowar in the data collection and Dr. Gordon Huang.

References
Antoine J, Fischer G, Makowski M (1997) Multiple criteria land use analysis. Appl Math Comput 83:195–215

Cheng S, Chan CW, Huang GH (2003) An integrated multi-criteria decision analysis and inexact mixed integer linear programming approach for solid waste management. Eng Appl Artif Intel 16(5):543–554

Collins M, Steiner F, Rushman M (2001) Land-use suitability analysis in the United States: historical development and promising technological achievements. J Environ Manage 28(5):611–621

Dai F, Lee C, Zhang X (2001) GIS-based geo-environmental evaluation for urban land-use planning: a case study. Eng Geol 61(4):257–271

Gomes E, Lins M (2002) Intergrating geographical information systems and multi-criteria methods: a case study. Ann Oper Res 116:243–269

Huang G, Xia J (2001) Barriers to sustainable water-quality management. J Environ Manage 61:1–23

Joerin F, Thériault M, Musy A (2001) Using GIS and outranking multicriteria analysis for land-use suitability assessment. Int J Geogr Inf Sci 15(2):153–174

Kiker G, Bridges T, Varghese A, Seager T, Linkov I (2005) Application of multicriteria decision analysis in environmental decision making. Integr Enviro Assess Manage 1(2):95–108

Kunwar P, Kachhwaha T, Kuma A, Agrawal AK, Singh A, Mendiratta N (2010) Use of high-resolution IKONOS data and GIS technique for transformation of landuse/landcover for sustainable development. Curr Sci India 98(2):204–212

Leung H (2003) Land use Planning Made Plain. University of Toronto Press, Toronto, Canada

Liu Y, Lv X, Qin X, Guo H, Yu Y, Wang J, Mao G (2007) An integrated GIS-based analysis system for land-use management of lake areas in urban fringe. Land Scape Urban Plan 82(4):233–246

Malczewski J (1999) GIS and multicriteria decision analysis. John Wiley & Sons.

Malczewski J (2004) GIS-based land-use suitability analysis: a critical overview. Pro Plann 62(1):3–65

Malczewski J (2006) GIS-based multicriteria decision analysis: a survey of the literature. Int J Geogr Inf Sci 20(7):703–726

Phua M, Minowa M (2005) A GIS-based multi-criteria decision making approach to forest conservation planning at a landscape scale: a case study in the Kinabalu Area, Sabah, Malaysia. Land Scape Urban Plan 71(2):207–222

Sharifi M, Boerboom L, Shamsudin K, Veeramuthu L (2006) Spatial Multiple Criteria Decision Analysis in Integrated Planning for Public Transport and Land use Development Study in Klang Valley, Malaysia. ISPRS Technical Commission 2nd Symposium, United States, pp 85–91

13

Groundwater remediation design using physics-based flow, transport, and optimization technologies

Larry M Deschaine[1,2]*, Theodore P Lillys[1] and János D Pintér[3]

Abstract

Background: The purpose of this work was to demonstrate an approach to groundwater remedial design that is automated, cost-effective, and broadly applicable to contaminated aquifers in different geologic settings. The approach integrates modeling and optimization for use as a decision support framework for the optimal design of groundwater remediation systems employing pump and treat and re-injection technologies. The technology resulting from the implementation of the methodology, which we call Physics-Based Management Optimization (PBMO), integrates physics-based groundwater flow and transport models, management science, and nonlinear optimization tools to provide stakeholders with practical, optimized well placement locations and flow rates for remediating contaminated groundwater at complex sites.

Results: The algorithm implementation, verification, and effectiveness testing was conducted using groundwater conditions at the Umatilla Chemical Depot in Umatilla, Oregon, as a case study. This site was the subject of a government-sponsored remedial optimization study. Our methodology identified the optimal solution 40 times faster than other methods, did not fail to perform when the physics-based models failed to converge, and did not require human intervention during the solution search, in contrast to the other methods. The integration of the PBMO and Lipschitz Global Optimization (LGO) methods with standalone physically based models provides an approach that is applicable to a wide range of hydrogeological flow and transport settings.

Conclusions: The global optimization based solutions obtained from this study were similar to those found by others, providing method verification. Automation of the optimal search strategy combined with the reliability to overcome inherent difficulties of non-convergence when using physics models in optimization promotes its usefulness. The application of our methodology to the Umatilla case study site represents a rigorous testing of our optimization methodology for handling groundwater remediation problems.

Keywords: Groundwater modeling, Contamination, Remediation, Optimization

Background

The increasing scarcity and degradation of potable water resources is an issue of global concern. Sources of water quality contamination include releases of chemical and radionuclide contaminants from point-source and non-point source origins. The costs of addressing water quality issues on a global scale are substantial. Regulatory agencies such as the U.S. Environmental Protection Agency (EPA)

and counterpart agencies in other countries must consider economic and human health costs with budgetary constraints in their efforts to clean up contaminated groundwater and restore water resources to beneficial reuse. The EPA documents this need to perform groundwater remediation in an optimal manner in the "*National Strategy to Expand Superfund Optimization Practices from Site Assessment to Site Completion*" (USEPA, 2012).

The objective of this research is to develop and demonstrate an automated, cost-effective, and broadly applicable approach to groundwater remedial design applicable to contaminated aquifers in different geologic settings. This paper presents the development, scope

* Correspondence: larry.m.deschaine@alum.mit.edu
[1]HydroGeoLogic, Inc., Reston, VA 20190, USA
[2]Department of Energy & Environment, Chalmers University of Technology, Göteborg, SE 412 96, Sweden
Full list of author information is available at the end of the article

and application of the simulation-optimization approach for remediating contaminated groundwater including reliability and efficiency verification to find globally optimized solutions to groundwater remediation problems. The approach is demonstrated using a well-studied, publically documented site example that was the subject of a government-sponsored remedial optimization study.

During the past two decades, researchers and engineers have been seeking methods for optimizing the design of ground water treatment systems. Table 1 provides an overview of existing algorithms and methods along with their salient features. Recognizing the limitations of the previously used optimization approaches and the need for efficient and effective groundwater remediation optimization tools motivates this research. This new approach combines flexible and efficient optimization techniques with commonly used subsurface groundwater flow and transport simulation models.

Table 1 Key features and limitations of previous optimization tools

Tool identification	Key features and limitations
GWM: Ground-Water Management Process for the U.S. Geological Survey (USGS) MODFLOW-2000 (Ahlfeld, et al., 2005)	• Performs optimization using Linear Programming (LP) or Sequential Linear Programming (SLP). • Tightly integrated to the MODFLOW code. • Handles only confined flow and mildly non-linear unconfined flow situations.
MGO: Modular Groundwater Optimizer (Zheng and Wang, 2003) based on MODFLOW and the MT3DMS code (Zheng and Wang, 1999) for contaminant transport simulation	• Performs optimization using heuristic global optimization methods, including Genetic Algorithm (GA) and Tabu Search (TS). • Tightly integrated to the MODFLOW and MT3DMS codes. • Computationally burdensome and cumbersome to use even for relatively straightforward practical situations.
SOMOS: Simulation/Optimization Modeling System (Peralta, 2004)	• Performs optimization using a combination of GA, TS, and Artificial Neuron Network (ANN) in conjunction with groundwater flow and solute transport modeling.
SEA: Successive Equimarginal Approach, a hybrid of the gradient-based method and the deterministic heuristic-based method (Guo, et al., 2007)	• Performs optimization using SEA to alleviate some of the computational burden of MGO. • Integrated with MODFLOW and MT3DMS. • Cumbersome to use requires frequent user intervention and may not lead to a global optimum.

DoD/ESTCP Simulation-optimization demonstration project

This approach is tested using a study problem posed as part of the joint U.S. Department of Defense (DoD)/Environmental Security Technology Certification Program (ESTCP) Groundwater Remediation Optimization Study (Minsker et al. 2004); the groundwater contamination remediation design at the Umatilla Chemical Depot in Umatilla, Oregon. The site has a pre-existing and operational remedy-in-place (RIP) installed to remediate the Royal Demolition Explosive (RDX) and 2,4,6-Trinitrotoluene (TNT) contaminated groundwater plumes. The RIP consists of a groundwater pump and treat remediation system.

The original research teams used three optimization approaches during the DoD/ESTCP study. The DoD/ESTCP report presents these approaches in their entirety in Appendix D, Volume II. The design approaches consisted of Subject Matter Expertise (SME); the Modular Groundwater Optimization (MGO) (Zheng and Wang 2002), and; the Simulation/Optimization Modeling Optimization Software System (SOMOS) (Systems Simulation/Optimization Laboratory SSOL 2002). The team from the University of Alabama applied the MGO approach, the team from Utah State University applied the SOMOS approach, and the group from GeoTrans applied an SME-based subjective engineering approach. The publically available project web site (http://www.frtr.gov/estcp) provides the DoD/ESTCP study reports, groundwater flow and transport models, and modeling files of the final solutions to this problem.

The Umatilla research problem formulation used for testing is ESTCP Formulation 1. The goal of the formulation aims to reduce the projected clean up times of RDX and TNT at minimal cost, subject to constraints on the total allowable pumping and injection, treatment capacity, and the number of new wells needed. The experimental design of the DoD/ESTCP study directed the investigators to consider the existing flow and transport models as "up-to-date and acceptable for design purposes". The groundwater flow model code is USGS MODFLOW-96 (McDonald and Harbaugh 1988; Harbaugh and McDonald 1996). The model code MT3DMS4 (Zheng and Wang 1999) for multispecies contaminant transport. The objective function calculator provided by ESTCP evaluates the cost associated with a remedial design and its performance. U.S. Army Corps of Engineers USACE (1996) developed and provided the MODFLOW-96 groundwater flow and MT3DMS4 solute transport models to the ESTCP study group.

The algorithm developed in this study - Physics-Based Management Optimization (PBMO) – is an automated simulation-optimization based method. Examination of the optimal design solution from PBMO with those from the ESTCP study demonstrates its effectiveness. The

PBMO solution acceptance metric is a cost equal to or lower than developed by the MGO team. MGO provided one of the lowest costs with the least amount of computational effort of the automated approaches. Using the same physically based models and objective function calculator in all the studies isolates the performance of the demonstrated optimization algorithms.

Umatilla chemical depot, Umatilla Oregon site background and description

Briefly, Umatilla is a 19,728-acre military reservation established in 1941 as an ordnance depot for the storage, renovation, and demilitarizing of conventional munitions, and for the storage of chemical munitions. As of 1994, the Umatilla site only stored chemical munitions awaiting destruction. A washout plant operated at the site in the 1950s and 1960s. Discharges to unlined lagoons consisted of an estimated 85 million gallons washout water laden with RDX and TNT. The water table is present about 47 feet beneath the bottoms of the lagoons. The resulting soil and groundwater contamination caused the placement of Umatilla on the EPA National Priorities List (NPL) in

1984. Section 3.1.1 of the study report (Minsker et al. 2004) provides additional description of the Umatilla site and historical summary.

The Record of Decision (U.S. Army Corps of Engineers USACE 1994) – the legal document governing the remediation approach - specified a pump and treat system with reinjection of treated groundwater as the remedial alternative. The treatment system commenced operations in January 1997; this action is the RIP. The system comprised three active extraction wells [EW-1, EW-3 and EW-4], three active infiltration/recharge basins [IF-1, IF-2 and IF-3], and a granular activated carbon (GAC) treatment system with a capacity of 1,300 gallons per minute (gpm). Figure 1 provides the locations of the system components. The extraction well (EW-2) installed 100 feet northwest of EW-4 is not part of the RIP, but is available for inclusion as merited.

The initial RIP included an existing industrial lagoon designated IFL. However, suspicions that infiltrating water could spread the TNT plume rather than flushing the unsaturated zone of contamination resulted in discontinuing its use. By mid-July 1999, the RIP system had

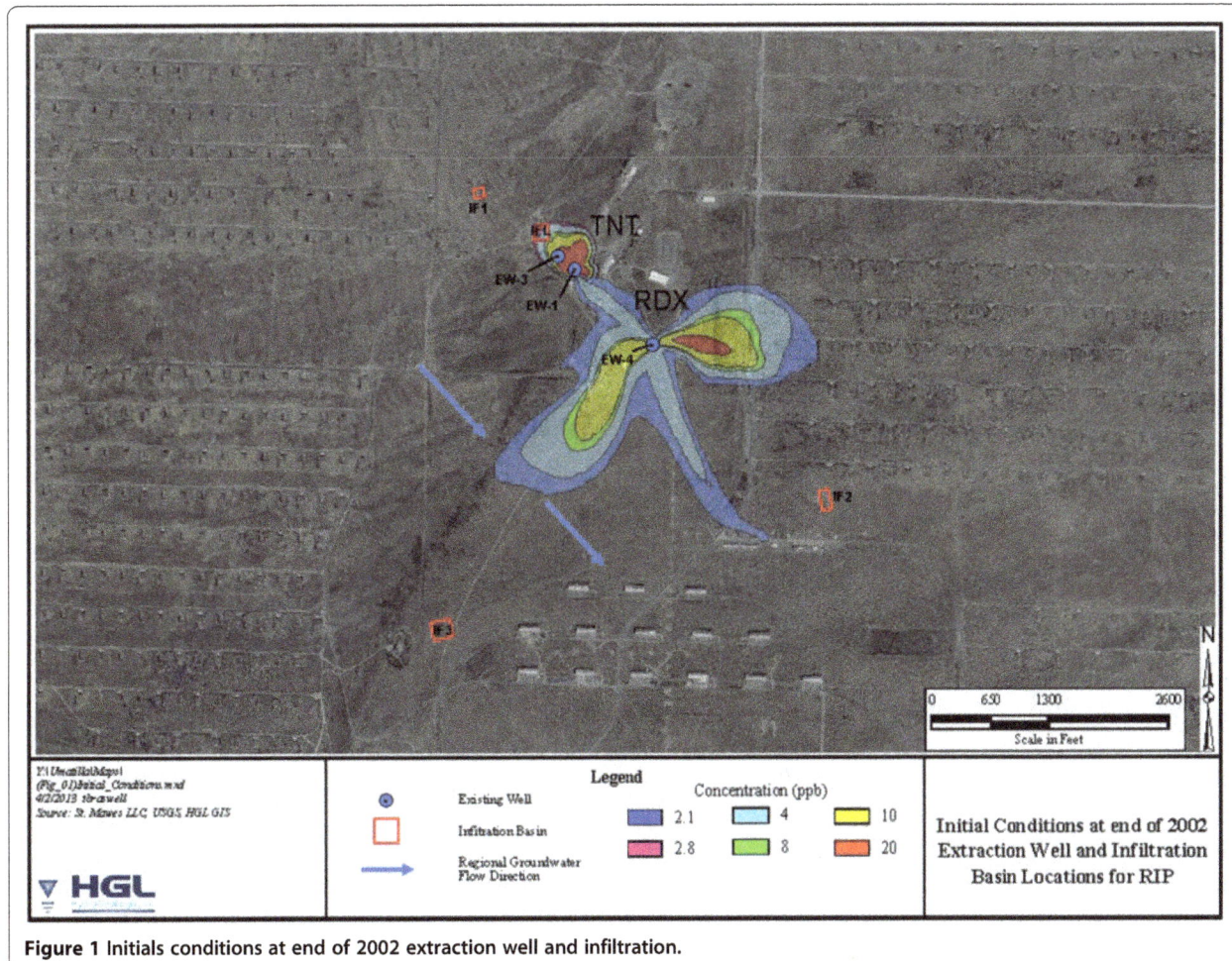

Figure 1 Initials conditions at end of 2002 extraction well and infiltration.

extracted, treated and recharged approximately 1.27 billion gallons of groundwater and removed an estimated 3,000 kilograms (kg) (6,614 pounds) of RDX and 400 kg (882 pounds) of TNT. The simulated RDX and TNT contaminant plumes, shown in Figure 2, represent the extent of contamination in the shallow aquifer at the end of 2002. The maximum RDX and TNT concentrations at that time were 28.2 and 86.7 micrograms per liter (μg/L), respectively.

Methods

Strategic planning for groundwater remediation, water resources planning and dewatering for mineral and resource mining requires tools that incorporate the complex constraints associated with the environment and support decision making with varying levels of uncertainty. This can include uncertainty in the conceptual site model, subsurface characterization (e.g. geologic material location and properties), chemical transport (e.g. reactions, natural attenuation, biodegradation), and funding levels (e.g. annual and total life cycle project) (ITRC, 2007).

Physics-based models: MODFLOW, MT3D, MODFLOW-SURFACT, and; MODHMS (HydroGeoLogic, Inc. HGL 2012) are highly effective at integrating remedial technology selection and application with competing remediation goals among regulators, site owners/custodians, and other stakeholders. The reasons for this are:

1. Comprehensive: Physics-based models incorporate realistic and measurable physical parameters required for accurately describing the fate and transport of dissolved contaminants within aquifers. Optimal solutions based on physics-based models are more reliable than those from "lumped" parameter or ad hoc approximations.
2. Efficient and Effective: Physics-based models are superior to lumped parameter or ad hoc models in accounting for changes in contaminant mass within aquifers. Increased accuracy yields solutions that better manage the use of remedial technologies and optimize costs and, therefore, be in complete control of the utilization of available funding.

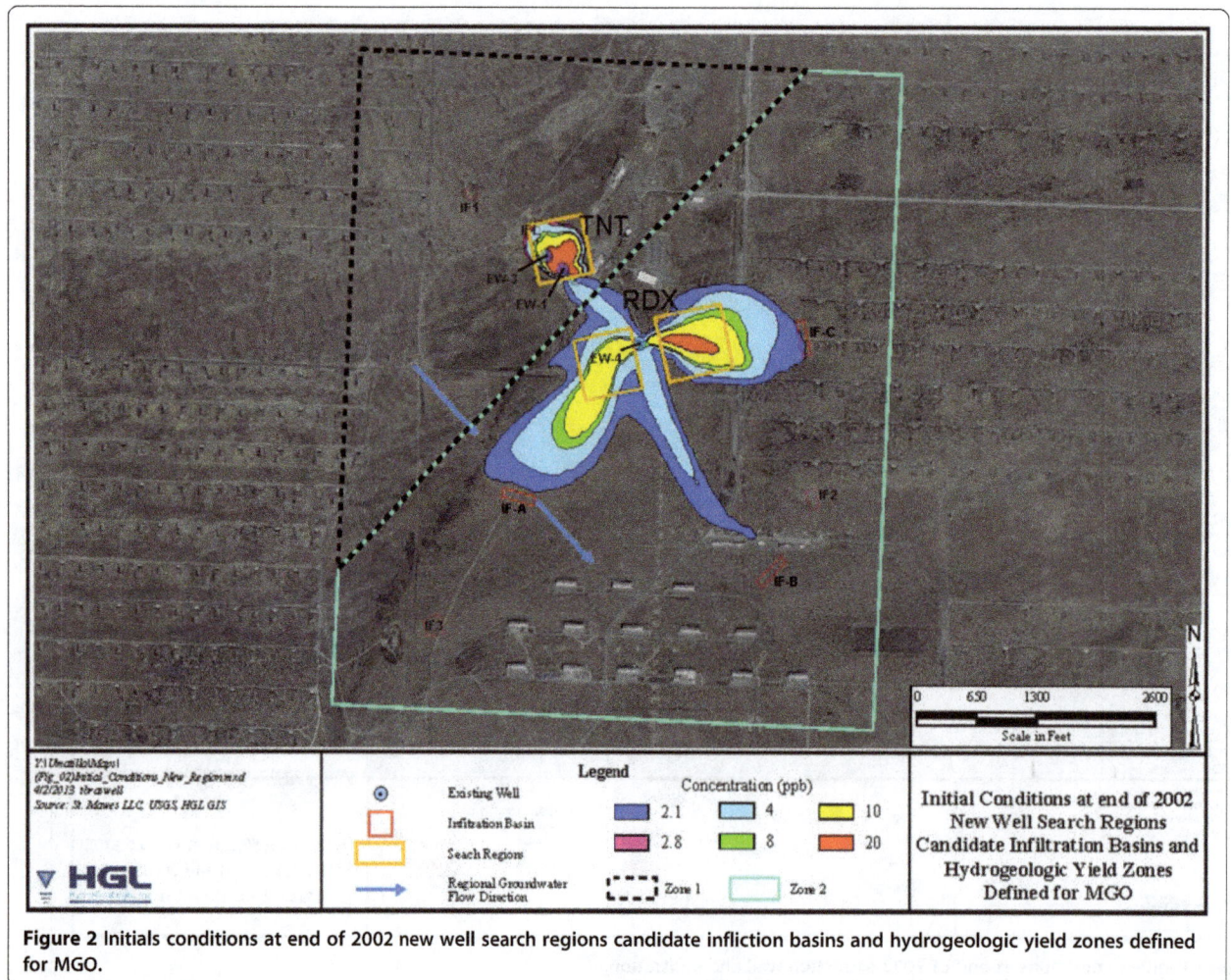

Figure 2 Initials conditions at end of 2002 new well search regions candidate infliction basins and hydrogeologic yield zones defined for MGO.

3. Flexible: Physics-based models can readily incorporate extended physical analysis to enable the development of optimal planning scenarios for processes that are outside of historical (data) observations, whereas models built on regression, interpolation, or extrapolation methods may not be representative.

Groundwater remediation problems amenable to physics-based decision optimization include the development of cost-effective and sustainable remedial designs; RIP evaluation and operation and maintenance (O&M) optimization; the development of optimized exit strategies to minimize life-cycle costs; the development of site completion/closure strategies; and water quality management issues for riparian and lacustrine settings, wetlands, and estuaries. This approach provides decision support to program and project managers regarding the best methods for remediating a site with contaminated groundwater.

Cost minimization is the overall optimization objective. In the case of groundwater remediation, well locations and their extraction or injection rates are the decision variables, and the simulated hydraulic heads and contaminant concentrations in groundwater are the state variables. *Decision variables* are the design elements of the problem studied. *State variables* represent the resultant simulated system. The *objective function* is a systematic accounting of costs which include the number, operation, and maintenance of the design elements (decision variables) and account for the changes in the modeled system (state variables) response from the candidate design. Computed over the life cycle of the simulation, the objective function represents the total remediation cost. The incorporation of *constraints* ensures the solution is practical and implementable. Constraints capture mechanical or physical process limitations (such as maximum pumping rates, and treatment train capacities) or interim and final values of state variables at discrete or continuous regions of the modeled domain. Assessment of *candidate solutions* is conducted by examining the cost value and the constraint compliance. A *single model function evaluation* is the cycle of the decision variable selection, process simulation, model objective function evaluation and constraint evaluation of a candidate design. *Optimization* is the automatic, guided process of the decision variable selection and application through repeated model function evaluations to arrive at the optimal objective function value that satisfies all constraints. The automatic adjustment of decisions variables terminates (ideally) at the global optimum (least cost).

PBMO's approach provides decision support for groundwater remediation, water resources planning and dewatering for mineral and resource mining via flexible integration of physics-based (groundwater flow and transport) models and optimization algorithms. Modular design promotes flexibility via independence of the physics model and optimization method. Linking the appropriate physics-based simulator with the best optimization algorithm(s) enables the solution to a wide range of problems types. Figure 3 shows how the modular design links the global optimization algorithms with the appropriate physics-based simulators to develop an optimal strategy. In this work, the first (top) half of the medallion represents the HGL_OPT optimization algorithms which include the Lipschitz Global Optimization (LGO©) solver suite (Pintér 1996, 2002, 2009, 2013). The second (bottom) half represents the physically based numerical groundwater flow and transport simulators USGS MODFLOW-96 and MT3DMS4B (in this study); however any process simulator with a text interface is implementable.

Subsurface simulators represent the physical processes of flow and transport in a mathematical form. Problem-specific, physics-based models represent the groundwater processes/conditions under investigation. From an optimization perspective, these simulation models can be *"black-boxes"*, that is models that receive input to produce an output without revealing the process. Since these groundwater flow and transport processes can occur in saturated or variably saturated conditions, in porous or fractured media, the numerical model used for the simulation will have a linearity *category*; the simulation can be linear, mildly non-linear (assumedly or provably convex), or highly non-linear (assumedly or provably non-convex). The system under consideration can be either single phase (water) or multiphase (water-NAPL-gas). The transport processes can be represented simply by using advective particle tracking, as a single non-reactive solute that undergoes dispersion and diffusion, or as a multicomponent system whose constituents interact with each other and the surrounding environment and are subject to hysteresis. The equations that govern these processes are Lipschitz-continuous, whose state-space representation can range from elliptic to parabolic or nearly hyperbolic when advection dominated. When no or low diffusion and dispersion are present relative to advection, steep fronts or shocks occur in the solutions. A complication for efficient optimization approach design arises when that the decision variables used in a candidate model evaluation during optimization change the model linearity category. Typically, the simulation of groundwater flow with slight drawdown is linear to mildly non-linear (because the saturated aquifer thickness does not change appreciably), whereas the simulation of significant groundwater drawdown or coupled groundwater flow and transport are most often highly non-linear. For the model scenario considered in this study, the flow system will range from linear far from the remediation system and contaminant plumes to highly non-linear in the contaminant plumes and near the remediation wells. The objective function is non-linear.

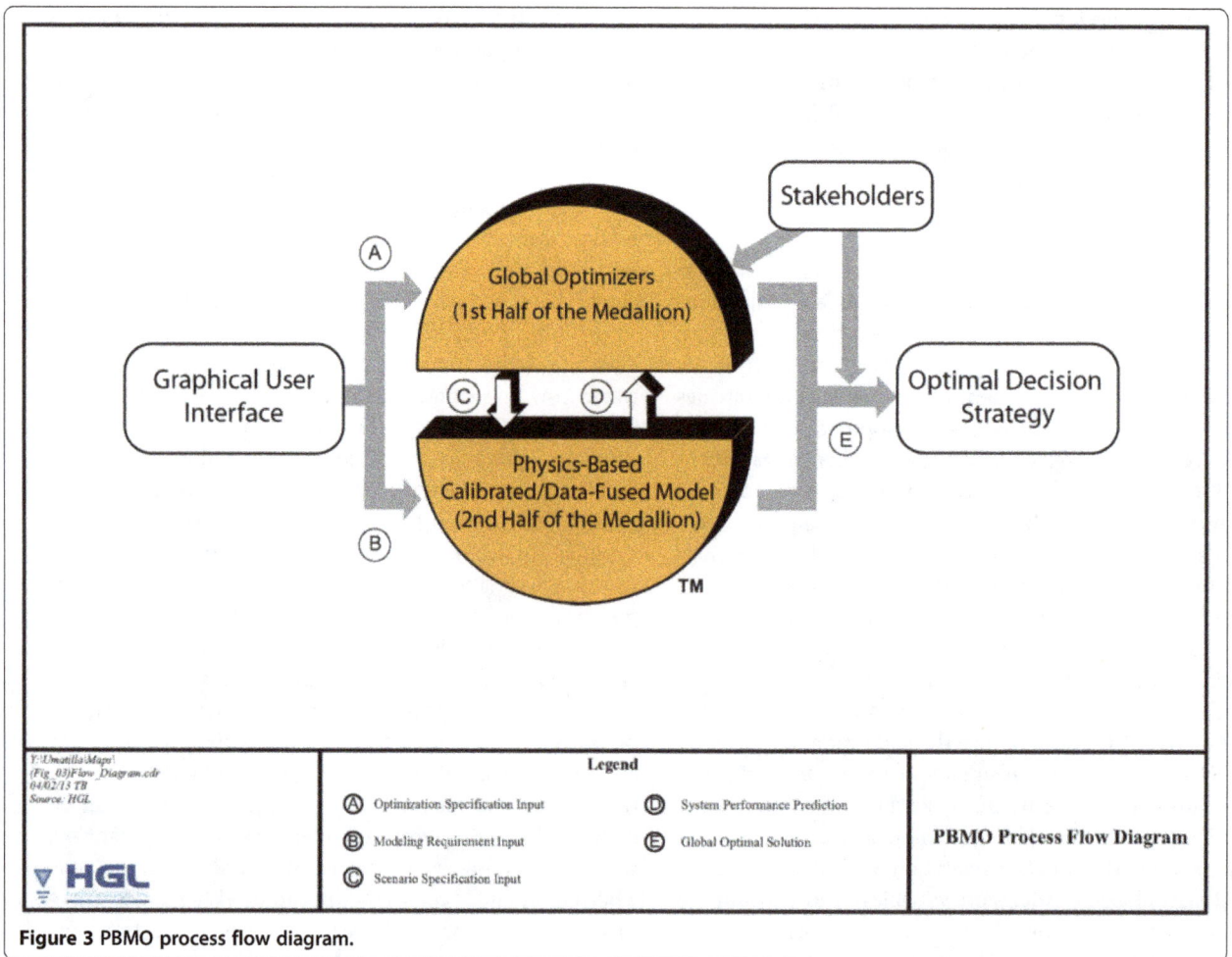

Figure 3 PBMO process flow diagram.

The USGS numerical simulator, MODFLOW-96, solves the following 3-D governing partial-differential equation for transient, saturated groundwater movement through porous media to yield a spatial distribution of the potentiometric head as a function of time:

$$\frac{\partial}{\partial x}\left(K_{xx}\frac{\partial h}{\partial x}\right) + \frac{\partial}{\partial y}\left(K_{yy}\frac{\partial h}{\partial y}\right) + \frac{\partial}{\partial z}\left(K_{zz}\frac{\partial h}{\partial z}\right) - q_s$$
$$= S_s\frac{\partial h}{\partial t} \tag{1}$$

Where K_{xx}, K_{yy} and K_{zz} are the values of hydraulic conductivity [L/T] along the x, y, and z coordinate axes, respectively, assumed to be parallel to the major axes of hydraulic conductivity; h is the potentiometric head [L]; q_s is a volumetric flux per unit volume of the aquifer and represents sources (injection wells) and/or sinks (extraction wells) of water [1/T]; Ss is the specific storage of the aquifer material [1/L]; and t is time (T).

Injection wells represent the infiltration/recharge basins. The initial ESTCP study used some terms interchangeably: injection, infiltration, and recharge. For clarity with the initial study language, we maintain that terminology in this paper to refer to the return of treated groundwater to the subsurface. The distribution of heads (one of the state variables), the source/sink data and, the candidate decision variables used or generated from the MODFLOW-96 simulation provide input to MT3DMS to solve the partial differential equation describing the fate and transport of contaminant species k in 3-D, transient groundwater flow systems. The governing equation is:

$$\frac{\partial(\theta C^k)}{\partial t} = \frac{\partial}{\partial x_i}\left(\theta D_{ij}\frac{\partial C^k}{\partial x_j}\right) - \frac{\partial}{\partial x_i}\left(\theta v_i C^k\right) + q_s C_s^k$$
$$+ \sum R_n \tag{2}$$

Where θ is the porosity of the aquifer material [-]; C^k is the dissolved concentration of species k [M/L^3]; $x_{i,j}$ is the distance along the respective coordinate axis, x, y or z [L]; D_{ij} is the hydrodynamic dispersion coefficient tensor [L^2/T]; v_i is the seepage or linear pore water velocity [L/T]; C_s^k is the concentration of the source or sink flux for species k [M/L^3], and; ΣR_n is the chemical reaction term [M/(L^{3}T)].

Darcy's Law couples the transport and flow equations.

$$v_i = -\frac{K_i}{\theta}\frac{\partial h}{\partial x_i} \tag{3}$$

McDonald and Harbaugh (1988) provide details on MODFLOW. Zheng and Wang (1999) provide details for MT3DMS.

To provide an optimization solver with the capability and efficiency to address this range of model types, PBMO is developed and implemented leveraging a suite of optimization algorithms that efficiently solve the various formulations expected to occur in the optimal decision support approaches presented here.

HGL_OPT is the master optimization driver. It comprises machine learning, objective function estimating methods, and SME-developed heuristics specifically useful for quickly developing high quality solutions to complex problems such as contaminant flow and transport remedial design (Deschaine 1992 and Deschaine 2003; Deschaine and Pintér 2003; and Deschaine et al. 1998, 2001, and Deschaine et al. 2011). These solutions can be used to focus (or initialize) the LGO-specific global and local optimization solvers.

The optimization algorithms included in the optimizer suite include linear programming (LP), sequential linear programming (SLP), sequential linear approximation (SLA), sequential quadratic programming (SQP), generalized reduced gradient (GRG), outer approximation (OA), Branch and Bound (BB), Globally Adaptive Random Search (GARS), and Multi-start Random Search (MS). Specifically, LP solves linear models; OA, SLP, SLA, SQP, and GRG serve to deal with mildly non-linear models; BB, GARS, and MS—in proper combination with SLA, SQP, and GRG—serve to handle highly nonlinear models.

Currently, the optimization system can handle model formulations with up to a few hundred binary (yes/no) decision variables, several thousand continuous variables, and a several thousand general constraints (in addition to variable bound constraints). The actual configuration can be adjusted to user demands. For example, it is possible to optimize models with 200 decision variables and 5,000 general constraints, or vice versa. Even larger model sizes can also be handled if the computer random-access memory (RAM) and the compiler used support this. The BB solver of LGO employs a mild assumption of Lipschitz continuity, which allows an efficient and a robust search for a global optimal value through a systematic partitioning and exploration of the entire feasible region. The GARS and MS solvers assume

basic model function continuity. The optimization suite solves the following general model:

i. Minimize the [arbitrary linear or nonlinear] objective function $f(x)$
ii. Subject to bound constraints and [arbitrary linear or nonlinear] constraints:

$$x \in D := \{x_l \le x \le x_u; g(x) \le 0\} \tag{4}$$

where x is a decision vector, an element of the real n-space $\boldsymbol{R}^n$; $f(x)$ is a continuous objective function, $f: \boldsymbol{R}^n \ \boldsymbol{R}$, ($\boldsymbol{R}=\boldsymbol{R}^1$); D is a non-empty set of admissible decisions, a subset of $\boldsymbol{R}^n$. As shown by (4), the set D is defined by l, u which is an explicit, finite n-vector bound of x (a "box") in $\boldsymbol{R}^n$; and $g(x)$ which is an m-vector of additional continuous constraint functions, $g: \boldsymbol{R}^n \ \boldsymbol{R}^m$.

The assumption is that Lipschitz continuity holds as expressed by the inequality

$$\left| f_j(x_1) - f_j(x_2) \right| \le L_j \| x_1 - x_2 \| \tag{5}$$

where $[L_j]$ is the Lipschitz constant. This equation means that the variability of the values computed by the objective function calculator is bounded with respect to the variability of the system input variables $[x]$ by the Lipschitz constant. This condition is an expected property of all physically based models.

Effective development and specification of a problem for groundwater remediation projects requires SME understanding of the physical processes and their constraints. The options under consideration include selecting which of the physics-based processes that need modeled (such as aquifer quality remediation via bioremediation or pump-and-treat) along with the constraints that the solution must satisfy. The formulation below is directly applicable and extensible to a wide range of groundwater dewatering and remediation problems. The basic elements of the mathematical formulation consist of an objective function and constraints.

$$\text{Minimize}: \ f(x) = \sum_{i=1}^{N} (\delta_i, \alpha_i, q_i) \tag{6a}$$

subject to:

$$\sum_{i=1}^{N} q_i \ge Q^* \ \forall i \in I \tag{6b}$$

$$q_i \le q_i^* \quad \forall i \in I \tag{6c}$$

$$c_j \le c_j^* \quad \forall j \in J \tag{6d}$$

$$h_k \geq h_k^* \quad \forall k \in K \qquad (6e)$$

$$t \leq T_{max} \qquad (6f)$$

Here, the objective function $f(x)$ is a simplified version of life cycle cost for illustrative purposes. It consists of the flow rate from extraction wells $[q_i]$, a unit cost to process and treat the water $[\alpha_i]$; and it assesses if an extraction well is pre-existing or not by examining for $\forall i$ locations and assigning a (0,1) multiplier if a well installation required ($\delta_i=1$) or is pre-existing ($\delta_i=0$). The actual formulations of the objective function for optimal design problems often involve quite extensive and detailed cost information and functions such as used in this study. The constraint $\sum_{i=1}^{N} q_i \geq Q^* \; \forall i \in I$ requires the installed system pumps at least some minimum volume of water Q. The $q_i \leq q_i^*$ constraint sets upper limits on a flow at each of the ith candidate well. The $c_j \leq c_j^*$ constraints ensure aquifer remediation to a certain acceptable residual level for all j locations. The $h_k \geq h_k^*$ constraints set minimum allowable water table elevations in the aquifer at all k locations. The $t \leq T_{max}$ constraint limits the allowable time for the activity to achieve the specified goals. These general constraints, like the cost function, can be modified and adapted as dictated by the needs of the project. For example, one can use this framework to incorporate constraints for differential land subsidence due to dewatering by using differencing constraints and adding constraints on the maximum slope of the water table or land surface. Similar approaches are effective for assessing depressurization and geotechnical stability.

The goal of constrained global optimization is to find at least one point x^* within the feasible region that satisfies $f(x^*) \leq f(x)$ for all x or to show that such a point does not exist. If no solution exists, then the decision makers realize it is an infeasible problem formulation. This enables the problem be reformulated. It is critical to determine if a feasible solution does not exist when solving a complex decision problem. This is a valuable capability of the physics-based optimization methodology. Many other optimization methods, specifically including heuristic techniques, cannot determine that a problem is infeasible: as a result, users can waste precious time and money searching for solutions with no hope of finding a feasible solution.

Umatilla optimization problem formulation
The three optimization problem formulations considered in the DoD/ESTCP study were:

Formulation 1: Minimize the cost to remediate RDX and TNT in 20 years or less using the current treatment plant maximum operating flow rate of 1,300 (gpm) as an upper limit on the total groundwater extraction rate;

Formulation 2: Same as Formulation 1, except the maximum treatment flow rate increased to 1,950 (gpm);

Formulation 3: Minimize the aggregate remaining mass of RDX and TNT in 20 years using the current treatment plant flow rate.

The PBMO approach addresses all these types of optimization formulations. This exercise focuses on finding a solution to the problem as stated by Formulation 1. Appendix D, Volume II of the DoD/ESTCP report (Minsker et al. 2004) provides the mathematical formulation of the problem statement in detail.

The objective function for Formulation 1 is specified as follows:

$$MINIMIZE \left(C_{CW} + C_{CB} + F_{CL} + F_{CE} + V_{CE} \right. \\ \left. + V_{CG} + V_{CS} \right) \qquad (7)$$

where:

C_{CW}: Capital costs of new wells

$$C_{CW} = (25 \times I_{EW2})^d + \sum_{i=1}^{N_Y} (75 \times N_{W_i})^d \qquad (8)$$

N_y is the modeling year when cleanup is achieved [yr] as defined by $[C_{RDX}] \leq 2.1 \; \mu g/L$ and $[C_{TNT}] \leq 2.8 \; \mu g/L$ as measured by the nodal concentration value in the top layer.

N_{Wi} is the total number of new extraction wells (except well EW-2) installed in year i. New wells may only be installed in years corresponding to the beginning of a 5-year management period. Capital costs do not apply to pre-existing extraction wells.

I_{EW2} is a flag indicator; 1 when well EW-2 first comes into service, 0 otherwise.

75 is the cost of installing a new well [K\$].

25 is the cost of putting existing well EW-2 into service [K\$].

d indicates the application of the discount function to yield Net Present Value (N_{PV}) defined as

$$N_{PV} = \frac{c}{(1+r)^{y-1}} \qquad (9)$$

c is the cost
r is the annual discount rate [1/yr]
y is the value of i in the summation

Table 2 Number of candidate extraction well locations in each of the three search boxes

Location	Coordinates	Size	Number of potential extraction well locations[1]
Box 1	(40,53) to (63,73)	24×21	127,260
Box 2	(82,63) to (91,83)	10×21	22,155
Box 3	(82,90) to (90, 96)	9×7	2,016

Includes single and double well location combinations per box.

C_{CB}: Capital costs of new infiltration basins

$$C_{CB} = \sum_{i=1}^{ny} \left(25 \times N_{B_i}\right)^d \tag{10}$$

N_{Bi} is the total number of new infiltration basins installed in year i. New recharge basins may only be installed in years corresponding to the beginning of a 5-year management period. The infiltration flux is evenly distributed throughout the basin.

25 is the cost of installing a new recharge basin independent of its location [K$/yr].

F_{CL}: Fixed cost of labor

$$F_{CL} = \sum_{i=1}^{N_Y} (237)^d \tag{11}$$

237 is the fixed annual O&M labor cost [K$/yr].

F_{CE}: Fixed cost of electricity (lighting, heating, and the like).

$$F_{CE} = \sum_{i=1}^{N_Y} (3.6)^d \tag{12}$$

3.6 is the fixed annual electric cost [K$/yr].

V_{CE}: Variable electric costs of operating wells (extraction and injection)

$$V_{CE} = \sum_{i=1}^{N_Y} \sum_{j=1}^{N_{wel_i}} \left(C_{W_{ij}} \times I_{W_{ij}}\right)^d \tag{13}$$

N_{wel_i} is the total number of extraction/injection wells active in year i.

Table 3 Optimal pumping strategies found using SME, MGO and PBMO for formulation 1 at Umatilla compared with existing RIP design

Name	Location (Layer, Row, Column)	RIP	SME design Stress period 1	SME design Stress period 2	MGO Stress period 1	PBMO Stress period 1
EW-1	(1,60,65)	−128	−280	−350	−307.5	−292.5
EW-2	(1,83,84)					
EW-3	(1,53,59)	−105		−360	−219.5	−292.5
EW-4	(1,85,86)	−887	−660			
New-1 (T&E)	(1,48,57)			−100		
New-2 (T&E)	(1,49,58)		−230	−360		
New-3 (MGO)	(1,48,59)				−360	
New-4 (MGO)	(1,48,55)				−283	
New-5 (PBMO)	(1,48,57)					−292.5
New-6 (PBMO)	(1,52,61)					−292.5
IF-L	*					
IF-1	*	233	282	585		
IF-2	*	405	405		380	390
IF-3	*	483	482		790	780
IF-4	*			585		
IF-A	*					
IF-B	*					
IF-C	*					
Total cost in net present value ($)		**$3,836,285**	**$2,230,905**		**$1,664,395**	**$1,644,085**

*See Figure 2 for location.
(a negative flow rate indicates pumping; positive indicates injections).

Table 4 Breakdown of the capital and O&M costs of RIP, SME, MGO and PBMO designs

Cost components	RIP existing design	SME strategy	MGO optimal strategy	PBMO optimal strategy
Capital costs of new wells (C_{CW})	$ -	$ 133,764	$ 150,000	$ 150,000
Capital costs of new infiltration basins (C_{CB})	$ -	$ 19,588	$ -	$ -
Fixed costs of labor (F_{CL})	$ 2,805,552	$ 1,263,086	$ 882,410	$ 882,410
Fixed costs of electricity (F_{CE})	$ 42,616	$ 19,186	$ 13,404	$ 13,404
Variable costs of electricity for operating wells (V_{CE})	$ 251,405	$ 91,952	$ 48,394	$ 48,402
Variable costs of changing GAC units (V_{CG})	$ 16,338	$ 14,301	$ 11,700	$ 11,382
Variable costs of sampling (V_{CS})	$ 720,374	$ 689,028	$ 558,487	$ 558,487
Objective function value	*$ 3,836,285*	*$ 2,230,905*	*$ 1,664,395*	*$1,644,085*

C_{Wij} is the electrical cost for well j in year i. Costs differ for wells depending on the extraction rates of well j in year i, Q_{ij}:

$$C_{W_{ij}} = 0.01\left(Q_{ij}\right) \quad \text{for } 0 \text{ gpm} < Q_{ij} \leq 400 \text{ gpm}$$
$$C_{W_{ij}} = 0.025\left(Q_{ij}\right) - 6 \quad \text{for } 400 \text{ gpm} < Q_{ij} \leq 1000 \text{ gpm}$$

I_{Wij} is a flag indicator; 1 if well j is active in year i, 0 otherwise.

V_{CG}: Variable costs of changing GAC units

$$V_{CG} = \sum_{i=1}^{N_Y} [\gamma(\bar{c}_i) \times m_i]^d \tag{14}$$

$\bar{c}_i$ is the average influent concentration [μg/L] (RDX plus TNT) into the treatment plant from all of the extraction wells in the year i calculated as:

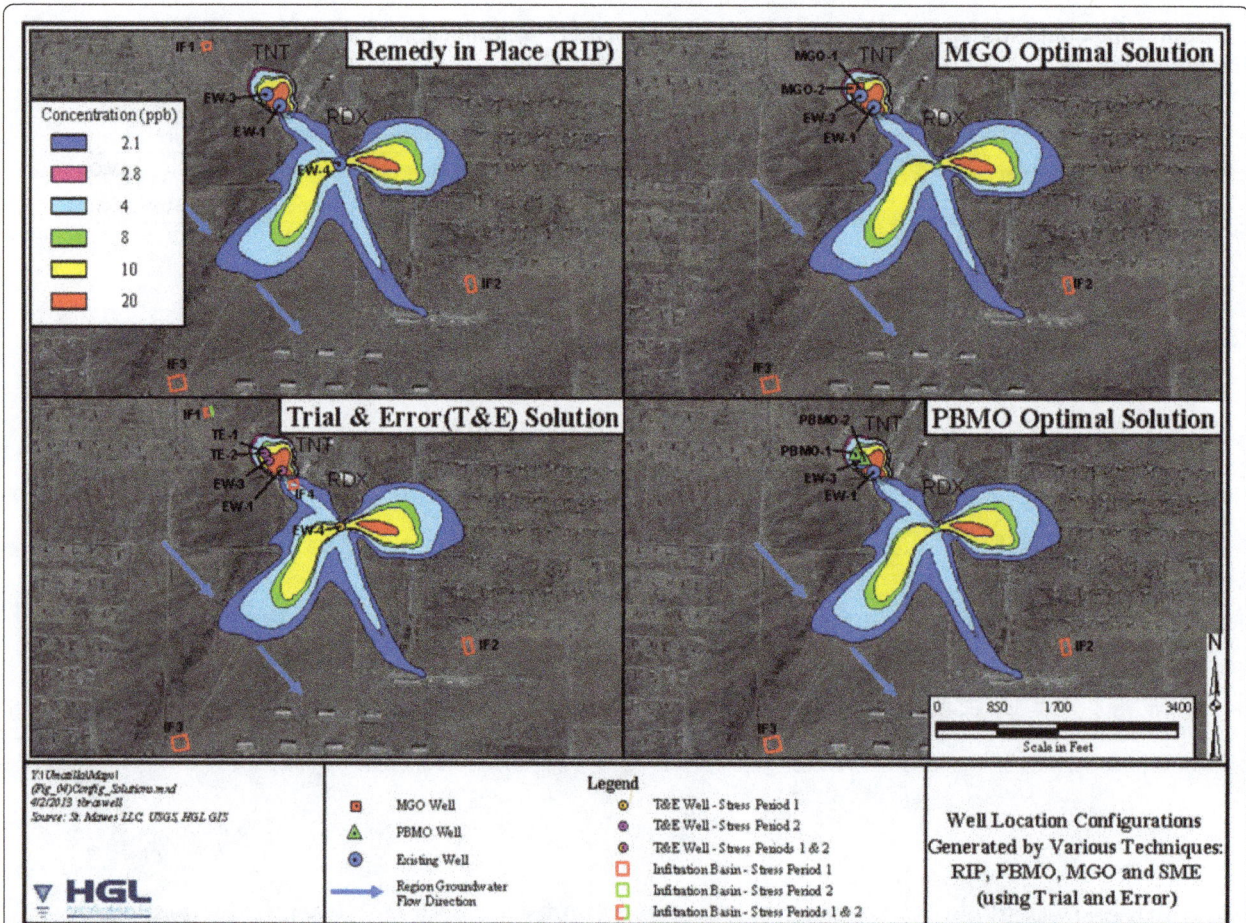

Figure 4 Well location configurations generated by various techniques: RIP, PBMO, MGO and SME (using trial and error).

$$\bar{c}_i = \frac{\sum_{j=1}^{N_{wel_i}} Q_{ij}\bar{c}_{ij}}{\sum_{j=1}^{N_{wel_i}} Q_{ij}} \tag{15}$$

$\bar{c}_{ij}$ is the average influent concentration [μg/L] (RDX plus TNT) from well j in the year i

$\gamma(\bar{c}_i)$ is the cost of mass removed [K$/kg] as a function of average influent concentration into the treatment plant in year i, calculated as:

$$\gamma(\bar{c}_i) = \frac{-0.5(\bar{c}_i) + 225}{1000} \tag{16}$$

m_i is the mass of contaminant removed [kg] during year i calculated as:

$$m_i = \sum_{j=1}^{N_{wel_i}} Q_{ij}\bar{c}_{ij} \times \beta \tag{17}$$

β is a conversion factor to produce the result in units of [kg/yr].

V_{CS}: Variable cost of sampling.

$$V_{CS} = \sum_{i=1}^{N_Y} [150 \times (A_i/I_A)]^d \tag{18}$$

I_A is the initial plume area in layer 1 of the model based on the extent of RDX and TNT as measured in January 2003 where RDX and TNT exceeded their respective cleanup goals (2.1 and 2.8 μg/L, respectively) [m²]

150 is the annual sampling cost (as of January 2001) and considers both labor and analysis costs [K$/yr]

A_i is the modeled plume area in layer 1 in year i. This is evaluated at the beginning of a 5 year management period [m²]. A_i is defined as:

$$A_i = \sum_{j=1}^{N_{col}} \sum_{k=1}^{N_{row}} [\Delta x_j \Delta y_k \times I_{C_{jk}}] \tag{19}$$

N_{col} is the number of grid cell columns in the x direction

N_{row} is the number of grid cell rows in the y direction

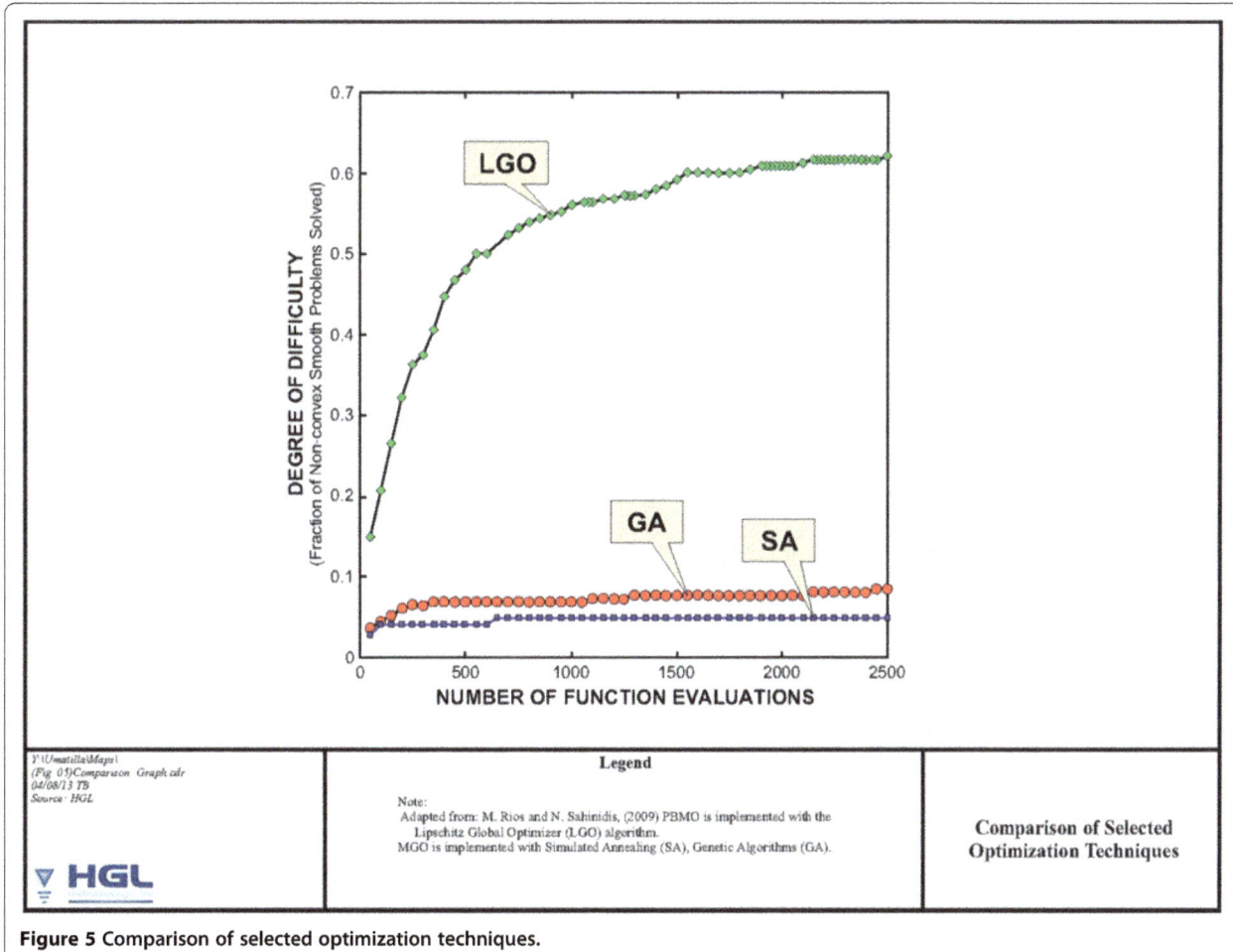

Figure 5 Comparison of selected optimization techniques.

Δx_j is the width of the jth grid cell column [m]

Δy_k is the width of the kth grid cell row [m]

$I_{C_{jk}}$ is a flag where:

C_{RDX}^{jk} is the concentration of RDX in the grid cell with indices j and k

C_{TNT}^{jk} is the concentration of TNT in the grid cell with indices j and k

The Formulation 1 constraints are:

1) The modeling period consists of four 5-year management periods beginning with January 2003 (i or $year = 1$).
2) Modifications to the system may only occur at the beginning of each management period.
3) Remediation in the top layer of the model must be achieved within 20 years (e.g., RDX ≤ 2.1 μg/L and TNT ≤ 2.8 μg/L everywhere in top model layer).
4) The total pumping rate, adjusted for the average amount of uptime, cannot exceed the treatment capacity of 1,300 (gpm) in any stress period. Evaluation of this constraint occurs at the beginning of each

5-year *management period*. It is computed as:

$$Q^*/_\alpha \leq 1300 \text{ gpm} \tag{20}$$

Here

α is a coefficient that accounts for the amount of average uptime ($\alpha=0.9$)

Q^* is the total modeled flow rate during a 5-year management period.

5) The hydrology dictates the upper sustainable flow limit on extraction wells. Extraction wells in Zone 1 may pump at a maximum rate of 400 (gpm), whereas extraction wells in Zone 2 may operate to a maximum of 1,000 (gpm). See Figure 2 for definitions of Zones 1 and 2:

$$\text{If Zone}(j,k) = 1,$$
$$\text{then } q_{jk}^*/_\alpha \leq 400 \text{ (gpm)} \tag{21}$$
$$\text{else } q_{jk}^*/_\alpha \leq 1,000 \text{ (gpm)}$$

where:

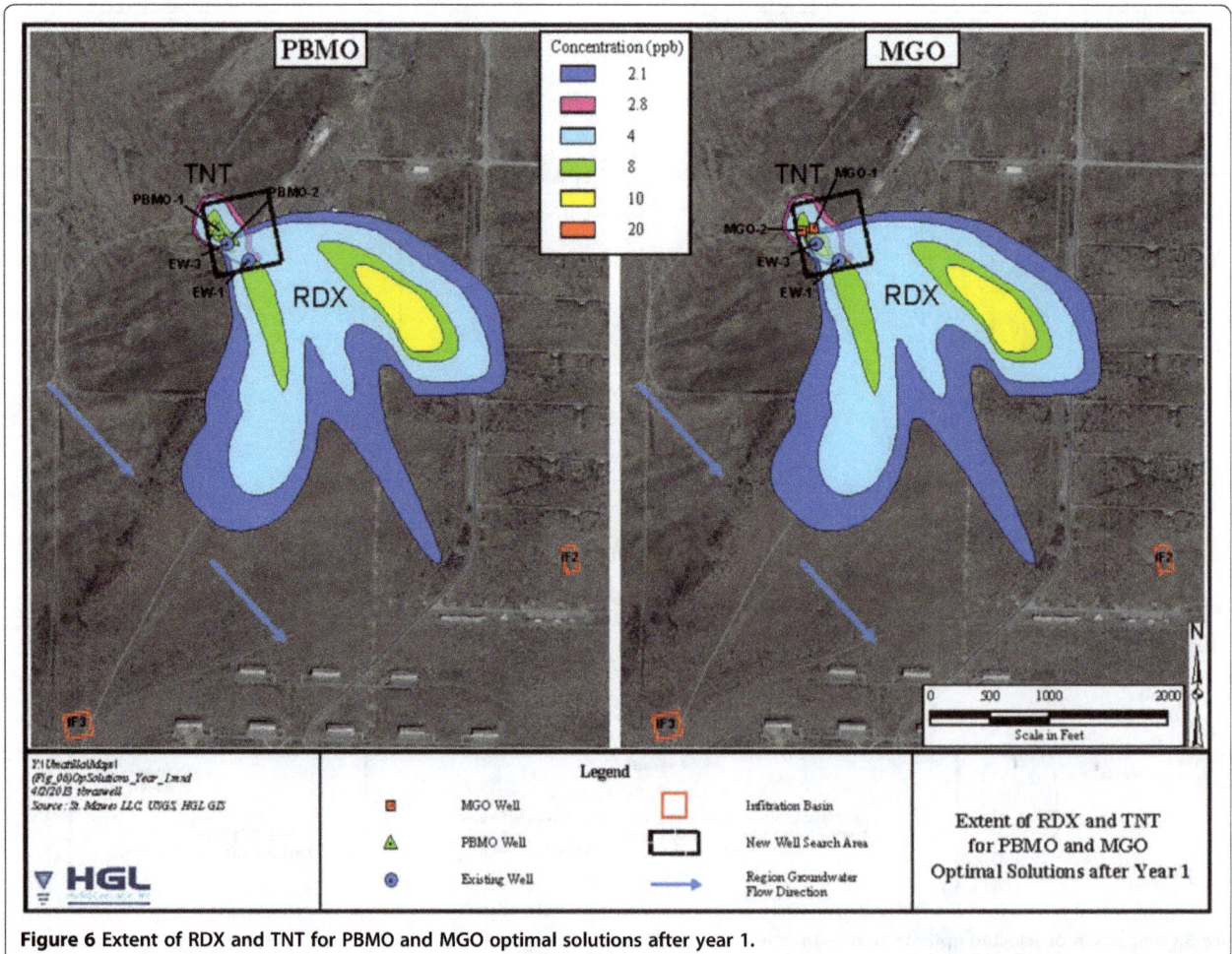

Figure 6 Extent of RDX and TNT for PBMO and MGO optimal solutions after year 1.

Zone(j,k) is a function of the jth grid cell column and k^{th} grid cell row that returns 1 if model grid (j,k) corresponds to Zone 1, and returns 2 if (j,k) corresponds to Zone 2

q_{jk}^{*} is the modeled extraction rate at model grid location (j,k).

6) It is unallowable for the extent of groundwater contamination to increase beyond initial conditions at any time during the remediation.

7) Total pumping and infiltration rates must be balanced at the beginning of every management period

$$\left| \sum_{j=1}^{N_{wel_i}} Q_{ij} - \sum_{k=1}^{N_{rech_i}} Q_{kj} \right| \leq 1 \text{ gpm}, \forall i \in \{1, 6, 11, 16\} \quad (22)$$

where:

N_{rech_i} is the number of injection wells operating in year i.

In summary, the optimization problem can be stated as follows: find the combination of simulated extraction and injection well locations and their operating rates that

minimize the cost of reducing RDX and TNT concentrations within a 20 year time horizon while satisfying all the constraints on well, remedy operations, and plume behavior. The remedial system designs use the calendar year 2003 as the starting point.

The groundwater flow and transport models provided for the study simulate 20 years with four management periods of five years each. The extraction and injection flow rates can vary across the management periods, but not within a management period. The locations of the extraction and injection infrastructure can only vary between candidate solutions, not between management periods.

Optimization solution approach

The optimization problem is solved by defining the regions to search for the globally optimal settings of the decision variables, and prescribe how to conduct the search. The decision variable search regions mimicked, as closely as possible, the regions established for new well locations and infiltration basins by the MGO team. The MGO team confined their search for new extraction well locations to regions where the plume densities were highest as shown

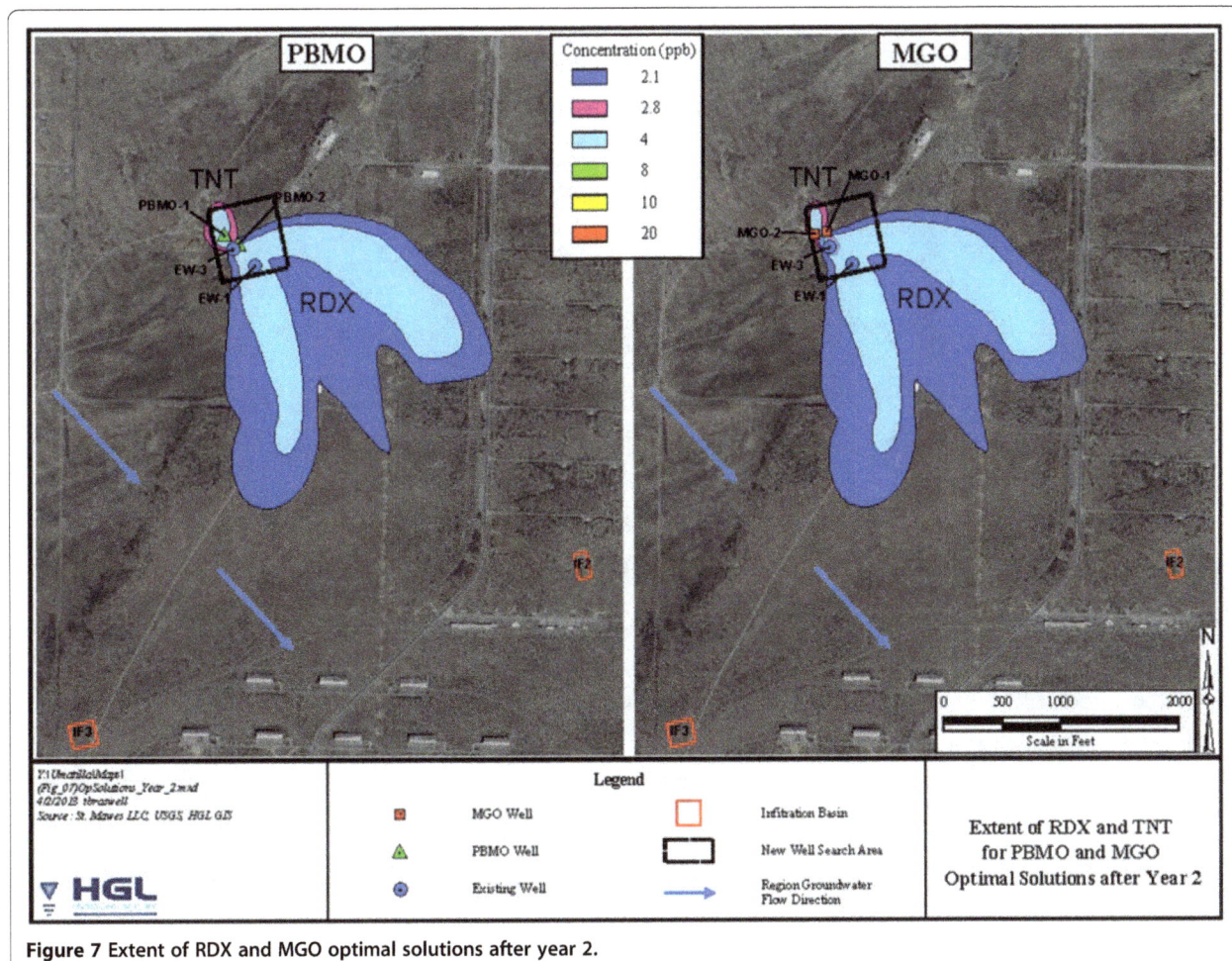

Figure 7 Extent of RDX and MGO optimal solutions after year 2.

on Figure 2. Each region contains the areas of the highest concentration for the three principal lobes of the two contaminant plumes. In addition to these search areas for extraction wells, locations for three new infiltration basins, IF-A, IF-B, and IF-C in the original study at the extent of the RDX plume to the east, southeast, and southwest as alternatives to the four pre-existing recharge basins. In all, a total of 11 candidate regions exist for locating decision variables: four extraction areas and seven injection/recharge areas; the locations as defined by the MGO team. These 11 regions make up the infrastructure search areas of the remedy used to alter the distribution of RDX and TNT in groundwater. Active remediation solutions require non-zero total extraction and injection fluxes; hence, there must be at least one extraction location and one injection location active for all viable candidate solutions.

Within the extraction locations, a well(s) can be located anywhere in the search box (we restrict the well position to a model node). Table 2 provides the box location and number of candidate position locations for one or two wells in each singular box.

Allowing an extraction well(s) to be located in any of the three search boxes results in 302,253 potential extraction well location combinations. The 128 different candidate location configurations for the infiltration basins results in a total of 38,688,384 candidate infrastructure system designs. Simultaneously, when determining the infrastructure configuration design, the water flow rates are optimized. This magnitude of options illustrates the difficulty of finding the optimal solution either by random searching or by using the SME subjective engineering judgment.

The approach used to solve the Umatilla problem consisted of the following generalized automated search strategy:

1) Evaluate the cost of the RIP. The RIP consists of three extraction wells and three infiltration basins. Store these results as a current minimum.
2) Begin evaluation of different combinations of extraction wells and infiltration basins.
   a. Set number of evaluation epochs equal to one.

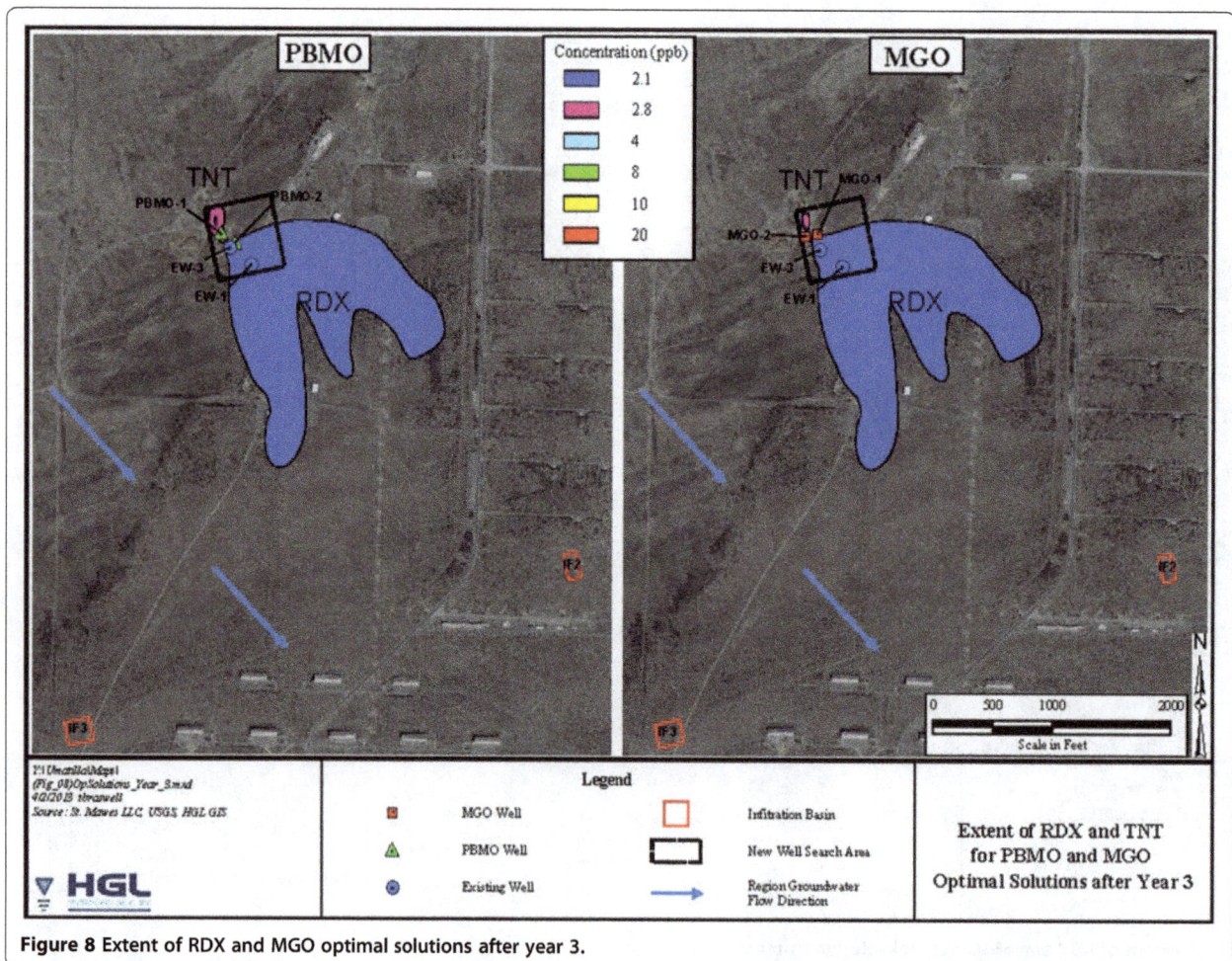

Figure 8 Extent of RDX and MGO optimal solutions after year 3.

b. Initial extraction location: Initiate the solution using the study maximum number of new wells (two), located in Box 1 (which contains the RDX and TNT plumes).

c. Initial injection location: IFL located in Box 1, the innermost infiltration basin.

d. Initiate search strategy. Begin to cycle through and test the 4 pumping areas and 7 injection areas for quality of solution, use extraction rate total = system maximum capacity of 1170 (gpm) (which is 1,300 (gpm) * 90% uptime) as done by the MGO team.

e. Set search cycle a minimum number of evaluations (12).

f. Upon finding a cost lower than current minimum cost:
   i. Select the solutions with the two lowest costs.
   ii. Explore these regions by conducting brief local random searches (LRS) on each solution to (statistically) determine the most promising configuration of wells and basins. Perform

initial analysis using the physically-based model. The optimization search progresses and generates an increasing number of function evaluations. Machine learning is invoked to provide candidate solution objective function evaluation estimation which reduces computation burden when model evaluations models are extensively time consuming.

g. Store these results (configuration, rates, cost).

h. Evaluate best current solution.
   i. If solution improves, replace previous remedial design configuration and objective function value.
   ii. Else, continue.

3) Initiate a global optimization analysis using the LGO search options of GARS followed by GRG; initialize using currently best identified solution.

a. Upon finding a cost lower than current minimum cost:
   i. Store these results.

b. Else, continue.

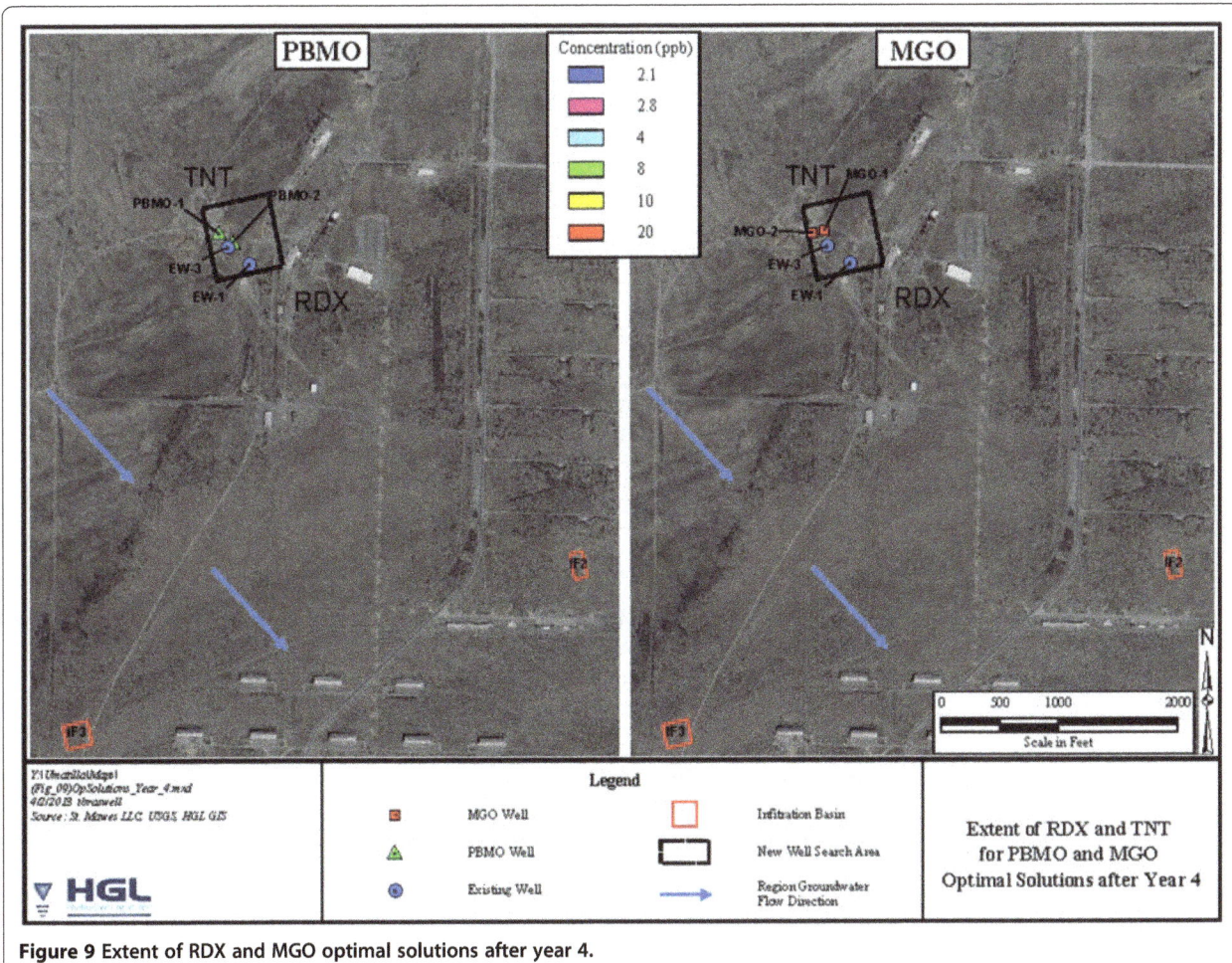

Figure 9 Extent of RDX and MGO optimal solutions after year 4.

4) If termination criteria met,
   a. Stop.
   b. Else, increment number of evaluation epochs by 1 until the user-specified maximum number reached. Store and print results.
   c. Go to step 2.

PBMO's partition and exploration approach enables implementation of the search strategy be conducted on multiple central processing units (CPUs). However, this test used a sequential implementation to mimic ESTCP test conditions. The search continues until the termination criterion satisfied. The termination criterion can either be the targeted optimal cost, the total number of flow and transport simulations, the number of simulations since the optimal value was last improved, or the total simulation (CPU) time consumed. In this case, the termination criterion was the optimal total remedial cost as published by the MGO team. The initial starting point for the total flow rate is the maximum treatment plant flow rate 1,170 (gpm) and mimics the MGO team. Initially, all extraction wells in Box 1 equally pumped; all

extracted water injected into the single infiltration gallery IFL, and new candidate extraction wells placed randomly during the infrastructure search phases.

Results and discussion

The total net present value cost from PBMO ($1,664,085) significantly improved upon the costs committed for remediation at the site by implementing the RIP ($3,836,285) and the design achievable via SME ($2,230,905). The PBMO results mimicked MGO ($1,664,395) and SOMOS ($1,663,841). Table 3 presents the well location strategies and cost results for RIP, SME, MGO and PBMO (since MGO and SOMOS were similar).

The termination criterion in the test was the MGO cost value. PBMO found a lower cost that MGO in fewer than 120 model evaluations. Additional optimization analysis performed during reliability testing produced multiple solutions with lower costs - as low as $1,663,240 – and with different new well extraction locations and different rates in the extraction wells. These subsequent findings illustrate the multi-extremal structure of this design problem,

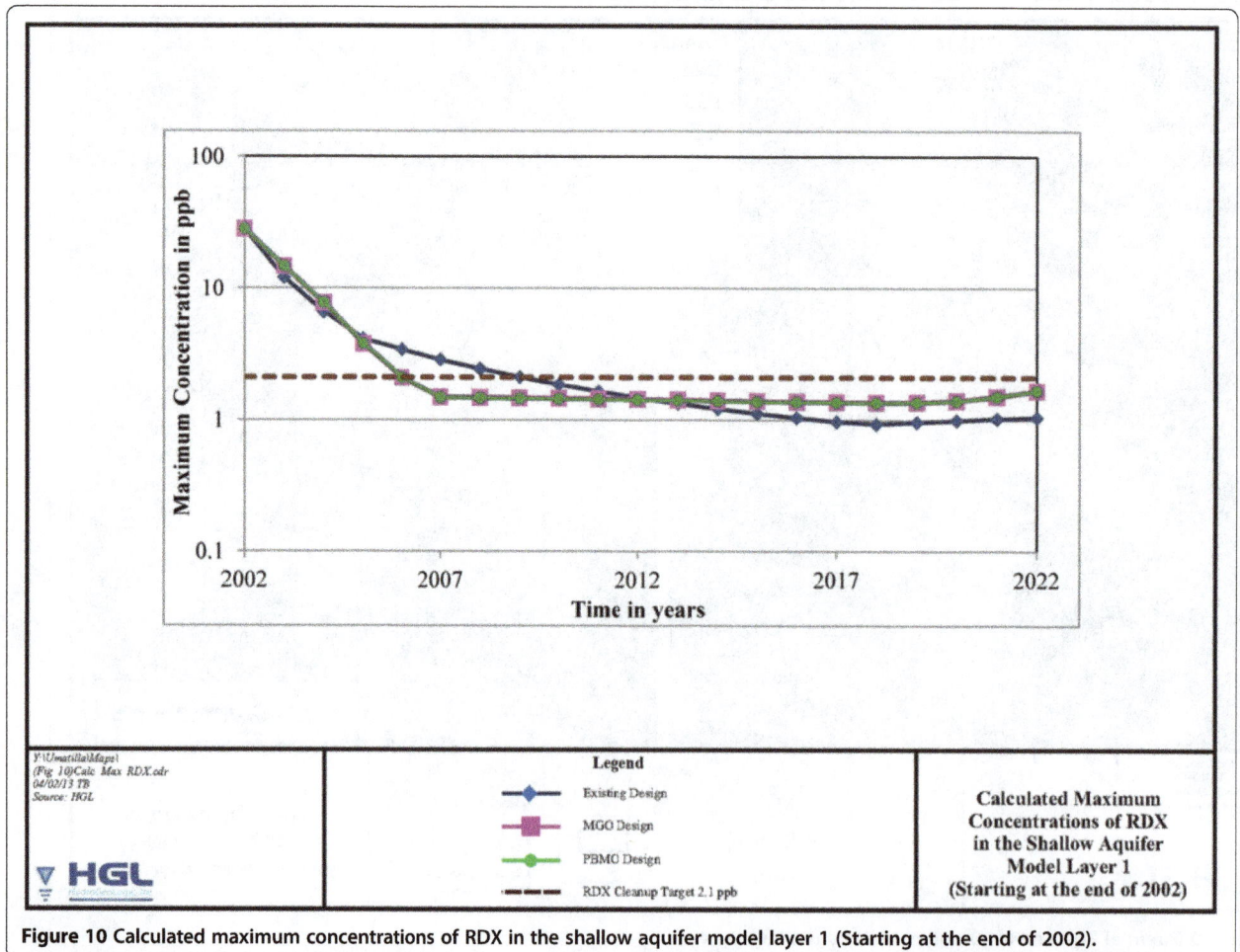

Legend
- Existing Design
- MGO Design
- PBMO Design
- RDX Cleanup Target 2.1 ppb

Calculated Maximum Concentrations of RDX in the Shallow Aquifer Model Layer 1 (Starting at the end of 2002)

Figure 10 Calculated maximum concentrations of RDX in the shallow aquifer model layer 1 (Starting at the end of 2002).

thereby calling for global optimization based solution approaches.

Simulation model reliability over the range of feasible inputs is essential for determining the globally optimal variable values. We observed that 10.7% of the viable candidate designs simulated in the groundwater flow and transport models during the optimization search failed to converge. This primarily occurred when the extraction well flow rates were set near the high end of the acceptable range and which caused flow model cells to dewater. In instances where model cells become dewatered, the groundwater flow simulator could not converge to a solution without the application of a non-physical lower bound on the head in the dewatered cell[a]. Given that the two simulators execute sequentially during a function evaluation (groundwater flow followed by fate and transport), the failure of the flow simulator to provide a convergent solution prevents the transport simulator from predicting contaminant distributions resulting in a lost candidate design evaluation cycle.

PBMO addresses issues that arise in model simulations due to these harsh modeling scenarios imposed by formal optimization. PBMO examines the model simulation solution time and the flow and transport mass balance errors. Automated solver parameter adjustments can take place if the solution becomes unstable or inefficient. If a model nevertheless fails to converge after a user-specified maximum number of numerical solver parameter setting attempts, penalty functions divert the optimal search from exploring this solution region. Hence, while the algorithm handles the non-convergent model issue, should the simulations be noted to fail to converge either at an appreciable rate or in regions of the search space where the suspected location of the optimal value, a more robust model code should be used. This will alleviate the risk of not locating the true globally optimal solution. This study design necessarily used the modeling system used by the other teams in spite of the 10.7% model simulation failure rate.

Table 4 presents a breakdown of the capital, and operations and maintenance costs for each of the four strategies. Figure 4 shows the well locations for RIP and T&E conducted by the SME's, and the optimal well locations determined by MGO and PBMO The PBMO solution places two new wells into the TNT plume, as did MGO.

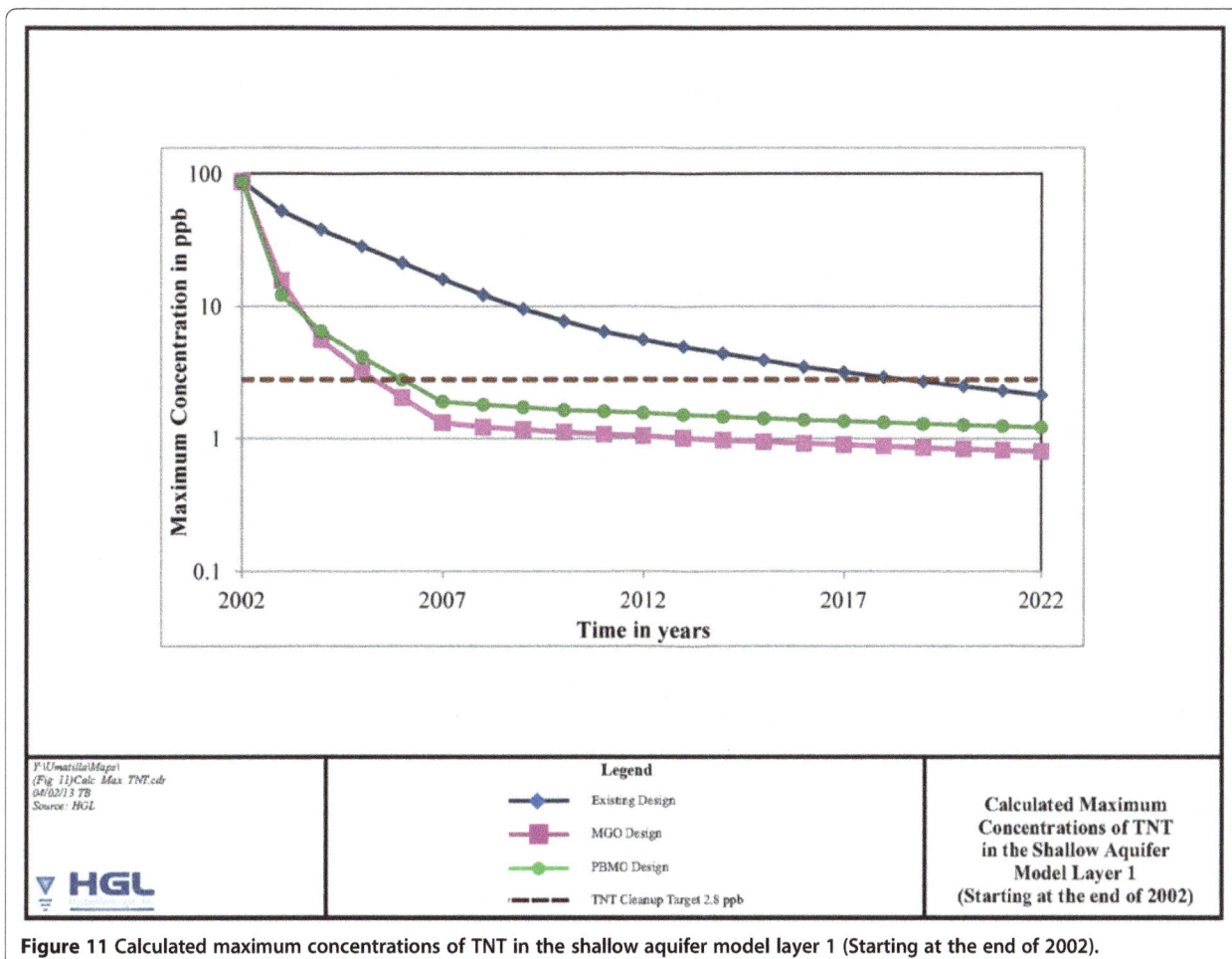

Figure 11 Calculated maximum concentrations of TNT in the shallow aquifer model layer 1 (Starting at the end of 2002).

Examining the performance comparisons of the algorithms[b] reported in the ESTCP study for the number of flow and transport simulations, PBMO found its solution using under 120 flow and transport simulations as compared to the estimated 5,000 simulations reported by the MGO investigators [see Volume II of the DoD/ESTCP report (Minsker et al., 2004)]. This is a significant improvement in efficiency, by an estimated factor of over 40. Furthermore, PBMO ran unattended, whilst MGO required numerous human interventions.

The authors believe that the observed increased efficiently lies in the integrated optimization search strategy. The PBMO approach supports rapid determination of "good" solution spaces that work synergistically with LGO, the core global optimization algorithm. The investigation of (Rios and Sahinidis 2012) supports this position via the results of testing 23 optimization algorithms, including LGO. The results of that comparative study indicate LGO to be orders of magnitude more efficient and reliable than the techniques selected by the original ESTCP study teams (see Figure 5, adapted from that study). Regarding the final solution to the Umatilla problem, the solutions generated

by the optimization techniques are conceptually similar: all three utilize the same existing extraction wells, EW-1 and EW-3; both use the same infiltration basins, IF-2 and IF-3; and both locate new extraction wells near to the center of mass of the TNT plume, and remediate the site in 4 years. Figures 6, 7, 8, 9 show the RDX and TNT plumes at the end of Year 1 through Year 4— the time of remediation completion. Figure 10 and Figure 11 illustrate the maximum concentration of both contaminants over time in the top layer of the model. The (slight) over design by MGO compared with PBMO is evidenced regarding the TNT remediation results. The TNT remediation did not occur at the same time as the RDX remediation and the water quality was remediated cleaner than required by the project specifications.

Remediation well starting position: sensitivity tests

The algorithm's reliability test consisted of assessing its ability to converge on the optimal result from different ill-placed starting locations. The well location search region used was the same search region defined for the northernmost box, the one that contains the initial RDX

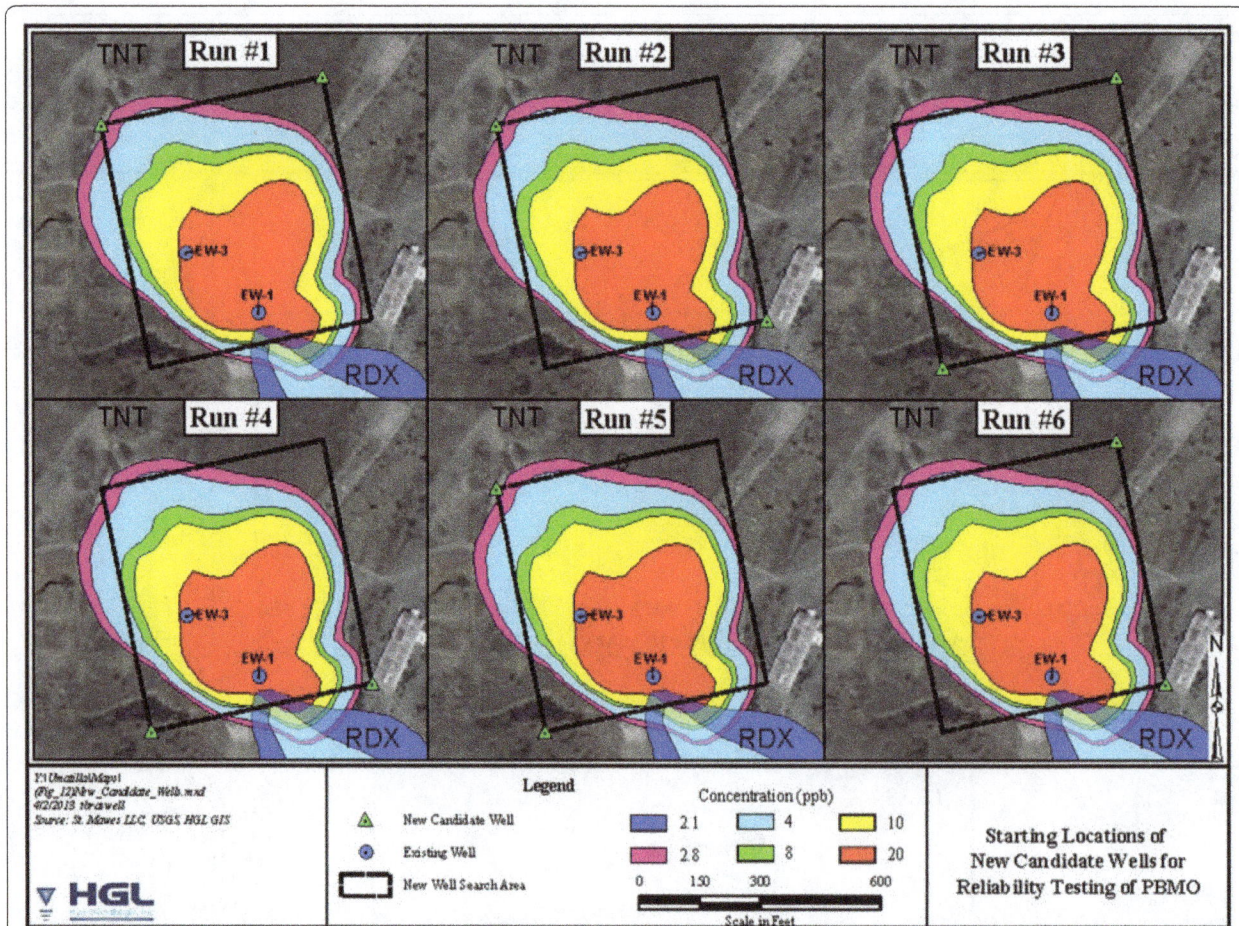

Figure 12 Starting locations of new candidate wells for reliability testing of PBMO.

Table 5 PBMO computational performance during robustness testing

Run #	Starting locations for new wells (L,R,C)	Optimal locations (L,R,C)	Optimal cost ($K)	# Flow /transport iterations to optimal[3]	CPU time to optimal (min)[1,2]
1	(1,40,53)	(1,48,57)	1,664.1	124	164.3
	(1,40,73)	(1,52,61)			
2	(1,40,53)	(1,48,57)	1,664.1	124	164.5
	(1,63,73)	(1,52,61)			
3	(1,63,53)	(1,48,57)	1,664.1	127	177.8
	(1,40,73)	(1,52,61)			
4	(1,63,53)	(1,48,57)	1,664.1	127	178.3
	(1,63,73)	(1,52,61)			
5	(1,40,53)	(1,48,57)	1,664.1	124	161.7
	(1,63,53)	(1,52,61)			
6	(1,40,73)	(1,48,57)	1,664.1	124	161.0
	(1,63,73)	(1,52,61)			

[1]Intel Core i5 CPU 760 @ 2.80 GHz, 8 GB RAM, Win7 Pro.
[2]Intel Core i7 CPU 870 @ 2.93 GHz, 8 GB RAM, Win7 Pro.
[3]The number of iterations for each optimization run includes 1 evaluation for RIP and 24 evaluations for the optimal infrastructure configuration search.

and TNT plumes as shown in Figure 2. However, instead of using the best configuration of the extraction well locations determined thus far (i.e. from the infrastructure search), the search box corners specify the initial new extraction well locations. The test scenarios consisted of six combinations of initial locations for the two new extraction wells at the four corners. Figure 12 shows the six different initial starting configurations for the two new extraction wells represented as the green triangles on the various corners. The results of interest are the final solution, the optimal placement of two new extraction wells to accompany EW-1 and EW-3, the number of simulation/optimization iterations required, and the elapsed time needed to achieve the solution. Table 5 presents the test results. Each of the six test cases found the same optimal locations. The number of model simulations varies between 124 and 127. In every scenario, identification of the optimal solution occurred in less than 3 hours of CPU time. In nearly 100 iterations, the GARS optimization algorithm found the optimal solution regardless that the search initiates from locations outside the contaminant plume. Reliability and minimal dependence on the starting solution initialization are fundamental and desirable features of a high-quality global solver.

Conclusions

The study demonstrates the effectiveness of an automated, cost-effective, and broadly applicable approach to groundwater remedial design. The application of the PBMO methodology to the Umatilla case study site represents a rigorous testing and validation exercise. This numerical analysis efficiently and automatically identified the best solution found by others. Extension of the approach for evaluation and optimal aquifer remediation management for different sites in different geologic settings is accomplished by using a site specific calibrated groundwater flow and transport model. The approach identified the optimal solution about 40 times faster than any of the other methods by reducing the number of time consuming flow and transport model evaluations. Comprehensive automation promotes efficiency, effectiveness and usability. Merit for reliably optimizing complicated systems is achieved through the ability to overcome the inherent non-convergence difficulties that occur when using numerical models for process simulation.

Endnotes

[a]This approach was explicitly applied to one of the other study sites in the DoD/ESTCP study – see Section 3.2.3.4 in Volume I of Minsker et al. (2004).

[b]The model files representing the optimal solution identified by MGO and SOMOS are on the web site, however, the input files to MGO or SOMOS to recreate the optimization study and generate the optimal solution are not available. Therefore, it was not possible to independently verify the reported operational performance of MGO or SOMOS. Hence, we use the reported values for the number of model runs, with MGO being the lesser number of ~5,000. To put this efficiency result in perspective, it would take nearly 5 days of clock time on a modern computer (~1.4 minutes per flow and transport simulation) to solve the problem using MGO, whereas PBMO found a somewhat better solution in less than 3 hours. In addition, the MGO solution required several stops and starts as well as other interceding actions by the investigators to reach the final answer presented in the study. PBMO required a single execution without any human intervention: the optimization process was completely automated.

Abbreviations

BB: Branch and Bound (optimization method in LGO); CCW: Capital Costs of New Wells; CCB: Capital Costs of New Infiltration Basins; CPU: central processing unit; DoD: U.S. Department of Defense; EPA: U.S. Environmental Protection Agency; ESTCP: Environmental Security Technology Certification Program; EW: extraction well; FCL: Fixed Costs of Labor; FCE: Fixed Costs of Electricity; GA: Genetic algorithm; GAC: Granular activated carbon; GARS: Globally Adaptive Random Search (optimization method in LGO); gpm: Gallons per minute; GRG: Generalized reduced gradient (optimization method in LGO); HGL: HydroGeoLogic, Inc.; kg: Kilograms; LGO: Lipschitz Global Optimization software package; LP: Linear programming; LRS: Local random search; µg/L: Micrograms per liter; MGO: Modular Groundwater Optimization; MODFLOW: Modular Finite-Difference Ground-Water Flow ModelMS, multi start random search (optimization method in LGO); MT3DMS: Modular 3-D Multi-Species Transport Model; NPL: National Priorities List; NPV: Net present value; OA: Outer approximation; O&M: Operations and maintenance; PBMO: Physics-Based Management Optimization; ppb: Part per billion; RAM: Random-access memory; RDX: Royal Demolition Explosive; RIP: Remedy In Place; SA: Simulated annealing; SLA: Sequential linear approximation; SLP: Sequential linear programming; SME: Subject matter expert; SOMOS: Simulation/Optimization Modeling Optimization Software System; SQP: Sequential quadratic programming; TNT: 2,4,6-Trinitrotoluene; TS: Tabu search; Umatilla: Umatilla Chemical Depot; VCE: Variable Costs of Electricity for Operating Wells; VCG: Variable Costs of Changing GAC Units; VCS: Variable Costs of Sampling.

Competing interests

The authors declare that they have no competing interest.

Authors' contributions

The work presented here was carried out in collaboration between all authors. LD defined the research theme, designed and implemented the overall optimization approach. JP provided LGO optimization component and TL coded the objective function calculator. All authors have contributed to the paper preparation, and have seen and approved the manuscript.

Authors' information

Larry M Deschaine PE is a graduate of MIT (1984), the University of Connecticut (1992) and is completing his PhD work in Complex Systems at Chalmers University of Technology in Göteborg, Sweden (expected 2013). He is a Principal Engineer at HydroGeoLogic and a recognized expert in the development and application of simulation and optimization techniques to large-scale energy and environmental engineering problems. His work covers the range of surface water, ocean and groundwater characterization and simulation; remedial design and construction; monitoring optimization; detection of unexploded ordnance, and; the development of self-learning adaptive algorithms which support optimization of electric power distribution networks including grid integration of renewables. His work has received numerous awards, including a US Vice-Presidential Hammer Award, and he has written more than 100 journal articles, book chapters, and other technical documents.

János D. Pintér is a researcher and practitioner with four decades of experience in the area of systems modeling and optimization, including algorithm and software development. He holds MSc, PhD and DSc degrees in Mathematics, with specialization in the Operations Research area. He has authored and edited books including an award-winning monograph, and has written more than 200 journal articles, book chapters, and other technical documents. Dr. Pintér is an active member and officer of several professional organizations, and he serves on the editorial board of several professional journals.

Theodore P. Lillys PE is a Senior Engineer at HydroGeoLogic, Inc. with over fifteen years of experience in research and applications in groundwater modeling, optimization, statistics, and software development. He holds and MS (1997) in civil and environmental engineering specializing in subsurface hydrology, numerical methods, and mathematical optimization techniques. Mr. Lillys' areas of expertise lie in groundwater modelling, management optimization of subsurface remedial designs and resource management, optimization of long term monitoring networks, and software development.

Acknowledgements

This research was conducted in concert with LD's PhD studies in Complex Systems at Chalmers University, with financial support by HydroGeoLogic, Inc, 11107 Sunset Hills Road, Suite 400, Reston, VA 20190 (www.hgl.com). An animation of this case study has been prepared; see (http://www.hglsoftware.com/cleanup.cfm).

Author details

[1]HydroGeoLogic, Inc., Reston, VA 20190, USA. [2]Department of Energy & Environment, Chalmers University of Technology, Göteborg, SE 412 96, Sweden. [3]PCS Inc, Halifax, NS, Canada.

References

Ahlfeld DP, Barlow PM, Mulligan AE (2005) GWM – a ground-water management process for the US Geological Survey modular ground-water model (MODFLOW – 2000). US Geological Survey, Open-File Report 2005-1072, 124 Pages, 2005. https://water.usgs.gov/nrp/gwsoftware/mf2005_gwm/OFR2005_1072.pdf

Deschaine LM (1992) Cost evaluation and optimization of ground water pump and treat programs, MSCE thesis. University of Connecticut, Storrs, Connecticut

Deschaine LM, Ahlfeld DP, Ades MJ, O'Brien D (1998) An optimization algorithm to minimize the life cycle cost of implementing an aquifer remediation project - theory and case history. Society for Modeling and Simulation International, Simulators International XV, Simulation Series 30(3):53–58

Deschaine LM, Regmi S, Patel JJ, Fox TA, Ades MJ, Katyal A (2001) Design optimization of groundwater quality management challenges using the outer approximation method, The Society for Modeling and Simulation International. Advanced Simulation Technology Conference, Seattle, WA, USA, pp pages 88–93, April 2001

Deschaine LM (2003) Simulation and optimization of large-scale subsurface environmental impacts; investigations, remedial design and long term monitoring. Journal of Mathematical Machines and Systems, Kiev 3(4):Pages 201–218

Deschaine LM, Pintér JD (2003) A comparison of the outer approximation method and lipschitz global optimization on optimal groundwater quality management. Presented at the INFORMS Annual Meeting Conference, Atlanta, Georgia, October 19-22, 2003

Deschaine LM, Nordin JP, Pintér JD (2011) A computational geometric / information theoretic method to invert physics-based mec-model attributes for mec discrimination. Journal of Mathematical Machines and Systems, National Academy of Sciences of Ukraine, Kiev 2:50–61

Guo X, Zhang CM, Borthwick JC (2007) Technical Report. Water Resources Research Journal 43(No. 8):WO8416–14 Pages, August. http://onlinelibrary.wiley.com/doi/10.1029/2006WR004947/abstract

Harbaugh AW, McDonald MG (1996) User's documentation for modflow-96, an update to the u.s. geological survey modular finite-difference ground-water flow model. U.S. Geological Survey Open-File Report 96–485, Reston VA, pp 56–485

HydroGeoLogic, Inc. (HGL) (2012) MODFLOW-SURFACT™ Version 3.0; User's manual and guide. VA, USA, 20190

ITRC (2007) In-Situ Bioremediation of Chlorinated Ethene DNAPL Source Zones: Case Studies. Prepared by The Interstate Technology and Regulatory Council, Bioremediation of Dense Non-Aqueous Phase Liquids (Bio DNAPL) Team. Refer to Chapter 9 "Simulation and Optimization of Subsurface Environmental Impacts; Investigations, Remedial Design and Long Term Monitoring of BioNAPL Remediation Systems"., pp 128–147, April 2007. http://www.itrcweb.org/Guidance/GetDocument?documentID=11

McDonald MG, Harbaugh AW (1988) A modular three-dimensional finite-difference groundwater flow model, Techniques of Water Resources Investigations Book 6. U.S. Geological Survey, Washington, D.C

Minsker B, Zhang Y, Greenwald R, Peralta R, Zheng C, Harre K, Becker D, Yeh L, Yager K (2004) Final technical report for application of flow and transport optimization codes to ground water pumping and treat systems, Technical Report to the Environmental Security Technology Certification Program. Volumes I-III. TR-2237-ENV. Engineering Service Center, Port Hueneme, California, http://www.frtr.gov/estcp/estcp.htm

Peralta RC (2004) "SOMOS: Simulation/Optimization Modeling System". User's Manual, Software Engineering Division, Department of Biological and Irrigation Engineering. Utah State University, Logan, UT, p 48 Pages, April 2004. http://www.frtr.gov/estcp/source_codes.htm

Pintér JD (1996) Global optimization in action. Kluwer Academic Publishers, Dordrecht Boston London, 1996. Now distributed by Springer Science + Business Media, New York

Pintér JD (2002) Global optimization: software, test problems, and applications. In: Pardalos PM, Romeijn HE (eds) Handbook of Global Optimization, Volume 2. Kluwer Academic Publishers, Dordrecht, pp 515–569

Pintér JD (2009) In: Pardalos PM, Coleman TF (eds) Software development for global optimization. American Mathematical Society, Providence, RI

Pintér JD (2013) LGO - A model development and solver system for global-local nonlinear optimization user's guide. Published and distributed by Pintér Consulting Services, Inc, Canada, www.pinterconsulting.com. First edition: June 1995; Current edition: March 2013

Rios M, Sahinidis N (2012) "Derivative-free optimization: A review and comparison of software implementations". J Glob Optim. doi:10.1007/s10898-012-9951-y, http://link.springer.com/content/pdf/10.1007%2Fs10898-012-9951-y.pdf

Systems Simulation/Optimization Laboratory (SSOL) (2002) Simulation/Optimization Modeling System (SOMOS) Users Manual. SS/OL, Biological & Irrig. Eng. USU, Logan, UT, p 457

U.S. Army Corps of Engineers (USACE) (1994) Defense Environmental Restoration Program, Final Record of Decision, Umatilla Depot Activity Explosives Washout Lagoons Ground Water Operable Unit., June 7, 1994

U.S. Army Corps of Engineers (USACE) (1996) Final Remedial Design Submittal, Contaminated Groundwater Remediation, Explosives Washout Lagoons, Umatilla Depot Activity, Hermiston Oregon, January 1996

USEPA (2012) National strategy to expand superfund optimization practices from site assessment to site completion., (September 2012) (PDF) OSWER 9200.3-75 http://www.epa.gov/oerrpage/superfund/cleanup/postconstruction/optimize.htm

Zheng C, Wang PP (1999) MT3DMS: A modular three-dimensional multispecies transport model for simulation of advection, dispersion and chemical reactions of contaminants in groundwater systems; documentation and user's guide, Contract Report SERDP-99-1. U.S. Army Engineer Research and Development Center, Vicksburg, MS, available at http://hydro.geo.ua.edu/mt3d

Zheng C, Wang PP (2002) MGO – A modular groundwater optimizer incorporating modflow/mt3dms, documentation and user's guide. Draft., April 2002. http://www.frtr.gov/estcp/source_codes.htm

Zheng C, Wang P (2003) "Application of flow and transport optimization codes to groundwater pump-and-treat systems: Umatilla Army Depot, OR". Technical Report, Revised version 2/2003. University of Alabama, Tuscaloosa, AL, p pp. 41, http://www.frtr.gov/estcp/demonstration_sites.htm

14

A simulation-optimization approach for assessing optimal wastewater load allocation schemes in the Three Gorges Reservoir, China

Zheng Wang[1,5], Shuiyuan Cheng[1*], Lei Liu[2*], Xiurui Guo[1], Yuan Chen[1], Cuihong Qin[1], Ruixia Hao[3], Jin Lu[4] and Jijun Gao[4]

Abstract

Background: The Three Gorges Reservoir (TGR) has been facing deteriorated water quality issues since the construction of the Three Gorges Dam (TGD) in 1994. However, no previous studies have used a simulation-optimization assessment framework to examine the waste-load allocation patterns in the TGR area for alleviating its water pollution problem. In this study, a simulation-optimization modeling approach was developed for addressing this issue, through combining an environmental fluid dynamic code (EFDC)-based water quality simulation model and a waste-load allocation optimization model into a general framework.

Results: The approach was applied to a TGR section (Changshou-Fuling section) for identifying the optimal waste-load allocation schemes among its 11 wastewater discharge outlets. Firstly, the EFDC model was run to simulate the water quality response in the receiving water body under a single discharge load scenario, and the simulated COD and NH_4^+-N concentrations were used to calculate the pollution mixing zone (PMZ), the pollution mixing zone per unit load (PMZPL), and sensitivity index (SI) pertaining to that outlet. These values were then used in the formulation of the waste-load allocation optimization model, with its objective being to maximize the environmental performance under constraints that existing waste discharge loads in terms of total wastewater amount, total pollutant mass, and existing PMZ size can't be exceeded.

Conclusions: Modeling results give an optimal waste-load allocation ratio for each discharge outlet within the study section, and its implications to the reservoir water quality management were analyzed. It is anticipated that the develop approach can be extended to the entire TGR area for better water quality management studies and practices.

Keywords: Environmental fluid dynamic code (EFDC); Water quality simulation; Waste-load allocation optimization; Simulation-optimization approach; Three Gorges Reservoir

Background

The Three Gorges Dam (TGD) spans the Yangtze River by the town of Sandouping, located in the Yiling District of Yichang City, in Hubei province, China. It is a hydroelectric dam and the world's largest power station in terms of installed capacity (21,000 MW). The TGD construction started in 1994 and the dam body was completed in 2006. Since then, the water level behind the dam has gradually increased and reached its designed maximum level of 175 m in October 2010. When the water level is at its maximum, the upstream of the Three Gorges Reservoir (TGR) along the Yangtze River is about 660 km in length and has an average of 1.12 km in width. The TGR contains 39.3 billion m^3 of water with a total surface area of 1,045 km^2, and the reservoir watershed has a total area of 58,000 km^2 (CWRC Changjiang Water Resource Commission 1997). The TGR is situated within an attitude of E106°–115° 50′ and N29°16′–31°25′, as shown in Figure 1. It encompasses 26 towns or counties under the jurisdiction of Metro Chongqing Municipality and Hubei Province, with a total population of 20.1 million.

* Correspondence: sycheng@bjut.edu.cn; Lei.Liu@Dal.Ca
[1]College of Environmental & Energy Engineering, Beijing University of Technology, Beijing 100022 China
[2]Department of Civil and Resource Engineering, Dalhousie University, Halifax, NS B3H 4R2 Canada
Full list of author information is available at the end of the article

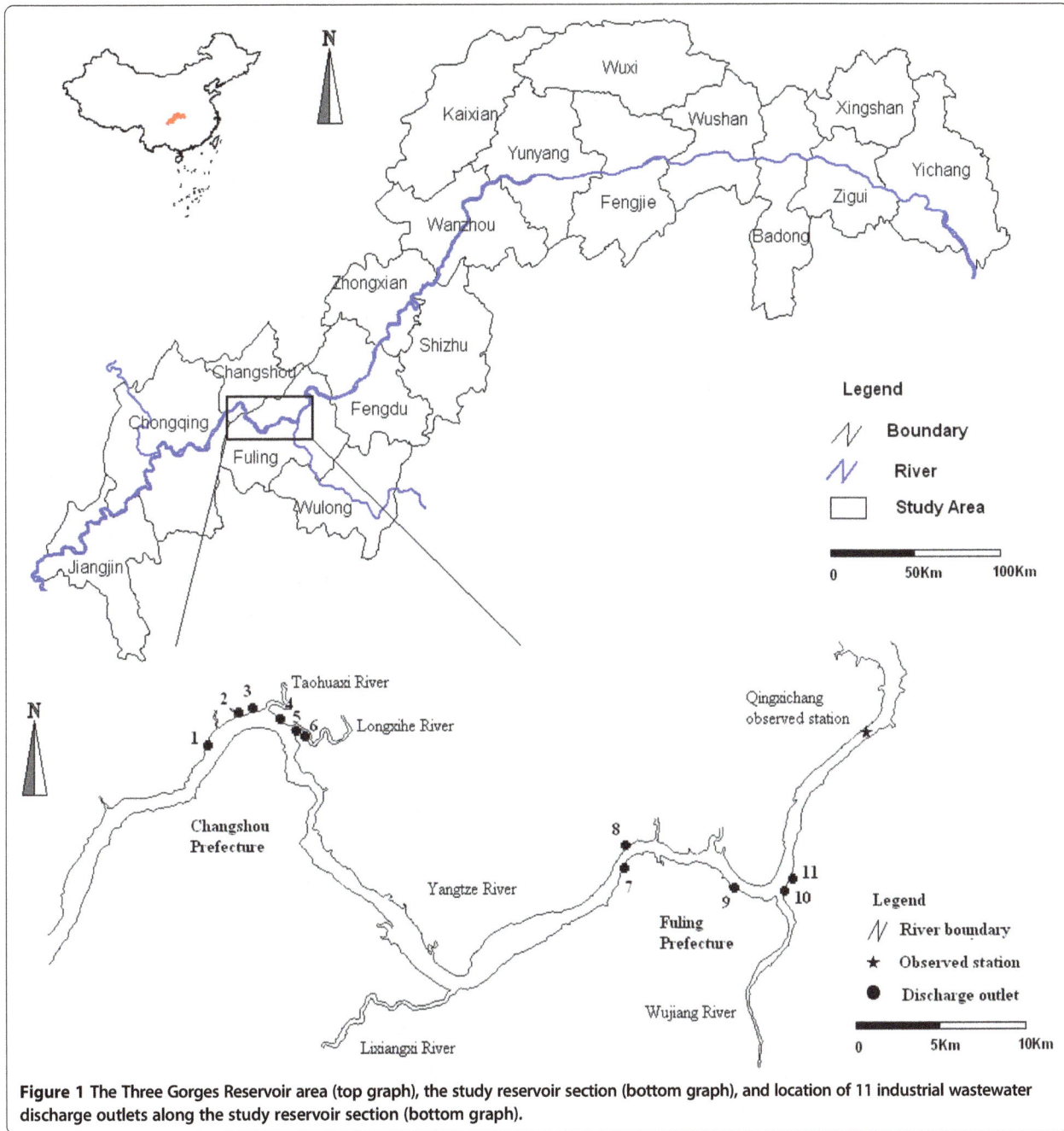

Figure 1 The Three Gorges Reservoir area (top graph), the study reservoir section (bottom graph), and location of 11 industrial wastewater discharge outlets along the study reservoir section (bottom graph).

As the consequences of fast population growth and rapid economic development in the past two decades in the TGR watershed, more and more industrial wastewater and domestic sewage are generated and discharged into the TGR, leading to the deterioration of water quality in the reservoir. Meanwhile, the construction and completion of TGD has resulted in significant changes of hydraulics of water flow in the reservoir, which further affects the transport and fate of various contaminants discharged into the reservoir water body. For example, the advective and diffusive transport of the chemicals in the reservoir water body has been significantly affected due to the slow-down of water flow. As a result, local authorities have been undertaking enhanced stresses in order to effectively respond to these concerns. In general, one reliable path is through designing effective environmental management schemes for dealing with this dilemma, and this requires a sound understanding of the significant contributors to the reservoir water pollution problem, and of the way the reservoir system will react to particular schemes or policies. It becomes obviously desired to study sound allocation schemes for better water quality management in the region.

Generally, the waste-load allocation problems are to determine the allowed discharge levels or required removal levels from a number of point pollution sources in a basin to achieve satisfactory (or aspired) water-quality responses in the receiving water body. Previously, the optimal allocation of waste loads has been typically solved by developing various forms of optimization models Mujumdar and Vemula 2004; Yang et al. 2011; Qin and Xu 2011). The decision variables in the optimization models are the discharge (or removal) levels of wastewater or pollutants at each of the point pollution sources. The objective function is often to maximize economic return or to minimize the treatment cost while the constraints are to ensure that the resulting water-quality responses in the receiving water body are satisfactory. Since the resulting water-quality responses in the receiving water body can only be quantified by water quality simulation models, the waste-load allocation problems are essentially required to conduct a simulation-optimization assessment of levels of waste load reductions and allocations from various sources while ensuring the water quality standards being satisfied. Examples of such studies include a waste load allocation model developed by Cho et al. (2003) for the heavily polluted Gyungan River in South Korea, where a modified QUAL2E model was used for the water quality simulation, and a simulation–optimization analysis of waste load allocation for a river water quality management by Mujumdar and Vemula (2004). In the TGR area, previously, many optimization modeling studies have been conducted but mainly focused on Yangtze River flood control (Cai et al. 2010), hydropower generation (Guo et al. 2011), Yangtze River watershed navigation (Wang and Ruan 2011), and regional water resources allocation and supply (Sun and Lv 2010).

The resulting water-quality response in the receiving water body depends on the total waste load discharged and allocation pattern among different sources, as well as the various hydraulic conditions within the water body. Surface water quality models have been deemed as sound engineering tools for rationalizing water quality management (Chapra 2003; Igbinosa and Okoh 2009; Chen et al. 2010). Various surface water quality modeling studies and software have been carried out and developed for supporting the management of surface water systems, and examples include the river and stream water quality model (QUAL2K) (Chapra et al. 2007), water quality analysis simulation program (WASP) (Wool et al. 2001), environmental fluid dynamics code (EFDC) (Tetra Tech I 2007) and MIKE11 (DHI 2003). Some models can tackle the problems related to chemical and biological processes through deterministic partial differential equations (Igwe et al. 2008; Shah et al. 2009). All the models include two essential components for executing two core tasks: (1) a hydrodynamic sub-model to simulate the water flow circulation and behavior, and (2) a

mass transport and kinetics sub-model to simulate the fate and movement of pollutants in the target water bodies. For the TGR area, previous water quality modeling studies have been focused on the simulation of 1D and 2D hydraulic flow in the mainstream of Yangtze River (Huang 2006), analysis of the natural assimilative capacity and water environmental capacity, definition and calculation of pollution mixing zones (Jiang et al. 2005), and some others (Zhao et al. 2011).

Literature survey show that, although the water pollution issue in the TGR has become more and more serious since the construction of the TGD, no previous studies have used a simulation-optimization assessment framework to examine the waste-load allocation patterns in the TGR area for alleviating TGR's water pollution problem. The objective of the present work is thus to combine a water quality simulation model and waste-load allocation model into a general methodology framework, and aims to address the water pollution issue in the TGR area. In this study, the Changshou-Fuling section is a typical mainstream section along the Yangtze River and was selected as the target study section for implementing the modeling runs and identifying the optimal waste-load allocation schemes. The EFDC model was used as the surface water quality model to simulate the flow hydrodynamics and chemical transport. It helps rank and screen the major waste discharge outlets within the study reservoir section. The pollutant mixing zone (PMZ) formed by one single outlet for different pollutants was calculated to establish the linkage between the pollution plume size with the strength of each discharge outlet, and a sensitivity index can then be calculated based on the calculated PMZ. The spatial and temporal distribution of PMZ was calculated by the EFDC model. The waste-load allocation patterns were obtained through optimizing the sensitivity index of each waste discharge outlet. According to the modeling results, the schemes for optimal waste-load allocations among 11 different discharge outlets were analyzed. The implications to formulate the strategies for improving the water quality for the entire region were also discussed. It is anticipated that the methodology developed in this study can be extended to other TGR areas for better water quality management studies and practices.

The Changshou-Fuling reservoir section

The Changshou-Fuling section of the TGR was selected as the study area, as shown in Figure 1. It is subject to a tropical monsoon climate of Northern Asia and has an annual mean temperature of 18°C, an average precipitation of 1170 mm per year, and an average evaporation of 1300 mm per year, and an average wind speed of 1.4 m/s (Zhao et al. 2011). This reservoir section has an annual mean flow velocity of 0.8 m/s, a mean flow rate of 12,000 m³/s, and a mean water depth of 50 m (MWRC Ministry Of Water Resources of China 2008). Changshou and Fuling Districts are two major industrial bases of ChongQing Municipality. They have a total of 11 industrial wastewater discharge outlets located along the study section, discharging a large amount of untreated industrial wastewater into the TGR. Among them, 6 discharge outlets are located in Changshou District and the other 5 are located in Fuling District (as indicated in Figure 1, bottom graph, and Figure 2). The study section is approximately 90 km

long, with 18 km located in Changshou District in the upper reach and 72 km in Fuling District in the lower reach. The Wujiang River flows into this section right before the discharge outlet #10 and is the second largest inflow tributary of the TGR. Figure 3 gives a digital elevation map of the study reservoir section. With the rapid population growth and fast economic growth in the past two decades, the overall water quality of in the

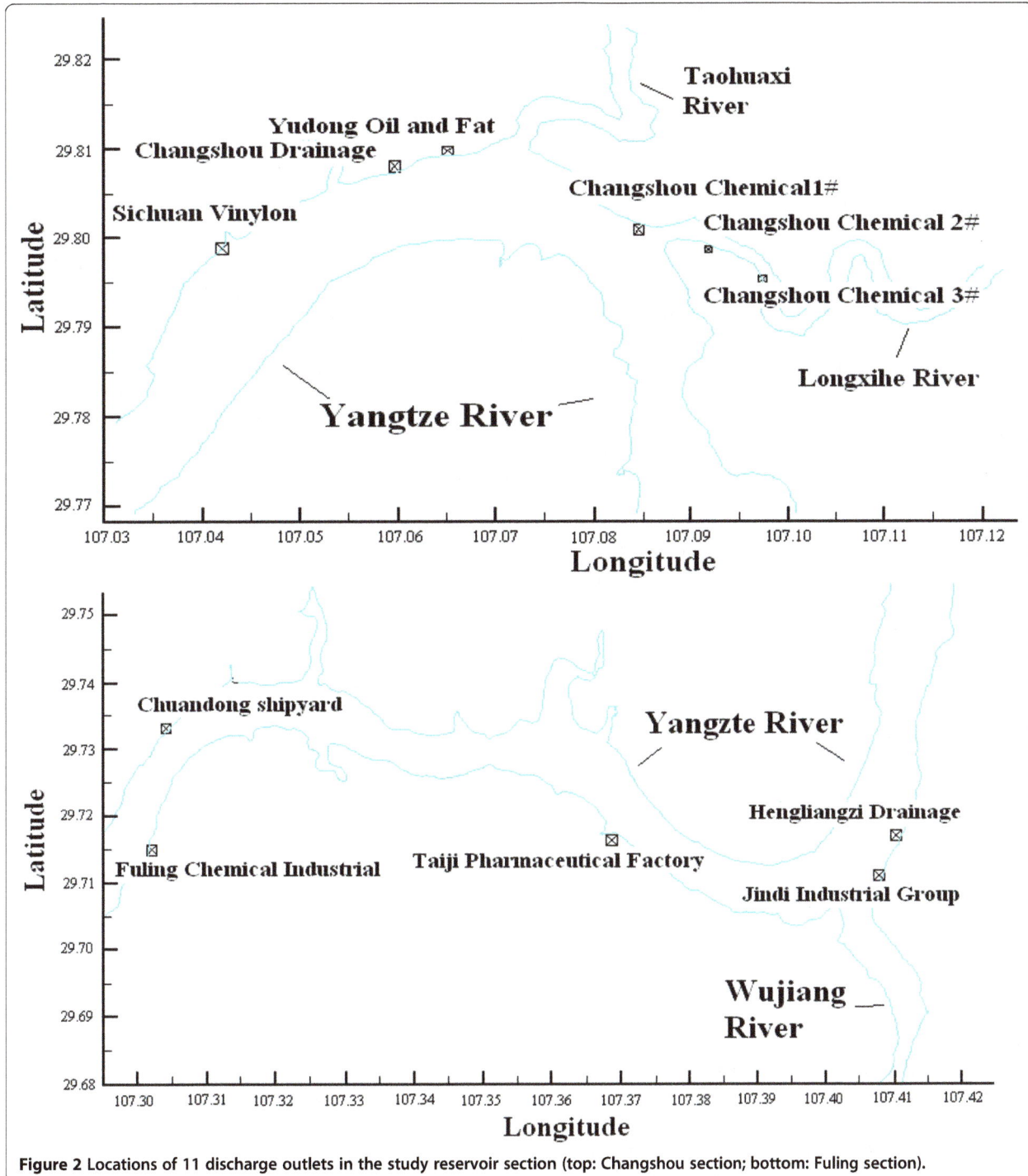

Figure 2 Locations of 11 discharge outlets in the study reservoir section (top: Changshou section; bottom: Fuling section).

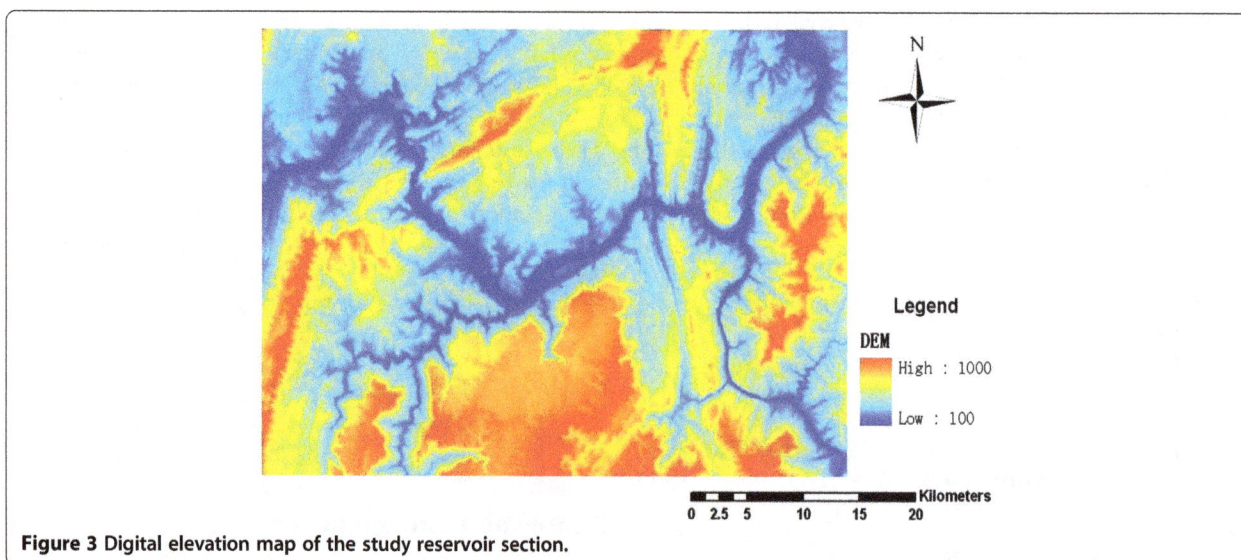

Figure 3 Digital elevation map of the study reservoir section.

Changshou-Fuling section has been gradually deteriorated and has been in very poor conditions (Yibing et al. 2007). Furthermore, the construction and completion of the TGD have made the situation even worse.

In this study, a topographic map with a resolution of 1:25,000 was used to delineate the river channel characteristics, and the water depth along the study section was sourced from the Department of Hydrology, Changjiang Water Resources Commission (Changjiang WRC). Hydrological data were obtained from the Bureau of Hydrology, the Ministry of Water Resources of China (MWRC). Water quality data include the concentrations of DO, COD, NH_4^+-N, observed at the Qingxichang monitoring station in 2008, and were provided by the ChongQing Environmental Science Research Institute. Meteorological data were downloaded from the website of the China Meteorological Data Sharing Service System. Industrial and municipal point pollution source data for 2008 for Changshou and Fuling Districts were provided by the Changjiang Water Resources Protection Institute. Data of nutrients (N, P) from non-point sources within the study section were provided by Chongqing Environmental Science Research Institute. The Level-II water quality standard specified in the National Environmental Quality Standards for Surface Water (GB3838-2002) was applied, and the standards for NH_4^+-N and COD are 0.5 mg/L and 15 mg/L, respectively. The discharge load from each discharge outlet in the year of 2008 is given in Table 1.

Methods

The simulation-optimization assessment approach
Definitions of pollutant mixing zone (PMZ) and sensitivity index (SI)
Before we discuss how the surface water quality simulation model and waste-load allocation model was combined into

a general assessment framework, a few concepts need to be defined. A primary concept is the pollution (or pollutant) mixing zone (PMZ), and it was defined as the area of a contamination plume caused by the wastewater discharge where the water quality within the zone exceeds the Level-II national surface water quality standards. The area of the PMZ for each discharge outlet is affected by their respective factors, mainly including the discharge load from the outlet, hydro-dynamic flow feature around the outlet, natural assimilative capacity of local water body, and background concentrations of the pollutants of interest. It was calculated by the EFDC model. The pollutant mixing zone per unit load (PMZPL) can then be defined and calculated by:

$$PMZPL_i = \frac{PMZ_i}{DL_i} \tag{1}$$

where PMZ_i is the surface area of the pollutant mixing zone formed by the ith discharge outlet (in m²); DL_i

Table 1 Discharge load from 11 discharge outlets in 2008

Number	Discharge outlet	District	Discharge load (kg/d)	
			COD	NH_4^+-N
1	Sichuan Vinylon	Changshou	6959	316
2	Changshou Drainage	Changshou	1200	100
3	Yudong Oil and Fat	Changshou	1225	82
4	Changshou Chemical #1	Changshou	1927	551
5	Changshou Chemical #2	Changshou	628	338
6	Changshou Chemical #3	Changshou	1372	677
7	Fuling Chemical Plant	Fuling	7778	1399
8	Chuandong shipyard	Fuling	295	38
9	Taiji Pharmaceutical Factory	Fuling	562	15
10	Jindi Industrial Group	Fuling	522	68
11	Hengliangzi Drainage	Fuling	2055	205

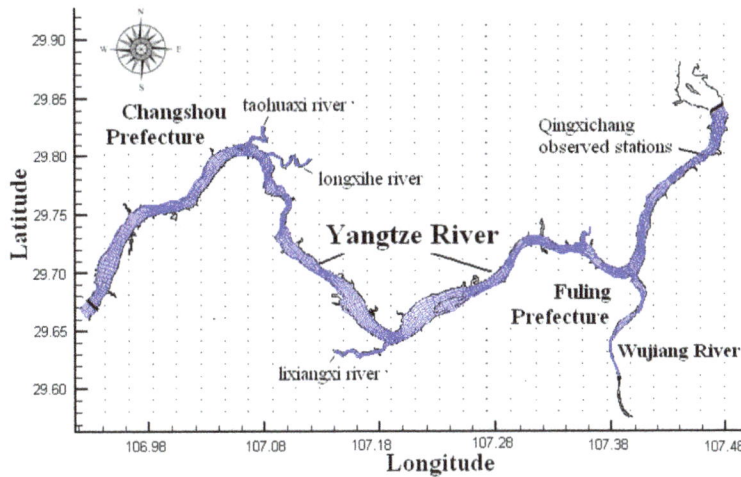

Figure 4 The computation grid designed for the Changshou-Fuling reservoir section.

refers to the discharge load from the ith discharge outlet (in kg/day).

In this study, the monthly and annually average values of PMZ and PMZPL for each individual discharge outlet were calculated by the EFDC-based water quality simulation model. In a specific PMZ calculation, only the discharge load from the outlet of interest was considered and used in the EFDC simulation, while the discharge loads from all other outlets were assumed to be zero. As a result, the calculated PMZ and MPZPL only reflect the pollution contribution from each individual outlet. A sensitivity index can then be defined to indicate how sensitive or important each individual discharge outlet could be among all the outlets in terms of its impact on water pollution. The formula to calculate the sensitivity index for each outlet is given below:

$$SI_i = \frac{PMZR_i}{DLR_i} \tag{2}$$

Where SI_i is the sensitivity index for the ith discharge outlet (dimensionless); $PMZR_i$ is the ratio or share of the

PMZ formed by the ith outlet to the total PMZ area formed all the outlets (dimensionless); and DLR_i represents the ratio or share of the discharge load from the ith outlet to the total discharge load from all the outlets (dimensionless). It is indicated that a higher SI value means a bigger PMZ area influenced by a specific outlet and a stronger pollution impact from the same outlet. $PMZR_i$ and DLR_i can be calculated by the following two formulas:

$$PMZR_i = \frac{PMZ_i}{\sum\limits_{i=1}^{n} PMZ_i} \tag{3}$$

$$DLR_i = \frac{DL_i}{\sum\limits_{i=1}^{n} DL_i} \tag{4}$$

In Equations (3) and (4), the subscript i indicates the number of discharge outlet along the study reservoir section. In this study, there are a total of 11 outlets ($n = 11$).

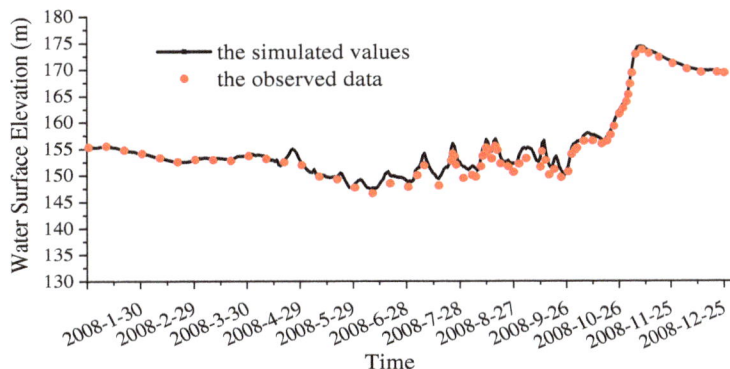

Figure 5 Comparison of simulated and observed water levels at Qingxichang monitoring station in 2008.

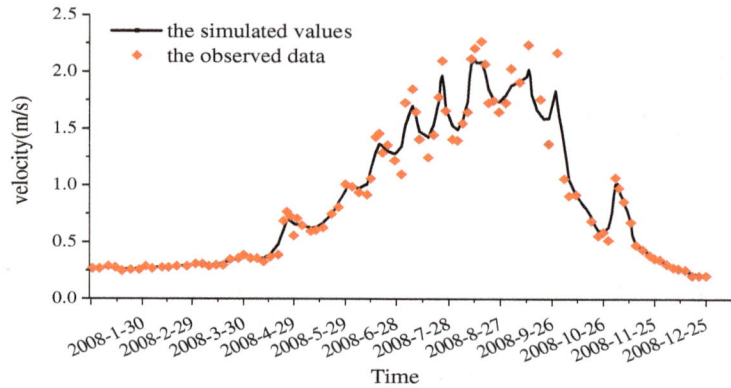

Figure 6 Comparison of simulated and observed flow velocity at Qingxichang monitoring station in 2008.

Simulation-optimization assessment framework

In this study, a hydro-dynamic water quality simulation model was developed using the environmental fluid dynamics code (EFDC) (Tetra Tech I 2007). The EFDC-based simulation model was used to calculate the monthly and annually average value of PMZ formed by each individual discharge outlet. The discharge load from one individual outlet was used as the single load input in each run of the EFDC simulation. Also, a linear optimization model was formulated to seek the optimal waste-load allocation schemes among different discharge outlets within the study section. In this study, the water quality simulation and waste-load allocation model were then combined into a general assessment framework. The assessment process was carried out in a sequential manner. It starts with running the EFDC simulation model to provide the values of parameters used in the optimization model, and then the optimization model was run to search for the best waste-load allocation schemes. The details of the modeling and assessment processes are proved in the following context.

EFDC-based water quality simulation modeling
Model description

The EFDC is a public domain, open source, surface water modeling system, and includes hydrodynamic, sediment and contaminant, and water quality modules which are fully integrated in a single source code implementation. EFDC has been applied to over 100 water bodies including rivers, lakes, reservoirs, wetlands, estuaries, and coastal ocean regions in support of environmental management and regulatory requirements. The EFDC was originally developed at the Virginia Institute of Marine Science (VIMS) and the School of Marine Science of The College of William and Mary, by Dr. John M. Hamrick in 1988 (Tetra Tech I 2007). The hydrodynamics of the EFDC is based on the Princeton Ocean Model; the physics of the two models are the same (Blumberg and Mellor 1987).

For simulating the water quality, the EFDC solves the hydrodynamics and drives the transport of chemicals by using the hydrodynamic results. The water quality module in the EFDC solves the phytoplankton kinetics, which includes nutrient uptake, growth, respiration, discretion,

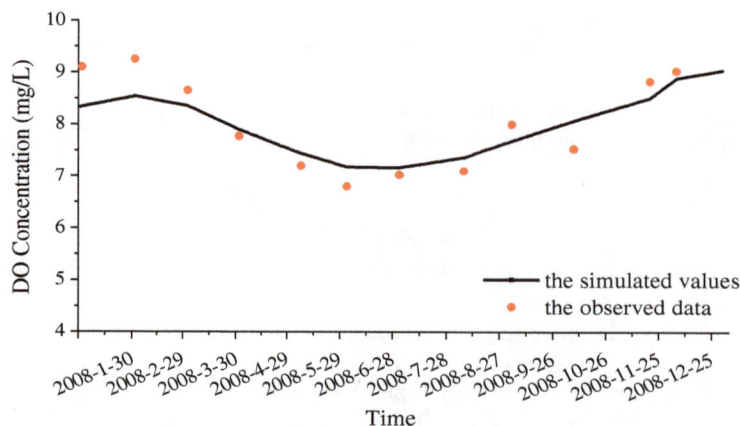

Figure 7 Comparison of simulated and observed DO concentration at Qingxichang monitoring station in 2008.

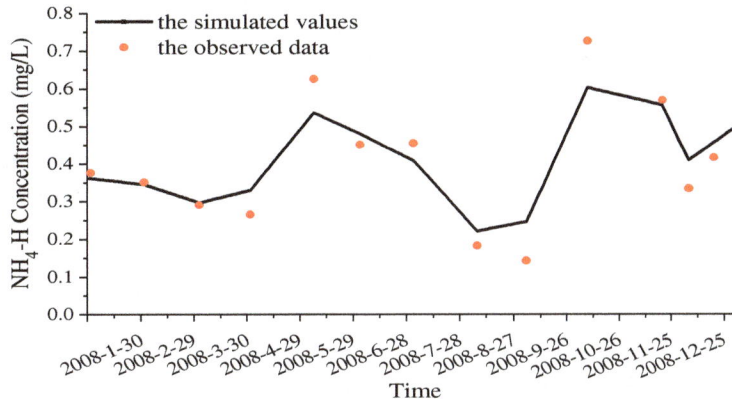

Figure 8 Comparison of simulated and observed NH_4^+-N concentration at Qingxichang monitoring station in 2008.

settling processes and nutrient transformations. The EFDC water quality module is based on CE-QUAL-ICM and can simulate up to 22 state variables (Hamrick 1996). The governing mass-balance equation for each of the water quality state variables is expressed as in the Equation (5).

$$\frac{\partial(m_x m_y HC)}{\partial t} + \frac{\partial(m_y HuC)}{\partial x} + \frac{\partial(m_x HvC)}{\partial x} + \frac{\partial(m_x m_y wC)}{\partial x}$$
$$= \frac{\partial}{\partial x}\left(\frac{m_y HA_x}{m_x}\frac{\partial C}{\partial x}\right) + \frac{\partial}{\partial y}\left(\frac{m_x HA_y}{m_y}\frac{\partial C}{\partial y}\right) + \frac{\partial}{\partial z}\left(\frac{m_x m_y A_z}{H}\frac{\partial C}{\partial z}\right) + S_c$$

$$(5)$$

Where, C denotes the concentration of a water quality state variable; u, v, and w refer to the velocity components in x-, y-, and z-directions in generalized curvilinear and sigma coordinate systems; A_x, A_y, A_z are the turbulent diffusivities in the x-, y-, and z-directions, respectively; H is the water column depth; m_x, m_y are the horizontal curvilinear coordinate scale factors; S_c represents internal and external sources and sinks per unit volume, which are

either generated and/or consumed by kinetic processes. The kinetic formulations of water quality component in the EFDC are primarily from CE-QUAL (Hamrick 1996).

Model setup

A horizontal computation grid was designed for the study reservoir section and for running the EFDC model, as shown in Figure 4. In this study, the orthogonal curvilinear grids were used for setting up the EFDC model to fit the natural flow boundary. The model maps the orthogonal curvilinear grid with the software Delft3D (Zuo 2007). The designed grids consist of 4373 cells in the horizontal direction, with grid size ranging from 10 m to 300 m. A vertical sigma coordinate was evenly distributed by 5 layers for better simulating the bottom topography (Xie et al. 2010). Model simulation time step was set to 5 s. For the initial conditions, the water surface elevation for each active cell was set to 155 m and the water quality variables for each active cell were set same to the initial data (Wu and Xu 2011; Zhao et al. 2011; Parka et al. 2005; Li et al. 2011).

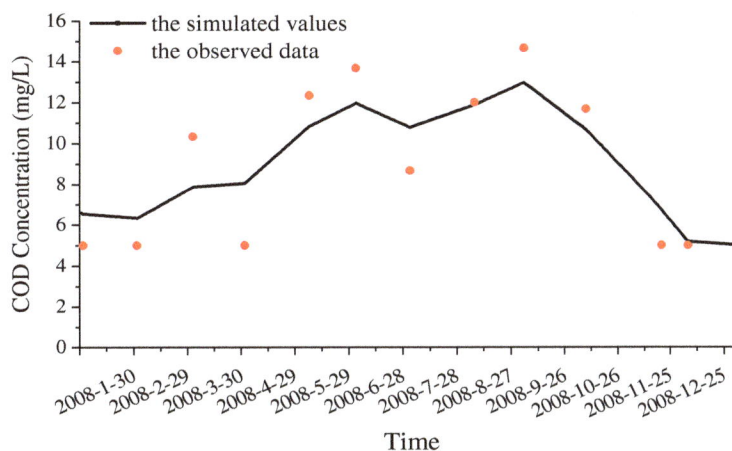

Figure 9 Comparison of simulated and observed COD concentration at Qingxichang monitoring station in 2008.

Table 2 Error analysis between simulated and observed verification parameters

Variable	Mean error	Absolute mean error	Absolute relative mean error
Water surface level (m)	0.081	1.004	0.65%
Velocity (m/s)	0.048	0.058	5.53%
DO (mg/L)	0.073	0.353	4.34%
NH_4^+-N (mg/L)	0.008	0.051	16.51%
COD (mg/L)	0.233	1.537	20.94%

Model verification

To verify the EFDC model, the simulated results were compared with the field water quality data observed in the year of 2008 at the Qingxichang Monitoring Station. The parameters used for model verification include two hydrodynamic variables (i.e., water level and flow velocity) and three water quality variables (i.e., concentrations of DO, NH_4^+-N and COD), all of which are on a daily average basis. The field data were collected on a monthly basis and the daily average was recorded for each monitored parameter.

Figures 5, 6, 7, 8 and 9 present the model verification results in terms of water levels, flow velocity, DO concentration, NH_4^+-N concentration and COD concentration, respectively. The mean error, absolute mean error and absolute relative mean error for each comparison are also calculated and given in Table 2. The model verification results indicate that the simulated and observed hydrodynamic variables are in excellent agreements. For three water quality variables as indicated in Figures 7, 8 and 9 and Table 2, bigger discrepancies were observed, particularly for NH_4^+-N concentration and COD concentration. However, considering the inherent uncertainties and complexities involved in water quality simulation process, this level of discrepancy are still acceptable and are

regarded within the satisfactory range. According to the model verification results, it is believed that the developed EFDC-based model can be applied to the hydrodynamic and water quality simulation for the study reservoir section.

Waste-load allocation optimization modeling

In order to achieve a better environmental performance through optimal waste-load allocation among different outlets, the natural assimilative capacities of the receiving water body near each outlet should be taken into consideration. For example, the field tests show that, the receiving water bodies near the discharge outlets at Changshou Chemical Plant #1, Hengliangzi Drainage, and Fuling Chemical plant, has high capacities in diluting COD by physical advection–diffusion mechanisms and chemical and biological degradation; while the receiving water bodies near the other 8 discharge outlets have higher capacities in diluting the ammonia-nitrogen (NH_4^+-N). For formulating the optimization model, a concept of environmental performance index (EPI) was defined to reflect combined effects of the waste load level from each outlet and the assimilative capacity of the receiving water body near the outlet. Its magnitude represents the relative importance of each outlet in improving water quality with its discharge load being optimally allocated.

The values of the EPI were determined by the AHP (analytic hierarchy process) approach. This approach decomposes the original problem into a hierarchy of more easily comprehended sub-problems, each of which can be analyzed independently. An AHP hierarchy generally consists of an overall goal, a group of options or alternatives for reaching the goal, and a group of factors or criteria that relate the alternatives to the goal. In this study, the overall goal is the overall water quality improvement and performance in the reservoir; the group of alternatives refers to the 11 discharge outlets and their calculated SI

Table 3 The simulated PMZ values and the calculated values of DLR, PMZR and SI for each discharge outlet

Discharge outlet	COD				NH_4–H			
	DLR	PMZ (m²)	PMZR	SI	DLR	PMZ (m²)	PMZR	SI
Sichuan Vinylon	28.38%	70331.72	32.29%	**1.14**	8.34%	8196.08	9.02%	**1.08**
Changshou Chemical #1	7.86%	15560.12	7.14%	**0.91**	14.54%	11215.92	12.34%	**0.85**
Changshou Chemical #2	2.56%	8882.03	4.08%	**1.59**	8.92%	12465.81	13.72%	**1.54**
Changshou Chemical #3	5.59%	17469.15	8.02%	**1.43**	17.87%	26242.91	28.88%	**1.62**
Changshou Drainage	4.89%	13042.31	5.99%	**1.22**	2.64%	2468.27	2.72%	**1.03**
Yudong Oil and Fat	5.00%	13513.16	6.20%	**1.24**	2.16%	1942.11	2.14%	**0.99**
Hengliangzi Drainage	8.38%	19770.31	9.08%	**1.08**	5.41%	5062.12	5.57%	**1.03**
Fuling Chemical Plant	31.72%	43934.76	20.17%	**0.64**	36.92%	20195.29	22.23%	**0.60**
Taiji Pharmaceutical Factory	2.29%	6234.39	2.86%	**1.25**	0.40%	398.97	0.44%	**1.11**
Jindi Industrial Group	2.13%	5812.61	2.67%	**1.25**	1.79%	1946.98	2.14%	**1.19**
Chuandong shipyard	1.20%	3269.58	1.50%	**1.25**	1.00%	728.92	0.80%	**0.80**

Table 4 The derived EPI values for each outlet by the APH approach

Discharge outlet	EPI values
Sichuan Vinylon	0.0843
Changshou Chemical #1	0.1068
Changshou Chemical #2	0.0593
Changshou Chemical #3	0.0581
Changshou Drainage	0.0868
Yudong Oil and Fat	0.0895
Hengliangzi Drainage	0.0885
Fuling Chemical Plant	0.1515
Taiji Pharmaceutical Factory	0.0811
Jindi Industrial Group	0.0766
Chuandong shipyard	0.1075

values (as presented in Table 3). These alternatives are related to the overall goal through two water pollutants, i.e., NH_4^+-N and COD. Once the hierarchy was built, various alternatives were systematically evaluated by comparing them to one another two at a time, with respect to their impact on a factor above them in the hierarchy. In making the comparisons, both concrete data about the alternative and any experience about the alternative' relative importance could be used. The AHP converts these evaluations and comparisons to numerical values that can be processed and compared over the entire range of the problem. In this study, the derived EPI values are presented in Table 4, and they represent a comprehensive and rational ranking for the discharge outlets in terms of their important and weight in contributing to water quality improvement in the study section.

With the EPI being defined, a linear programing model was then formulated to search for the optimal waste-load

allocation schemes. The decision variable is the ratio or percentage of wastewater discharge amount from one specific outlet over the total wastewater amount discharged from all the outlets. For this problem, our goal aims to achieve a best water quality improvement result, and thus, the objective function of the linear model is to maximize the overall environmental performance achieved by optimally allocating the wastewater discharge among all the outlets. The overall environmental performance is mathematically represented by the sum of the product of decision variable and the corresponding EPI for all the outlets. The optimal solutions are searched under a number of constraints that the existing pollution situations could be exceeded. The existing pollution situations were expressed in terms of total wastewater discharge, total pollutants discharge, and existing PMZ plume size. Also, the discharge ratio for each outlet is pre-determined in a certain range. The complete linear waste-load allocation model is presented below:

Objective function:

$$\text{Max } F = \sum_{i=1}^{11} \alpha_i x_i \tag{6a}$$

Subject to:
(1) PMZ constraint for pollutant COD

$$\sum_{i=1}^{11} \left(M\beta_{iCOD}\gamma_{iCOD} \right) x_i \leq S_0\_COD \tag{6b}$$

(2) PMZ constraint for pollutant NH_4^+-N

$$\sum_{i=1}^{11} \left(M\beta_{iNH_4^+-N}\gamma_{iNH_4^+-N} \right) x_i \leq S_0\_NH_4^+-N \tag{6c}$$

(3) Total COD discharge constraint

Table 5 Numeric values of the parameters used in the linear optimization model

Discharge outlet	γ_{iCOD} [m²/(kg/d)]	$\gamma_{iNH_4^+-N}$ [m²/(kg/d)]	β_{iCOD} [kg/m³]	$\beta_{iNH_4^+-N}$ [kg/m³]	Q_COD [kg/d]	Q_NH_4^+-N [kg/d]	Q_WW [m³/d]
Sichuan Vinylon	10.11	25.94	0.11	0.005	6959	316	63260.27
Changshou Chemical #1	8.07	20.36	0.158	0.045	1927	551	12196.16
Changshou Chemical #2	14.14	36.88	0.158	0.085	628	338	3972.6
Changshou Chemical #3	12.73	38.76	0.158	0.078	1372	677	8680.55
Changshou Drainage	10.87	24.68	0.06	0.005	1200	100	20000
Yudong Oil and Fat	11.03	23.68	0.158	0.085	1225	82	7753.42
Hengliangzi Drainage	9.62	24.69	0.05	0.005	2055	205	41095.89
Fuling Chemical Plant	5.65	14.44	0.8	0.126	7778	1399	652.05
Taiji Pharmaceutical Factory	11.09	26.60	0.215	0.028	562	15	1369.86
Jindi Industrial Group	11.14	28.63	0.354	0.064	522	68	21945.21
Chuandong shipyard	11.08	19.18	0.56	0.015	295	38	1002.74

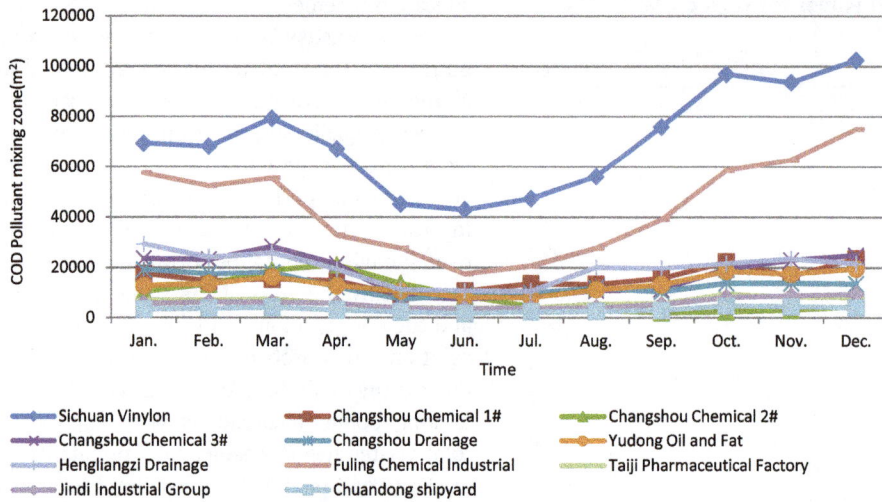

Figure 10 Monthly average size of COD-PMZ formed by each discharge outlet.

$$\sum_{i=1}^{11}\left(M\beta_{iCOD}\right)x_i \leq Q_{0\_COD} \qquad (6d)$$

(4) Total NH_4^+-N discharge constraint

$$\sum_{i=1}^{11}\left(M\beta_{iNH_4^+-N}\right)x_i \leq Q_{0N}H_4^+-N \qquad (6e)$$

(5) Total wastewater discharge constraint

$$\sum_{i=1}^{11}Mx_i \leq Q_{0\_WW} \qquad (6f)$$

(6) Discharge ratio constraints

$$x_i^{min} \leq x_i \leq x_i^{max} \qquad (6g)$$

$$\sum_{i=1}^{11}x_i = 1 \qquad (6h)$$

(7) Technical constraints

$$x_i \geq 0 \qquad (6i)$$

In Model (6), x_i represents the ratio or percentage of wastewater discharge amount over the total amount discharged from all the outlets, and the subscript i represent the 11 discharge outlets along the study section ($i = 1, 2, ..., 11$); x_i^{max} and x_i^{min} denote the upper and lower limits of x_i, respectively; α_i refers to the EPI for each discharge outlet (dimensionless); M is the total wastewater discharge amount from all the outlets (m³/d); β_{iCOD} and $\beta_{iNH_4^+-N}$ are the average COD and NH_4^+-N concentrations

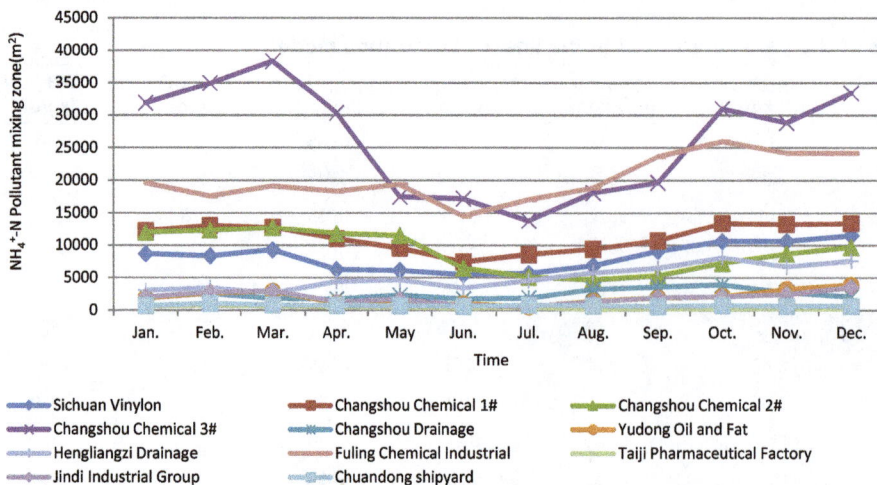

Figure 11 Monthly average size of NH_4^+–H-PMZ formed by each discharge outlet.

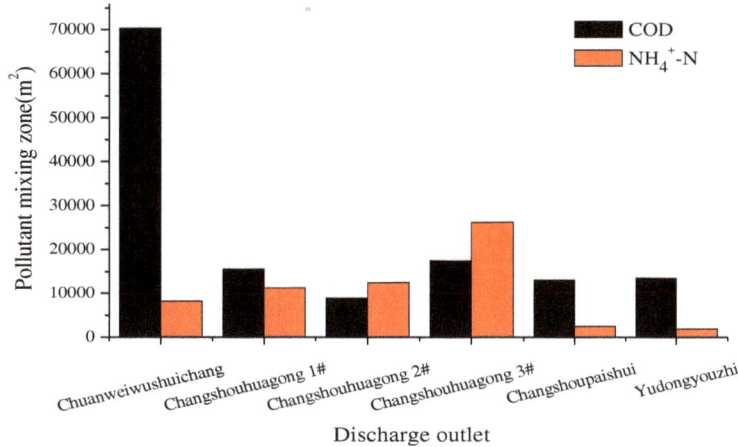

Figure 12 The simulated yearly average size of PMZ formed by 6 outlets in Changshou District.

(in kg/m^3) in the wastewater stream discharged from the ith outlet; γ_{iCOD} and $\gamma_{iNH_4^+-N}$ are the calculated numeric values of the pollutant mixing zone per unit load (in m^2/(kg/d)) for COD and NH$_4^+$-N for the ith outlet, respectively (i.e., γ_{iCOD} = $PMZPL_i$ for COD, and $\gamma_{iNH_4^+-N}$ = $PMZPL_i$ for NH$_4^+$-N); S_0_COD and $S_{0N}H_4^+-N$ represents the existing plume of the pollutant mixing zone (m^2) in the entire study section for COD and NH$_4^+$-N, respectively, calculated by the EFDC model; Q_0_COD and $Q_0\_NH_4^+-N$ represents total amount of pollutants COD and NH$_4^+$-N discharged currently from all the outlets (in kg/d); Q_0_WW is the total wastewater amount discharged currently from all the outlets (in m^3/d). Table 5 gives the numeric values of β_{iCOD}, $\beta_{iNH_4^+-N}$, γ_{iCOD}, and $\gamma_{iNH_4^+-N}$. Table 5 also presents the discharge amount of COD, NH$_4^+$-N and wastewater from each outlet (i.e., Q_i_COD, $Q_i\_NH_4^+-N$, and Q_i_WW, respectively), and sum of each column equals to their total, Q_0_COD, $Q_0\_NH_4^+-N$, and Q_0_WW, respectively.

Results and discussions

The verified EFDC model was run to calculate the spatial and temporal distributions of monthly average COD and NH$_4^+$-N concentrations, which were used to calculate the size of PMZ and the value of PMZPL for each outlet. The monthly average size of PMZ formed by each discharge outlet was calculated and presented in Figures 10 and 11, with Figure 10 for COD and Figure 11 for NH$_4^+$-N. It is apparent that the PMZ size fluctuates from month to month for a specific outlet and is also different from one outlet to another. The calculated PMZ values are the fundamental data for computing other parameters, such as those data presented in Table 5.

The EFDC model was then run to calculate the yearly average size of PMZ for each discharge outlet, and the yearly average PMZPL can then be calculated using Equation (1). They can then be used to calculate DLR and PMZR and SI using Equations (2), (3) and (4). Table 3 gives the simulated PMZ values and the calculated values of DLR, PMZR and SI for each discharge outlet. It is indicated

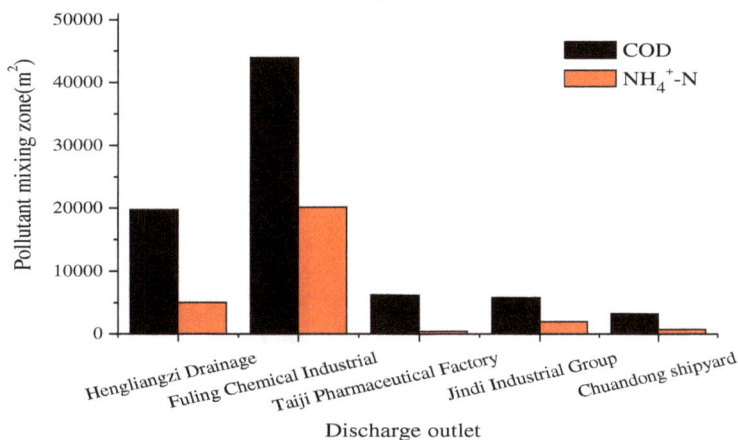

Figure 13 The simulated yearly average size of PMZ formed by 5 outlets in Fuling District.

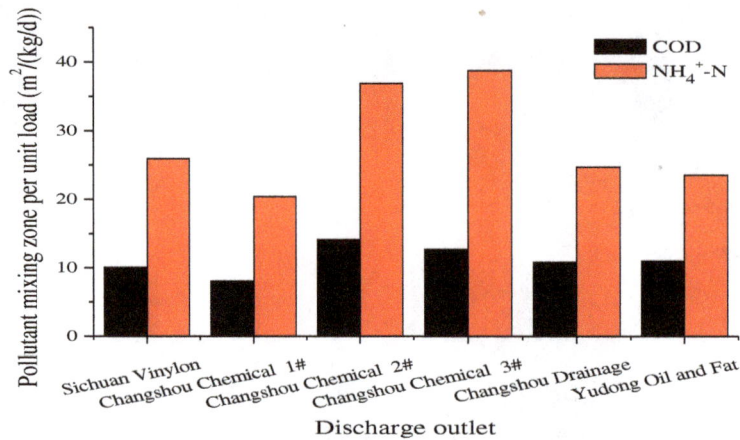

Figure 14 The calculated yearly average of PMZPUL for 6 discharge outlets in Changshou District.

that different discharge outlet has different level of sensitivity index. A higher SI value represents a bigger PMZ area influenced by a specific outlet and a stronger pollution impact from the same outlet. Therefore, the concept of sensitivity index can help rank the significance and influence of a specific discharge outlet on the formation of the pollutant mixing zone in its surrounding water body in comparison to all other outlets. From this sense, a discharge outlet with a high SI has a big potential for improving reservoir water quality if a same level of waste load is cut across the board from all outlets. In other words, this outlet will have a better environmental performance than others if appropriate action is taken for water pollution control purpose.

Based on the data presented in Table 3, the yearly average PMZ size of COD and NH_4^+-N formed by each outlet is plotted as bar graphs in Figures 12 and 13. Figure 12 is the plot for 6 outlets in Changshou District, and Figure 13 is the plot for 5 outlets in Fuling District. In Changshou District, the biggest COD-PMZ was formed near the

outlet of Sichuan Vinylon, with an area of 70331.72 m^2, and it account for 51% of the total COD-PMZ formed in the reservoir section in Changshou District. The biggest NH_4^+-N-PMZ was formed near the outlet of Changshou Chemical #3, with an area of 26242.91 m^2, which accounts for 42% of total NH_4^+-N-PMZ formed in Changshou. In Fuling District, the biggest COD-PMZ and NH_4^+-N-PMZ was both formed near the outlet of Fuling Chemical Industrial, with an area of 43934.76 m^2 and 20195.29 m^2, which accounts for 56% and 71% of the total formed in Fuling, respectively.

Similarly, the calculated yearly average of PMZPL for COD and NH_4^+-N is plotted as bar graphs in Figures 14 and 15, respectively. Figure 14 is the plot for 6 outlets located in Changshou District, and Figure 15 is the plot for 5 outlets located in Fuling District. Some interesting observations could be obtained. For example, the biggest yearly average PMZPL for COD appears near the outlet of Changshou Chemical #2 while that for NH_4^+-N is near Changshou Chemical #3. Both discharge outlets are located

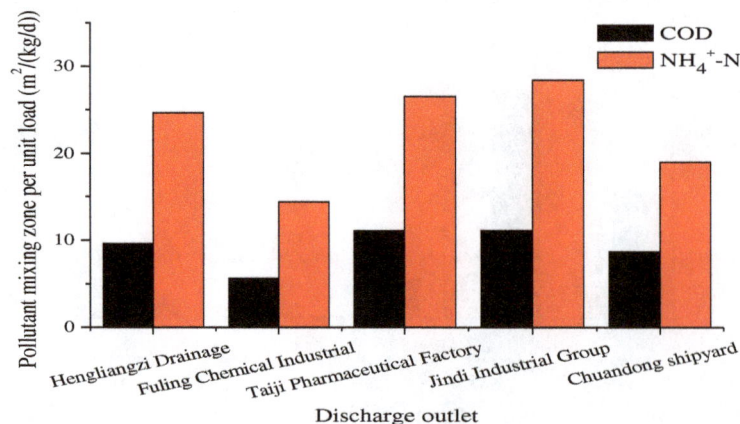

Figure 15 The calculated yearly average of PMZPUL for 5 discharge outlets in Fuling District.

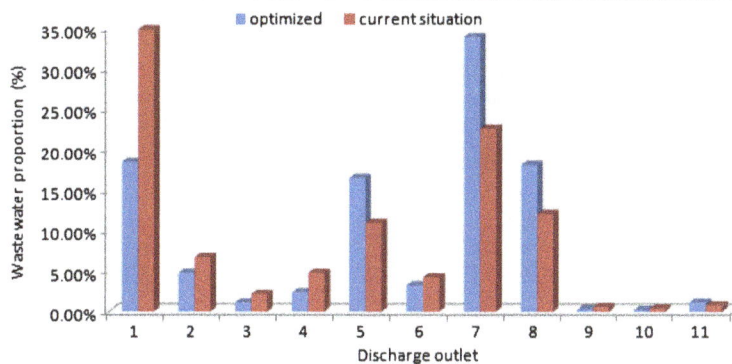

Figure 16 The optimal wastewater allocation result (blue bar graphs) and its comparison with existing discharge load level for each outlet in the study section.

at Longxihe River which is a tributary of the Yangtze River. The return water from the Yangtze River changes the hydraulics of the Longxihe River and might be the main cause of a large PMZ formation. It is also observed from Figure 14 that the PMZPL value of Chemical #2 almost doubles that of Chemical #1. According to the data presented in Tables 1 and 3, the PMZ sizes formed by Changshou Chemical #1 and #2 are very close (11216 m^2 for #1 and 12466 m^2 for #2), however, the NH_4^+-N discharge load from Chemical #1 is 551 kg/d while that from Chemical #2 is only 338 kg/d. There are two possible reasons to explain this: (1) the NH_4^+-N concentration from Chemical #2 is much higher (0.085 kg/m^3) than that from Chemical #1; (2) Chemical #1 is located at the estuary close to mainstream of Yangtze River which make the NH_4^+-N be easily dispersed and diluted.

In this study, the formulated optimization model (6) was solved using Lingo Software. The optimization results give an optimal allocation of wastewater discharge loads among 11 discharge outlets, as presented in Figure 16 and Table 5. A comparison between the existing discharge load level and the optimal load level is also provided. It is indicated that the outlets of Changshou Drainage, Hengliangzi Drainage, and Fuling Chemical Industrial need to increase their discharge load and ratio of the discharge load among the total while the allocation ratios for others need to reduced, for achieving the best environmental performance in terms of reservoir water quality improvement. It is suggested that the management policy and strategy should not only focus on the reduction of industrial wastewater discharge, but also on the increased treatment of domestic wastewater.

Conclusions

In this study, a simulation-optimization assessment framework was developed to help address the water pollution issue in the TGR area, through combining an EFDC-based water quality simulation model and a waste-load allocation optimization model into a general methodology framework.

The developed method was applied to the Changshou-Fuling reservoir section for ratifying the approach and identifying the optimal waste-load allocation schemes among its 11 wastewater discharge outlets. After being verified, the EFDC model was used to simulate the flow hydrodynamics and contaminant transport in the study reservoir section. The simulated COD and NH_4^+-N concentrations were used to calculated the size of pollution mixing zone formed at each discharge outlet. The pollution mixing zone per unit load as well as the sensitivity index for each discharge outlet was also calculated to reflect the connections between the pollution plume and the load strength from each outlet. These parameters were used for developing the waste-load allocation optimization model with an objective of maximized environmental performance under constraints that existing waste discharge loads in terms of total wastewater amount, total pollutant weight, and existing PMZ size formed can't be exceeded. Modeling results give an optimal waste-load allocation ratio for each discharge outlet within the study reservoir section, and its implications to regional water quality management were analyzed. It is anticipated that the methodology developed in this study can be extended to the entire TGR area for better water quality management studies and practices.

Competing interests
The authors declare that they have no competing interests.

Authors' contributions
ZW focused on EFDC model setup, formulation, verification, and water quality simulation, and he was also responsible for the waste load allocation optimization model formulation and implementation. SC and LL were the technical advisors for the project, and they were responsible for designing the methodology framework, drafting the manuscript with the help of ZW, participating in modeling studies, and finalizing the project and manuscript. YC focused mainly on the EFDC modeling studies. XG participated in the calculation of pollution mixing zone and sensitivity index. CQ participated in the system analysis for the study reservoir section. RH, JL and JG were on the field trip to the study section for data collection and analysis, and they were also responsible for the modeling result analysis and interpretation. All authors read and approve the final manuscript.

Acknowledgments

This research was supported by the Major Science and Technology Program for Water Pollution Control and Treatment (No. 2009ZX07104-006), Natural Sciences Foundation of China (No. 51178005), and NSERC. The authors are grateful to the editors and the anonymous reviewers for their insightful comments.

Author details

[1]College of Environmental & Energy Engineering, Beijing University of Technology, Beijing 100022 China. [2]Department of Civil and Resource Engineering, Dalhousie University, Halifax, NS B3H 4R2 Canada. [3]College of Architecture & Civil Engineering, Beijing University of Technology, Beijing 100022 China. [4]Department of Water Environment, China Institute of Water Resources and Hydropower Research, Beijing 100038 China. [5]China Institute of Water Resources and Hydropower Research, Beijing 100038 China.

References

Blumberg AF, Mellor G (1987) A description of a three dimensional ocean circulation. In: Heaps NS (ed) Three-dimensional coastal ocean models. Coastal and estuarine science, vol 4. American Geophysical Union, Washington, DC, pp 1–19

Cai ZG, Zhang AD, Zhang J (2010) Optimized control on water level between Three Gorges dam and Gezhouba dam and its benefit analysis. J Hydroelectric Eng 29(3):11–17

Chapra S, Pelletier G, Tao H (2007) QUAL2K: A modeling framework for simulating river and stream water quality, Version 2.07: Documentation and user's manual. Civil and Environmental Engineering Department, Tufts University, Medford, Massachusetts, USA

Chapra SC (2003) Engineering water quality models and TMDLs. J Water Res Pl-ASCE 129(4):247–256

Chen HW, Yu RF, Liaw SL, Huang WC (2010) Information policy and management framework for environmental protection organization with ecosystem conception. Int J Environ Sci Te 7(2):313–326

Cho JH, Ahn KH, Chung WJ, Gwon EM (2003) Waste load allocation for water quality management of a heavily polluted river using linear programming. Water Sci Technol 48(10):185–190

CWRC (Changjiang Water Resource Commission) (1997) Study on eco-environmental impacts of Three Gorges project. Hubei Science and Technology Press, Wuhan

DHI (2003) MIKE 11: A modeling system for rivers and channels: short introduction tutorial. DHI Software, Horsholm, Denmark

Guo SL, Chen JH, Li Y (2011) Joint operation of the multi-reservoir system of the Three Gorges and the Qingjiang cascade reservoirs. Energies 4(7):1036–1050

Hamrick JM (1996) A user's manual for the environmental fluid dynamics computer code (EFDC). The College of William and Mary. Virginia Institute of Marine Science, Special Report

Huang C (2006) An one-dimension water body eutrophication simulation model for Chongqing section of the Three Gorges Reservoir. Dissertation, Southwest University, Chongqing, China

Igbinosa EO, Okoh AI (2009) Impact of discharge wastewater effluents on the physico-chemical qualities of a receiving watershed in a typical rural community. Int J Environ Sci Te 6(2):175–182

Igwe JC, Abia AA, Ibeh CA (2008) Adsorption kinetics and intraparticulate diffusivities of Hg, As and Pb ions on unmodified and thiolated coconut fiber. Int J Environ Sci Te 5(1):83–92

Jiang CB, Zhang LM, Chen LQ (2005) Parallel numerical simulation of pollution mixing zones in Fuling reach of Three Gorges Reservoir. J Hydroelectric Eng 24:82–87

Li YP, Huang GH, Nie SL, Chen X (2011) A robust modeling approach for regional water management under multiple uncertainties. Agric Water Manage 98(10):1577–1588

Yibing LV, Zhengyu G, Jun L (2007) Status of Water Quality in the Three Gorge s after the Water Storage Period. Res Env Sci 20(1):1–6

Mujumdar PP, Vemula VRS (2004) Fuzzy Waste Load Allocation Model: Simulation-Optimization Approach. J Comput Civil Eng 18(2):120–131

MWRC (Ministry Of Water Resources of China) (2008) Hydrology annals of the People's Republic of China: hydrological data of Yangtze basin, vol 6., pp 9–355

Parka K, Jungb HS, Kimb HS, Ahnc SM (2005) Three dimensional hydrodynamic eutrophication model (HEM-3D): Application to Kwang-Yang Bay, Korea. Mar Environ Res 60(2):171–193

Qin X, Xu Y (2011) Analyzing urban water supply through an acceptability-index-based interval approach. Adv Water Resour 34(7):873–886

Shah BA, Shah AV, Singh RR (2009) Sorption isotherms and kinetics of chromium uptake from wastewater using natural sorbent material. Int J Environ Sci Te 6(1):83–92

Sun YT, Lv WG (2010) Optimization of technical water supply system in three gorges power station. East China Elect Power 38(8):1188–1191

Tetra Tech I (2007) The environmental fluid dynamics code user' manual, US EPA Version 1 .01. Tetra Tech Inc, Fairfax, Virginia, USA

Wang XP, Ruan Q (2011) Optimization model for ship lock scheduling plans using ant colony algorithm. J Huazhong Univ Sci Tech 39(8):100–103

Wool TA, Ambrose RB, Martin JL, Comer EA (2001) Water quality analysis simulation program (WASP) Version 6.0, User' s Manual. US EPA, Region 4, Atlanta, USA

Wu GZ, Xu ZX (2011) Prediction of algal blooming using EFDC model: Case study in the Daoxiang Lake. Ecol Model 222(6):1245–1252

Xie R, Wu DA, Yan YX (2010) Fine silt particle pathline of dredging sediment in the Yangtze River deepwater navigation channel based on EFDC model. J Hydrodyn B22(6):760–772

Yang CC, Chen CS, Lee CS (2011) Comprehensive river water quality management by simulation and optimization models. Environ Model Assess 16(3):283–294

Zhao X, Shen ZY, Xiong M (2011) Key uncertainty sources analysis of water quality model using the first order error method. Int J Environ Sci Te 8(1):137–148

Zuo SH (2007) Introduction of numerical simulation software Dleft3D and its application on offshore area of Aojiang Estuarine. J China Hydrol 27(6):55–58

15

A 2-D Barakat-Clark finite difference numerical method and its comparison for water quality simulation modeling

Lei Jin[1*], Aili Yang[1], Li Wang[2] and Haiyan Fu[1]

Abstract

Background: This paper provides an advance numerical algorithm to solve both ordinary and partial differential equations of surface water quality models. It uses finite difference methods and structures explicit or implicit or other process forms to solve the water quality model. This study also considers the stability of solutions to obtain more accurate results among those numerical algorithms.

Results: Water quality modeling commonly manifests itself in ordinary and partial differential equations in a realistic world. This study has applied numerical solutions to simulate the changing process of water quality in one and two dimensional spaces or in multiple dimensional spaces. The solutions of these analytical methods are provided in this paper to attest the justifiability of these numerical methods. It demonstrates that the 2-dimensional Barakat-Clark numerical method can be a highly efficient tool in obtaining approximate results of ordinary and partial differential equations, which may prove difficult in finding the accurate solution by using conventional methods. At the same time, the stability analysis corroborated the convergence of those numerical solutions.

Conclusions: This study is the first attempt to compare the multiple numerical methods with the 2-D Barakat-Clark method in the water quality modeling process. The results clearly suggest that the Barakat-Clark method is better in reflecting the accuracy of the water quality modeling with stability for hydrological systems.

Keywords: Water quality modeling; Numerical algorithm; 2-D barakat-clark method; Stability; Finite difference

Background

In the past few decades, the applications of surface water quality models have been intensely studied as population growth and economic development makes more increases pollution in water resources, particularly in the surface hydrological system. Streeter and Phelps have been progressed the water quality models in 1925. Since then, many scientists (Kellogg 1964; Shampine et al. 1979; Stasa 1985; Christie 1987; Hoffman 1992; Rudi et al. 1997; Cash 2003; Shawgfeh 2004; Walter 2006) have carried out rigorous research in this area. Many water quality models, for example, the BOD model, are formed generally by ordinary or partial differential equations (Davis 1962; Na 1979; Taylor 1982; Evans 1985; Noye Noye 1992; Richard 1997; Moiianty 2004; Liu et al. 2005; Munavalli 2004; Timo et al. 2013). For

solving those equations, the traditional methods are using the finite difference method and the finite element method.

In this paper, the application of an explicit and an implicit method to solve the water quality equation includes and tests the first order ordinary differential equation, the second order ordinary differential equation and the second order partial differential equation. To illustrate the reasonability of these solutions, the stability of diffusion equations are also provided in this article. The analytic solutions obtained from these diffusion equations are also compared with other numerical methods. The results suggest that the numerical method can be as good to reflect the accurate solution of water quality diffusion equation in the real world condition.

Methods

Fundamental background of water quality modelling

The transference, diffusion, and degradation mathematical equations of pollutants in the hydrological systems are

---

* Correspondence: jinlei777@gmail.com
[1]College of Environmental Science and Engineering, Xiamen University of Technology, Xiamen, Fujian Province 361024, China
Full list of author information is available at the end of the article

three-dimensional in form. Therefore, according to the mass balance equation, the initial water quality transfer equation is as follows:

$$\frac{\partial C}{\partial t} + \sum_i V_i \frac{\partial C}{\partial X_i} = \sum_i \frac{\partial}{\partial X_i}\left(\varepsilon_{T,i}\frac{\partial C}{\partial X_i}\right) \pm \sum_k S_k$$

(1a)

where

X_i = the three axial X_1, X_2, X_3;
V_i = the velocity of three axial V_1, V_2, V_3;
S_k = the source and sink of contaminants;
C = the concentrate of contaminants;
t = time;

The C, V_i and $\varepsilon_{T,i}$ are the function of X_i. In the left side of the equation (1a), the first item indicates the change rate of concentration of contaminants. The second item on the left side shows the transfer rate of contaminants along the stream axial. The first item, in the right hand side of the equation, presents the dispersion of pollutants along a different axial. There, we are assuming that there is no molecular diffusion and turbulent diffusion of pollutants. The second item in left side indicates the sources of contaminants. The symbol +/– states the contaminants increase or decrease in the hydrological systems. In (1a) is a three-dimensional variable coefficient of the second-order Partial Differential Equation (PDE). The parameters of modeling require much information; especially the variable S on the right side of equation, will cause a serious complication in the solution process. It is difficult to get the analytic a solution. Therefore, people usually simplified this modeling process then used the numerical method to generate the approximate solutions. This paper targets several simplified water quality models and uses finite difference method and structures simulate process forms to solve the problem. It also considers the stability of the solutions to obtain more accurate results among those numerical algorithms.

Numerical methods for water quality modeling
Application of least squares method
If the velocity of the stream is very high, the variable of dispersion will be extremely low, so it can be ignored. Thus, the water shift modeling is presented as follows:

$$V\frac{dC}{dX} = kC$$

(2a)

The equation (2a) is a first order ordinary differential equation. The analytic solution of it can be obtained as follow:

$$C = C_0 e^{\frac{k}{V}x}$$

(3a)

Some of the parameters in (3a) can be solved by multiple linear regression method. Concurrently it can use the least squares method to establish the error function. To consider the amount of biochemical oxygen demand (BOD) and disintegration, we use L to indicate BOD, and K_1 present disintegration coefficient of BOD. Then, equation (2a) can is presented as:

$$-\frac{dL}{dt} = K_1 L$$

(4a)

The solution of the above equation is $L(t) = L_0 e^{-K_1 t}$. Therefore the oxygen consumption is $y(t) = L_0 - L(t) = L_0\left(1 - e^{-k_1 t}\right)$. By using experimental data, applying the least squares method estimates the initial value of L_0 and K_1. There let $K_1 = K'_1 + h$ then,

$$y(t) = L_0\left[1 - e^{-(K'_1 + h)t}\right] = L_0\left(1 - e^{-K'_1 t}e^{-ht}\right)$$

(5a)

let $y(t) = af_1 + bf_2$, and $a = L_0, b = L_0 h, f_1 = 1 - e^{-K'_1 t}, f_2 = te^{-K'_1 t}$. According to the least squares method to minimize $\sum_i\left[y(t) - y(t)_{testing}\right]^2 = \sum_i\left[af_1 + bf_2 - y(t)_{testing}\right]^2$ then, we can conclusion those coefficient a and b. The a and b can be indicated as following,

$$a = \frac{\sum f_2^2 \sum f_1 y(t) - \sum f_1 f_2 \sum f_2 y(t)}{\sum f_1^2 \sum f_2^2 - \left(\sum f_1 \sum f_2\right)^2},$$

$$b = \frac{\sum f_1^2 \sum f_2 y(t) - \sum f_1 f_2 \sum f_1 y(t)}{\sum f_1^2 \sum f_2^2 - \left(\sum f_1 \sum f_2\right)^2}$$

(6a)

The L_0 and step-size h can be obtained by the value of a and b. If the h in the equation is larger than 0.001, the result are not stability. Otherwise, the solution is reliable.

Application of explicit and implicit methods
One-dimensional water quality modeling When the geometry of stream varies greatly, for example the river is very long but it is very shallow; the one-dimensional water quality modeling can be expressed as follows:

$$E\frac{\partial^2 C}{\partial X^2} - V\frac{\partial C}{\partial X} + kC = \frac{\partial C}{\partial t}$$

(7a)

Obviously, the function (8a) is a one-dimensional second-order Partial Differential Equation (PDE). It is solved by explicit and implicit methods as following.

Solution of explicit and implicit methods The equation (8a) can be represented by finite difference. Those equations are expressed as follows:

$$\frac{\partial^2 C}{\partial X^2} = \frac{C_{i+1}^l - 2C_i^l + C_{i-1}^l}{\Delta X^2}$$

(8a)

$$\frac{\partial C}{\partial X} = \frac{C_{i+1}^l - C_{i-1}^l}{2\Delta X} \tag{8b}$$

$$\frac{\partial C}{\partial t} = \frac{C_i^{l+1} - C_i^l}{\Delta t} \tag{8c}$$

then the explicit method can solve them as follows,

$$C_i^{l+1} = \left(E\frac{\Delta t}{\Delta X^2} + V\frac{\Delta t}{2\Delta X} \right) C_{i-1}^l + \left(k\Delta t - 2E\frac{\Delta t}{\Delta X^2} + 1 \right) C_i^l$$
$$+ \left(E\frac{\Delta t}{\Delta X^2} - V\frac{\Delta t}{2\Delta X} \right) C_{i+1}^l \tag{9a}$$

the solution of implicit methods can be shown as:

$$-\left(E\frac{\Delta t}{\Delta X^2} + V\frac{\Delta t}{2\Delta X} \right) C_{i-1}^{l+1} + \left(2E\frac{\Delta t}{\Delta X^2} + 1 - k\Delta t \right) C_i^{l+1}$$
$$-\left(E\frac{\Delta t}{\Delta X^2} - V\frac{\Delta t}{2\Delta X} \right) C_{i+1}^{l+1} = C_i^l \tag{10a}$$

Stability Analysis of explicit and implicit methods

In this section the stability of the equations, (10a) and (11a), are examined. The stability of those methods are tested by using the method of the Fourier stability analysis also known as the von Neumann stability analysis. Let $C_j^l = u^l e^{ij\theta}$, and then the equation (10a) can be written as:

$$u^{l+1} e^{ij\theta} = \left(E\frac{\Delta t}{\Delta X^2} + V\frac{\Delta t}{2\Delta X} \right) u^l e^{i(j-1)\theta}$$
$$+ \left(k\Delta t - 2E\frac{\Delta t}{\Delta X^2} + 1 \right) u^l e^{ij\theta}$$
$$+ \left(E\frac{\Delta t}{\Delta X^2} - V\frac{\Delta t}{2\Delta X} \right) u^l e^{i(j+1)\theta} \tag{11a}$$

The amplification factor λ equal to:

$$\lambda = \left(E\frac{\Delta t}{\Delta X^2} + V\frac{\Delta t}{2\Delta X} \right) (\cos\theta - i\sin\theta)$$
$$+ \left(k\Delta t - 2E\frac{\Delta t}{\Delta X^2} + 1 \right)$$
$$+ \left(E\frac{\Delta t}{\Delta X^2} - V\frac{\Delta t}{2\Delta X} \right) (\cos\theta + i\sin\theta)$$
$$= -2E\frac{\Delta t}{\Delta X^2}(1 - \cos\theta) - V\frac{\Delta t}{\Delta X} i\sin\theta + k\Delta t + 1 \tag{12a}$$

subject to

$$|\lambda|^2 = \left[k\Delta t + 1 - 2E\frac{\Delta t}{\Delta X^2}(1 - \cos\theta) \right]^2 + \left(V\frac{\Delta t}{\Delta X}\sin\theta \right)^2 \tag{13a}$$

Let $s = 1 - \cos\theta$, $s \in [0, 2]$ then the equation (13b) can be re-written as:

$$F(s) = 4E^2\frac{\Delta t^2}{\Delta X^2}s^2 - 2E\frac{\Delta t}{\Delta X^2}(k\Delta t + 1)s + k^2\Delta t^2$$
$$+ 2k\Delta t + V^2\frac{\Delta t^2}{\Delta X^2}\sin^2\theta \tag{14a}$$

Therefore

$$F'(s) = 8E^2\frac{\Delta t^2}{\Delta X^2}s - 2E\frac{\Delta t}{\Delta X^2}(k\Delta t + 1) \tag{15a}$$

$$F''(s) = 8E^2\frac{\Delta t^2}{\Delta X^2} \geq 0 \tag{16a}$$

$$F(s) \leq 0 \Leftarrow \begin{cases} F''(s) \geq 0 \\ F(0) \leq 0 \\ F(2) \leq 0 \end{cases} \Leftrightarrow \begin{cases} k^2\Delta t^2 - 2k\Delta t + V^2\frac{\Delta t^2}{\Delta X^2}\sin^2\theta \leq 0 \\ 16E^2\frac{\Delta t^2}{\Delta X^2} - 4E\frac{\Delta t}{\Delta X^2}(k\Delta t + 1) + k^2\Delta t^2 \\ + 2k\Delta t + V^2\frac{\Delta t^2}{\Delta X^2}\sin^2\theta \leq 0 \end{cases} \tag{17a}$$

The solution of the explicit is unstable.

On the other hand, the implicit equation (11a) is also tested using the this Fourier analysis as follows. Let $C_j^l = u^l e^{ij\theta}$, the equation can be re-written as:

$$-\left(E\frac{\Delta t}{\Delta X^2} + V\frac{\Delta t}{2\Delta X} \right) u^{l+1} e^{i(j-1)\theta} + \left(2E\frac{\Delta t}{\Delta X^2} + 1 - k\Delta t \right) u^{l+1} e^{ij\theta}$$
$$-\left(E\frac{\Delta t}{\Delta X^2} - V\frac{\Delta t}{2\Delta X} \right) u^{l+1} e^{i(j+1)\theta} = u^l e^{ij\theta}$$

the amplification factor λ is equal to:

$$\lambda = 1 / \left[-\left(E\frac{\Delta t}{\Delta X^2} + V\frac{\Delta t}{2\Delta X} \right)(\cos\theta - i\sin\theta) \right.$$
$$\left. + \left(2E\frac{\Delta t}{\Delta X^2} + 1 - k\Delta t \right) - \left(E\frac{\Delta t}{\Delta X^2} - V\frac{\Delta t}{2\Delta X} \right)(\cos\theta + i\sin\theta) \right]$$
$$= 1 / \left[2E\frac{\Delta t}{\Delta X^2}(1 - \cos\theta) + V\frac{\Delta t}{\Delta X} i\sin\theta - k\Delta t + 1 \right] \tag{18a}$$

The results show $|\lambda|^2 \leq 1$. It indicates that the solution of the implicit method is unconditionally stable. Therefore, through analyzing the stability of equation (12a) and (13a), it is can conclude that the solution of the implicit method will be more accuracate than explicit method in the water quality diffusion equation.

Application the explicit, ADI and Barakat-Clark methods

Two-dimensional water quality modeling The two-dimensional water quality modeling is represented as follows:

$$\frac{\partial C}{\partial t} + V_1\frac{\partial C}{\partial X} + V_2\frac{\partial C}{\partial Y} = E_1\frac{\partial^2 C}{\partial X^2} + E_2\frac{\partial^2 C}{\partial Y^2} - kC \tag{19a}$$

As it shows, this model is a second-order Partial Differential Equation (PDE). It will be solved using explicit, ADI and Barakat-Clark methods. The finite difference equation can substituted in the PDE as follows:

$$\frac{C_{i,j}^{l+1}-C_{i,j}^{l}}{\Delta t/2} + V_1 \frac{C_{i+1,j}^{l}-C_{i-1,j}^{l}}{2\Delta X} + V_2 \frac{C_{i,j+1}^{l}-C_{i,j-1}^{l}}{2\Delta Y}$$
$$= E_1 \frac{C_{i+1,j}^{l}-2C_{i,j}^{l}+C_{i-1,j}^{l}}{(\Delta X)^2} + E_2 \frac{C_{i,j+1}^{l}-2C_{i,j}^{l}+C_{i,j-1}^{l}}{(\Delta Y)^2} - kC_{i,j}^{l}$$

$$(20a)$$

Solution of explicit, ADI and Barakat-Clark methods
Firstly, equation (20a) is solved by the explicit method. While ΔX is equal to ΔY, the equation (20a) can be re-written as:

$$C_{i,j}^{l} = \left(E_1 \frac{\Delta t}{2\Delta X^2} + V_1 \frac{\Delta t}{4\Delta X}\right)C_{i-1,j}^{l}$$
$$+ \left(1 - E_1 \frac{\Delta t}{2\Delta X^2} - E_2 \frac{\Delta t}{2\Delta X^2} - \frac{k\Delta t}{2}\right)C_{i,j}^{l}$$
$$+ \left(E_1 \frac{\Delta t}{2\Delta X^2} - V_1 \frac{\Delta t}{4\Delta X}\right)C_{i+1,j}^{l} \qquad (21a)$$
$$+ \left(E_2 \frac{\Delta t}{2\Delta X^2} + V_2 \frac{\Delta t}{4\Delta X}\right)C_{i,j-1}^{l}$$
$$+ \left(E_2 \frac{\Delta t}{2\Delta X^2} - V_2 \frac{\Delta t}{4\Delta X}\right)C_{i,j+1}^{l}$$

Secondly, equation (20a) is solved by the ADI method, a two stage time method. At this point, each time step size will be divided into two steps for calculation. The first stage uses a half of step from t^l to $t^{l+1/2}$ for calculating the result; thus the equation (21a) can be written as:

$$-\left(V_2 \frac{\Delta t}{4\Delta X} + E_2 \frac{\Delta t}{2\Delta X^2}\right)C_{i,j-1}^{l+1/2} + \left(1 + E_2 \frac{\Delta t}{\Delta X^2}\right)C_{i,j}^{l+1/2}$$
$$+ \left(V_2 \frac{\Delta t}{4\Delta X} - E_2 \frac{\Delta t}{2\Delta X^2}\right)C_{i,j+1}^{l+1/2} = \left(V_1 \frac{\Delta t}{4\Delta X} + E_1 \frac{\Delta t}{2\Delta X^2}\right)C_{i-1,j}^{l}$$
$$+ \left(1 - E_1 \frac{\Delta t}{\Delta X^2} - \frac{k\Delta t}{2}\right)C_{i,j}^{l} + \left(E_1 \frac{\Delta t}{2\Delta X^2} - V_1 \frac{\Delta t}{4\Delta X}\right)C_{i+1,j}^{l}$$

$$(22a)$$

the second step use anther step size from $t^{l+1/2}$ to t^l for calculating; the equation (21a) can be written as,

$$\frac{C_{i,j}^{l+1}-C_{i,j}^{l+1/2}}{\Delta t/2} + V_1 \frac{C_{i+1,j}^{l+1}-C_{i-1,j}^{l+1}}{2\Delta X} + V_2 \frac{C_{i,j+1}^{l+1/2}-C_{i,j-1}^{l+1/2}}{2\Delta Y}$$
$$= E_1 \frac{C_{i+1,j}^{l+1}-2C_{i,j}^{l+1}+C_{i-1,j}^{l+1}}{(\Delta X)^2}$$
$$+ E_2 \frac{C_{i,j+1}^{l+1/2}-2C_{i,j}^{l+1/2}+C_{i,j-1}^{l+1/2}}{(\Delta Y)^2} - kC_{i,j}^{l+1}$$

$$(23a)$$

While ΔX is equal to ΔY, the equation (23a) can be re-written as,

$$-\left(E_1 \frac{\Delta t}{2\Delta X^2} + V_1 \frac{\Delta t}{4\Delta X}\right)C_{i-1,j}^{l+1} + \left(1 + E_1 \frac{\Delta t}{\Delta X^2} + \frac{k\Delta t}{2}\right)C_{i,j}^{l+1}$$
$$-\left(E_1 \frac{\Delta t}{2\Delta X^2} - V_1 \frac{\Delta t}{4\Delta X}\right)C_{i+1,j}^{l+1} = \left(E_2 \frac{\Delta t}{2\Delta X^2} + V_2 \frac{\Delta t}{4\Delta X}\right)C_{i,j-1}^{l+1/2}$$
$$+ \left(1 - E_2 \frac{\Delta t}{\Delta X^2}\right)C_{i,j}^{l+1/2} + \left(E_2 \frac{\Delta t}{2\Delta X^2} - V_2 \frac{\Delta t}{4\Delta X}\right)C_{i,j+1}^{l+1/2}$$

$$(24a)$$

Lastly, equation (20a) is solved using the Barakat-Clark method. According the H. Z. Barakat, the solution of (19a) can be written as (25a) and (25b),

$$\frac{u_{i,j}^{n+1}-u_{i,j}^{n}}{\Delta t} + V_1 \frac{u_{i+1,j}^{n}-u_{i,j}^{n}}{\Delta X} + V_2 \frac{u_{i,j+1}^{n}-u_{i,j}^{n}}{\Delta Y}$$
$$= E_1 \frac{u_{i+1,j}^{n}-u_{i,j}^{n}-u_{i,j}^{n+1}+u_{i-1,j}^{n+1}}{\Delta X^2}$$
$$+ E_2 \frac{u_{i,j+1}^{n}-u_{i,j}^{n}-u_{i,j}^{n+1}+u_{i,j-1}^{n+1}}{\Delta Y^2} - ku_{i,j}^{n}$$

$$(25a)$$

$$\frac{v_{i,j}^{n+1}-v_{i,j}^{n}}{\Delta t} + V_1 \frac{v_{i,j}^{n}-v_{i-1,j}^{n}}{\Delta X} + V_2 \frac{v_{i,j}^{n}-v_{i,j-1}^{n}}{\Delta Y}$$
$$= E_1 \frac{v_{i+1,j}^{n+1}-v_{i,j}^{n+1}-v_{i,j}^{n}+v_{i-1,j}^{n}}{\Delta X^2}$$
$$+ E_2 \frac{v_{i,j+1}^{n+1}-v_{i,j}^{n+1}-v_{i,j}^{n}+v_{i,j-1}^{n}}{\Delta Y^2} - kv_{i,j}^{n}$$

$$(25b)$$

While ΔX is equal to ΔY, the equations (25a) and (25b) can be re-written as,

$$u_{i,j}^{n+1} = \left[\left((V_1+V_2)\frac{\Delta t}{\Delta X} - (E_1+E_2)\frac{\Delta t}{\Delta X^2} - k\Delta t + 1\right)u_{i,j}^{n}\right.$$
$$+ \left(E_1 \frac{\Delta t}{\Delta X^2} - V_1 \frac{\Delta t}{\Delta X}\right)u_{i+1,j}^{n} + \left(E_2 \frac{\Delta t}{\Delta X^2} - V_2 \frac{\Delta t}{\Delta X}\right)u_{i,j+1}^{n}$$
$$\left. + E_1 \frac{\Delta t}{\Delta X^2}u_{i-1,j}^{n+1} + E_2 \frac{\Delta t}{\Delta X^2}u_{i,j-1}^{n+1}\right] / \left[1 + (E+E_2)\frac{\Delta t}{\Delta X^2}\right]$$

$$(26a)$$

$$v_{i,j}^{n+1} = \left[\left(-(V_1+V_2)\frac{\Delta t}{\Delta X} - (E_1+E_2)\frac{\Delta t}{\Delta X^2} - k\Delta t + 1\right)v_{i,j}^{n}\right.$$
$$+ \left(E_1 \frac{\Delta t}{\Delta X^2} - V_1 \frac{\Delta t}{\Delta X}\right)v_{i-1,j}^{n} + \left(E_2 \frac{\Delta t}{\Delta X^2} - V_2 \frac{\Delta t}{\Delta X}\right)v_{i,j-1}^{n}$$
$$\left. + E_1 \frac{\Delta t}{\Delta X^2}v_{i+1,j}^{n+1} + E_2 \frac{\Delta t}{\Delta X^2}v_{i,j+1}^{n+1}\right] / \left[1 + (E_1+E_2)\frac{\Delta t}{\Delta X^2}\right]$$

$$(26b)$$

The concentration $C_{i,j}$ of equation (19a) at any time level $(n + 1)$ may be given as,

$$C_{i,j}^{n+1} = \left(u_{i,j}^{n+1} + v_{i,j}^{n+1}\right)/2 \qquad (27a)$$

Stability analysis of the explicit, ADI and Barakat-Clark methods

According to the above Fourier stability analysis, the solution of the explicit method (21a) is unstable. For a solution using the ADI method, let $C_j^l = u^l e^{ij\theta}$ and substitute it into the equation (24a). This results in:

$$
\begin{aligned}
p = &-\left(V_2 \frac{\Delta t}{4\Delta X} + E_2 \frac{\Delta t}{2\Delta X^2}\right)(\cos\theta - i\sin\theta) + \left(1 + E_2 \frac{\Delta t}{\Delta X^2}\right) \\
&+\left(V_2 \frac{\Delta t}{4\Delta X} - E_2 \frac{\Delta t}{2\Delta X^2}\right)(\cos\theta + i\sin\theta) \\
= &\ E_2 \frac{\Delta t}{(\Delta X)^2}(1 - \cos\theta) + 1 + V_2 \frac{\Delta t}{2\Delta X} i\sin\theta
\end{aligned}
$$

(28a)

$$
\begin{aligned}
q = &\left(E_1 \frac{\Delta t}{2\Delta X^2} + V_1 \frac{\Delta t}{4\Delta X}\right)(\cos\theta - i\sin\theta) + \left(1 - E_1 \frac{\Delta t}{\Delta X^2} - \frac{k\Delta t}{2}\right) \\
&+\left(E_1 \frac{\Delta t}{2\Delta X^2} - V_1 \frac{\Delta t}{4\Delta X}\right)(\cos\theta + i\sin\theta) \\
= &-E_1 \frac{\Delta t}{(\Delta X)^2}(1 - \cos\theta) - V_1 \frac{\Delta t}{2\Delta X} i\sin\theta + 1 - \frac{k\Delta t}{2}
\end{aligned}
$$

(28b)

When the amplification factor λ IS equal to $\lambda = \frac{q}{p}$, $s = 1 - \cos\theta, \theta \in [0, 2]$ the equations of (28a) and (28b) can be written as:

$$
p = E_2 \frac{\Delta t}{\Delta X^2} s + 1 + V_2 \frac{\Delta t}{2\Delta X} i\sin\theta
$$

(30a)

$$
q = -E_1 \frac{\Delta t}{2\Delta X^2} s - V_1 \frac{\Delta t}{2\Delta X} i\sin\theta + 1 - \frac{k\Delta t}{2}
$$

(30b)

By considering $|q|^2 \le |p|^2$, it is concluded that the solution of the ADI method is stable. The solution stability of the Barakat-Clark method is unconditionally stable because it has been proven by J. A. Clark and H. Z. Barakat (1966).

A case study

The experimental data of 10 groups are showed in the following table for the least squares method. It considers the process which is mentioned in equation (6a) It describes the least squares method to estimate the initial value of BOD and other parameters (Table 1).

Here it assumes that the $h = 0.001$ and $K'_1 = 0.1$. The results show $K_1 = 0.3166$, $L_0 = 175.5$ by going through the process. The L_0 value is the initial value of BOD. Therefore, the water quality model is $L(t) = 175.5e^{-0.3166t}$. It is a one-dimensional BOD model.

Table 1 Experimental data of water samples

Time t (day)	1	2	3	4	5	6	7	8	9	10
BOD y (t) (mg/L)	50	85	107	125	138	148	155	161	167	170

Now, let's use those other numerical methods including: explicit, implicit, ADI and Barakat-clark to simulate the degradation of contaminants in a section of stream systems. Under natural conditions, the initial value of concentration of pollutants is assumed to be 0 in this stream section. The inflow from the upstream concentration of contaminants is 150 mg/L. At this point, the equation (8a) is examined. This river section, it is 1000 m long. The water velocity is 17 m/s. The diffusion rate of pollutants E is 360. The degradation rate of the pollutant k is 0.15 in this stream. There is a water quality monitor state in every 200 meters. At the same time, the concentration of pollution is consider to be 0 mg/L at the end of this river, the data is required by the downstream management agencies. Under these conditions, the Department of Environmental Protection wants to know the background concentration of contaminants in this river in order to satisfy the requirement of those downstream cities. Therefore, by using simulation methods, the concentration of pollutants is estimated using the following methods.

The simulation of explicit method

The water quality modeling can be re-written as a standard form of the explicit method.

$$
\begin{aligned}
C_i^{l+1} = &\left(E\frac{\Delta t}{\Delta X^2} + V\frac{\Delta t}{2\Delta X}\right)C_{i-1}^l \\
&+\left(-k\Delta t - 2E\frac{\Delta t}{\Delta X^2} + 1\right)C_i^l \\
&+\left(E\frac{\Delta t}{\Delta X^2} - V\frac{\Delta t}{2\Delta X}\right)C_{i+1}^l
\end{aligned}
$$

(31a)

By substituted the initial value, the solution of equations is as follow,

$$
E\frac{\Delta t}{\Delta X^2} = 0.009, \quad V\frac{\Delta t}{2\Delta X} = 0.0425, \quad k\Delta t = 0.15
$$

$$
E\frac{\Delta t}{\Delta X^2} + V\frac{\Delta t}{2\Delta X} = 0.0515, \quad -k\Delta t - 2E\frac{\Delta t}{\Delta X^2} + 1
$$
$$
= 0.832, E\frac{\Delta t}{\Delta X^2} - V\frac{\Delta t}{2\Delta X} = -0.0335
$$

then,

$$
C_1^1 = 0.0515C_0^0 + 0.832C_1^0 - 0.0335C_2^0 = 7.725
$$

$$
C_2^1 = 0.0515C_1^0 + 0.832C_2^0 - 0.0335C_3^0 = 0
$$

$$
C_3^1 = 0.0515C_2^0 + 0.832C_3^0 - 0.0335C_4^0 = 0
$$

$$
C_4^1 = 0.0515C_3^0 + 0.832C_4^0 - 0.0335C_5^0 = 0
$$

$$
C_1^2 = 0.0515C_0^1 + 0.832C_1^1 - 0.0335C_2^1 = 14.1522
$$

$$
C_2^2 = 0.0515C_1^1 + 0.832C_2^1 - 0.0335C_3^1 = 0.3978
$$

The simulation of implicit method

Also, the water quality modeling can be re-written as a standard form of the implicit method.

$$-\left(E\frac{\Delta t}{\Delta X^2} + V\frac{\Delta t}{2\Delta X}\right)C_{i-1}^{l+1} + \left(2E\frac{\Delta t}{\Delta X^2} + 1 + k\Delta t\right)C_i^{l+1}$$

$$-\left(E\frac{\Delta t}{\Delta X^2} - V\frac{\Delta t}{2\Delta X}\right)C_{i+1}^{l+1} = C_i^l$$

$$(32a)$$

Let substitute the value of initial condition. The equations of implicit method is as follow,

$$E\frac{\Delta t}{\Delta X^2} + V\frac{\Delta t}{2\Delta X} = 0.0515, 2E\frac{\Delta t}{\Delta X^2} + 1 + k\Delta t$$

$$= 1.168, E\frac{\Delta t}{\Delta X^2} - V\frac{\Delta t}{2\Delta X} = -0.0335$$

then

$$\begin{bmatrix} 1.168 & 0.0335 & 0 & 0 \\ -0.0515 & 1.168 & 0.0335 & 0 \\ 0 & -0.0515 & 1.168 & 0.0335 \\ 0 & 0 & -0.0515 & 1.168 \end{bmatrix} \begin{bmatrix} C_1^1 \\ C_2^1 \\ C_3^1 \\ C_4^1 \end{bmatrix}$$

$$= \begin{bmatrix} 7.725 \\ 0 \\ 0 \\ 0 \end{bmatrix}$$

then these C_1^2, C_2^2, C_3^2, C_4^2 values can be obtained by using above results.

$$\begin{bmatrix} 1.168 & 0.0335 & 0 & 0 \\ -0.0515 & 1.168 & 0.0335 & 0 \\ 0 & -0.0515 & 1.168 & 0.0335 \\ 0 & 0 & -0.0515 & 1.168 \end{bmatrix} \begin{bmatrix} C_1^2 \\ C_2^2 \\ C_3^2 \\ C_4^2 \end{bmatrix}$$

$$= \begin{bmatrix} 7.725 + C_1^1 \\ C_2^1 \\ C_3^1 \\ C_4^1 \end{bmatrix}$$

The simulation of Barakat-Clark method

The water quality modeling can be re-written as (33a) to (33c), as follows.

$$C_{i,j}^{n+1} = \left(u_{i,j}^{n+1} + v_{i,j}^{n+1}\right)/2 \qquad (33c)$$

The parameters are $E\frac{\Delta t}{\Delta X^2} = 0.009$ $V\frac{\Delta t}{\Delta X} = 0.0815$ $k\Delta t = 0.15$. For u and v, each parameter can be solved as follows:

$$V\frac{\Delta t}{\Delta X} - E\frac{\Delta t}{\Delta X^2} - k\Delta t + 1 = 0.926, E\frac{\Delta t}{\Delta X^2} - V\frac{\Delta t}{\Delta X}$$

$$= -0.076, 1 + E\frac{\Delta t}{\Delta X^2} = 1.009$$

$$u_1^1 = \frac{0.926u_1^0 - 0.076u_2^0 + 0.009u_0^1}{1.009} = 1.338$$

$$u_2^1 = \frac{0.926u_2^0 - 0.076u_3^0 + 0.009u_1^1}{1.009} = 0.012$$

$$u_3^1 = \frac{0.926u_3^0 - 0.076u_4^0 + 0.009u_2^1}{1.009} = 0.0001$$

$$u_4^1 = \frac{0.926u_4^0 - 0.076u_5^0 + 0.009u_3^1}{1.009} = 0.0000008$$

$$u_3^1 = \frac{0.926u_3^0 - 0.076u_4^0 + 0.009u_2^1}{1.009} = 0.0001$$

$$u_3^1 = \frac{0.926u_3^0 - 0.076u_4^0 + 0.009u_2^1}{1.009} = 0.0001$$

$$u_4^1 = \frac{0.926u_4^0 - 0.076u_5^0 + 0.009u_3^1}{1.009} = 0.0000008$$

$$1 + E\frac{\Delta t}{\Delta X^2} = 1.009, E\frac{\Delta t}{\Delta X^2} = 0.009, -V\frac{\Delta t}{\Delta X} - E\frac{\Delta t}{\Delta X^2}$$

$$-k\Delta t + 1 = 0.756, -E\frac{\Delta t}{\Delta X^2} - V\frac{\Delta t}{\Delta X} = -0.094$$

$$\begin{bmatrix} 1.009 & -0.009 & 0 & 0 \\ 0 & 1.009 & -0.009 & 0 \\ 0 & 0 & 1.009 & -0.009 \\ 0 & 0 & 0 & 1 \end{bmatrix} \begin{bmatrix} v_1^1 \\ v_2^1 \\ v_3^1 \\ v_4^1 \end{bmatrix} = \begin{bmatrix} -14.1 \\ 0 \\ 0 \\ 0 \end{bmatrix}$$

The simulation of the ADI Method for two-dimensional water quality modeling

The initial condition of the stream has been given as 50 meters long and 40 meters wide. There is no pollution at the beginning. The contaminants came from upstream. This stream section has 4 monitoring points in every 10 meters. The concentrations of pollutants are 150 mg/L at X direction and 150 mg/L at Y direction. The flow velocity

$$u_i^{n+1} = \left[\left(V\frac{\Delta t}{\Delta X} - E\frac{\Delta t}{\Delta X^2} - k\Delta t + 1\right)u_i^n + \left(E\frac{\Delta t}{\Delta X^2} - V\frac{\Delta t}{\Delta X}\right)u_{i+1}^n + E\frac{\Delta t}{\Delta X^2}u_{i-1}^{n+1}\right] / 1 + E\frac{\Delta t}{\Delta X^2} \qquad (33a)$$

$$\left(1 + E\frac{\Delta t}{\Delta X^2}\right)v_i^{l+1} - E\frac{\Delta t}{\Delta X^2}v_{i+1}^{l+1} = \left(-V\frac{\Delta t}{\Delta X} - E\frac{\Delta t}{\Delta X^2} - k\Delta t + 1\right)v_i^l - \left(E\frac{\Delta t}{\Delta X^2} + V\frac{\Delta t}{\Delta X}\right)v_{i-1}^l \qquad (33b)$$

is 10 m/s in X direction and 5 m/s in Y direction. The diffusion rate of pollutants E_1 and E_2 all are equaled to 100. The degradation rate of the pollutant k is 0.15 in this stream. The department of EPA wants to know this section's water quality every 10 meter long and every 8 meters wide. Therefore, the simulation methods of pollutants in two-dimensional water quality are tested as follows:

According to the ADI method, the two-dimensional water quality equation can be re-written as follows by considering the first half step as ($l = 0.5$).

$$-\left(V_2\frac{\Delta t}{2\Delta Y}+E_2\frac{\Delta t}{\Delta Y^2}\right)C_{i,j-1}^{l+1/2}+\left(2+2E_2\frac{\Delta t}{\Delta Y^2}\right)C_{ij}^{l+1/2}$$

$$+\left(V_2\frac{\Delta t}{2\Delta Y}-E_2\frac{\Delta t}{\Delta Y^2}\right)C_{i,j+1}^{l+1/2}=\left(V_1\frac{\Delta t}{2\Delta X}+E_1\frac{\Delta t}{\Delta X^2}\right)C_{i-1,j}^{l}$$

$$+\left(2-E_1\frac{2\Delta t}{\Delta X^2}-k\Delta t\right)C_{ij}^{l}+\left(E_1\frac{\Delta t}{\Delta X^2}-V_1\frac{\Delta t}{2\Delta X}\right)C_{i+1,j}^{l}$$

$$(34a)$$

when $i = 1, j = 1, 2, 3, 4$

$$-1.197C_{1,0}^{0+1/2}+3.56C_{1,1}^{0+1/2}-0.363C_{1,2}^{0+1/2}$$
$$=0.75C_{0,1}^{0}+0.925C_{1,1}^{0}+0.25C_{2,1}^{0}$$

$$-1.197C_{1,1}^{0+1/2}+3.56C_{1,2}^{0+1/2}-0.363C_{1,3}^{0+1/2}$$
$$=0.75C_{0,2}^{0}+0.925C_{1,2}^{0}+0.25C_{2,2}^{0}$$

$$-1.197C_{1,2}^{0+1/2}+3.56C_{1,3}^{0+1/2}-0.363C_{1,4}^{0+1/2}$$
$$=0.75C_{0,3}^{0}+0.925C_{1,3}^{0}+0.25C_{2,3}^{0}$$

$$-1.197C_{1,3}^{0+1/2}+3.56C_{1,4}^{0+1/2}-0.363C_{1,5}^{0+1/2}$$
$$=0.75C_{0,4}^{0}+0.925C_{1,4}^{0}+0.25C_{2,4}^{0}$$

There $C_{1,1}^{0+1/2}, C_{1,2}^{0+1/2}, C_{1,3}^{0+1/2}, C_{1,4}^{0+1/2}$, they can be obtained by above equations. Then, it substituted $i = 2, j = 1, 2, 3, 4$, the equation (34a) can be written as following,

$$-1.197C_{2,0}^{0+1/2}+3.56C_{2,1}^{0+1/2}-0.363C_{2,2}^{0+1/2}$$
$$=0.75C_{1,1}^{0}+0.925C_{2,1}^{0}+0.25C_{3,1}^{0}$$

$$-1.197C_{2,1}^{0+1/2}+3.56C_{2,2}^{0+1/2}-0.363C_{2,3}^{0+1/2}$$
$$=0.75C_{1,2}^{0}+0.925C_{2,2}^{0}+0.25C_{3,2}^{0}$$

$$-1.197C_{2,2}^{0+1/2}+3.56C_{2,3}^{0+1/2}-0.363C_{2,4}^{0+1/2}$$
$$=0.75C_{1,3}^{0}+0.925C_{2,3}^{0}+0.25C_{3,3}^{0}$$

$$-1.197C_{2,3}^{0+1/2}+3.56C_{2,4}^{0+1/2}-0.363C_{2,5}^{0+1/2}$$
$$=0.75C_{1,4}^{0}+0.925C_{2,4}^{0}+0.25C_{3,4}^{0}$$

$C_{2,1}^{0+1/2}, C_{2,2}^{0+1/2}, C_{2,3}^{0+1/2}, C_{2,4}^{0+1/2}$, they can be obtained by above equations. Similarly, let substitute $i = 3, 4$, and $j = 1, 2, 3, 4$.

The second half of the ADI method is based on the results of the first step. By using $i = 1, j = 1, 2, 3, 4$, the equation (34a) can be written as,

$$-0.0165C_{0,1}^{1}+2.218C_{1,1}^{1}-0.0015C_{2,1}^{1}$$
$$=50.5C_{1,0}^{0+1/2}-98C_{1,1}^{0+1/2}-49.5C_{1,2}^{0+1/2}$$

$$-0.0165C_{1,1}^{1}+2.218C_{2,1}^{1}-0.0015C_{3,1}^{1}$$
$$=50.5C_{2,0}^{0+1/2}-98C_{2,1}^{0+1/2}-49.5C_{2,2}^{0+1/2}$$

$$-0.0165C_{2,1}^{1}+2.218C_{3,1}^{1}-0.0015C_{4,1}^{1}$$
$$=50.5C_{3,0}^{0+1/2}-98C_{3,1}^{0+1/2}-49.5C_{3,2}^{0+1/2}$$

$$-0.0165C_{3,1}^{1}+2.218C_{4,1}^{1}-0.0015C_{5,1}^{1}$$
$$=50.5C_{4,0}^{0+1/2}-98C_{4,1}^{0+1/2}-49.5C_{4,2}^{0+1/2}$$

These $C_{1,1}^{1}, C_{2,1}^{1}, C_{3,1}^{1}, C_{4,1}^{1}$ can be solved using above equations. Substituting $j = 1, i = 1, 2, 3, 4$, the equation (34a) can be written as,

$$-\left(E_1\frac{\Delta t}{\Delta X^2}+V_1\frac{\Delta t}{\Delta X}\right)C_{i-1,j}^{l+1}+\left(2+E_1\frac{2\Delta t}{\Delta X^2}+k\Delta t\right)C_{ij}^{l+1}$$

$$-\left(E_1\frac{\Delta t}{\Delta X^2}-V_1\frac{\Delta t}{2\Delta X}\right)C_{i+1,j}^{l+1}=\left(E_2\frac{\Delta t}{\Delta Y^2}+V_2\frac{\Delta t}{2\Delta Y}\right)C_{i,j-1}^{l+1/2}$$

$$+\left(2-E_2\frac{\Delta t}{\Delta Y^2}\right)C_{ij}^{l+1/2}+\left(E_2\frac{\Delta t}{\Delta Y^2}-V_2\frac{\Delta t}{2\Delta Y}\right)C_{i,j+1}^{l+1/2}$$

$$(34b)$$

$$-0.75C_{0,1}^{1}+3.075C_{1,1}^{1}-0.25C_{2,1}^{1}$$
$$=1.197C_{1,0}^{0+1/2}+1.22C_{1,1}^{0+1/2}+0.363C_{1,2}^{0+1/2}$$

$$-0.75C_{1,1}^{1}+3.075C_{2,1}^{1}-0.25C_{3,1}^{1}$$
$$=1.197C_{2,0}^{0+1/2}+1.22C_{2,1}^{0+1/2}+0.363C_{2,2}^{0+1/2}$$

$$-0.75C_{2,1}^{1}+3.075C_{3,1}^{1}-0.25C_{4,1}^{1}$$
$$=1.197C_{3,0}^{0+1/2}+1.22C_{3,1}^{0+1/2}+0.363C_{3,2}^{0+1/2}$$

$$-0.75C_{3,1}^{1}+3.075C_{4,1}^{1}-0.25C_{5,1}^{1}$$
$$=1.197C_{4,0}^{0+1/2}+1.22C_{4,1}^{0+1/2}+0.363C_{4,2}^{0+1/2}$$

Let substitute $j = 2, i = 1, 2, 3, 4$, the equation (34a) can be written as,

$$-0.75C_{0,2}^{1}+3.075C_{1,2}^{1}-0.25C_{2,2}^{1}$$
$$=1.197C_{1,1}^{0+1/2}+1.22C_{1,2}^{0+1/2}+0.363C_{1,3}^{0+1/2}$$

$$-0.75C_{1,2}^{1}+3.075C_{2,2}^{1}-0.25C_{3,2}^{1}$$
$$=1.197C_{2,1}^{0+1/2}+1.22C_{2,2}^{0+1/2}+0.363C_{2,3}^{0+1/2}$$

$$-0.75C_{2,2}^{1}+3.075C_{3,2}^{1}-0.25C_{4,2}^{1}$$
$$=1.197C_{3,1}^{0+1/2}+1.22C_{3,2}^{0+1/2}+0.363C_{3,3}^{0+1/2}$$

$$-0.75C_{3,2}^1 + 3.075C_{4,2}^1 - 0.25C_{5,2}^1$$
$$= 1.197C_{4,1}^{0+1/2} + 1.22C_{4,2}^{0+1/2} + 0.363C_{4,3}^{0+1/2}$$

Thus, $C_{1,2}^1 \cdot C_{2,2}^1, C_{3,2}^1, C_{4,2}^1$ can be obtained by above equations. Similarly, the other C value is calculated by substituting $i = 3, 4$.

The simulation of Barakat-Clark method for two-dimensional modeling

This model is similar to the one-dimensional water quality modeling. If it considers the two different directions (x,y). The group of equations can be written as,

$$u_{i,j}^{l+1} = \left\{ \left[\left(V_1 \frac{\Delta t}{\Delta X} + V_2 \frac{\Delta t}{\Delta Y} - E_1 \frac{\Delta t}{\Delta X^2} - E_2 \frac{\Delta t}{\Delta Y^2} - k\Delta t + 1 \right) u_{i,j}^l \right. \right.$$
$$+ \left(E_1 \frac{\Delta t}{\Delta X^2} - V_1 \frac{\Delta t}{\Delta X} \right) u_{i+1,j}^l + \left(E_2 \frac{\Delta t}{\Delta Y^2} - V_2 \frac{\Delta t}{\Delta Y} \right) u_{i,j+1}^l$$
$$\left. + E_1 \frac{\Delta t}{\Delta X^2} u_{i-1,j}^{l+1} + E_2 \frac{\Delta t}{\Delta Y^2} u_{i,j-1}^{l+1} \right] \right\} /$$
$$\left[1 + E_1 \frac{\Delta t}{\Delta X^2} + E_2 \frac{\Delta t}{\Delta Y^2} \right]$$

(35a)

in this case, the u values are as follows

$$u_{1,1}^1 = \left(0.312u_{1,1}^0 + 0.25u_{2,1}^0 + 0.363u_{1,2}^0 + 0.5u_{0,1}^1 + 0.78u_{1,0}^1 \right)/2.28$$
$$= 84.21$$

$$u_{1,2}^1 = \left(0.312u_{1,2}^0 + 0.25u_{2,2}^0 + 0.363u_{1,3}^0 + 0.5u_{0,2}^1 + 0.78u_{1,1}^1 \right)/2.28$$
$$= 61.7$$

$$u_{1,3}^1 = \left(0.312u_{1,3}^0 + 0.25u_{2,3}^0 + 0.363u_{1,4}^0 + 0.5u_{0,3}^1 + 0.78u_{1,2}^1 \right)/2.28$$
$$= 54$$

$$u_{1,4}^1 = \left(0.312u_{1,4}^0 + 0.25u_{2,4}^0 + 0.363u_{1,5}^0 + 0.5u_{0,4}^1 + 0.78u_{1,3}^1 \right)/2.28$$
$$= 51.37$$

$$u_{2,1}^1 = \left(0.312u_{2,1}^0 + 0.25u_{3,1}^0 + 0.363u_{2,2}^0 + 0.5u_{1,1}^1 + 0.78u_{2,0}^1 \right)/2.28$$
$$= 69.783$$

$$u_{2,2}^1 = \left(0.312u_{2,2}^0 + 0.25u_{3,2}^0 + 0.363u_{2,3}^0 + 0.5u_{1,2}^1 + 0.78u_{2,1}^1 \right)/2.28$$
$$= 37.4$$

$$u_{2,3}^1 = \left(0.312u_{2,3}^0 + 0.25u_{3,3}^0 + 0.363u_{2,4}^0 + 0.5u_{1,3}^1 + 0.78u_{2,2}^1 \right)/2.28$$
$$= 24.64$$

$$u_{2,4}^1 = \left(0.312u_{2,4}^0 + 0.25u_{3,4}^0 + 0.363u_{2,5}^0 + 0.5u_{1,4}^1 + 0.78u_{2,3}^1 \right)/2.28$$
$$= 19.695$$

The rest of u values can be calculated by substituting $i = 3, 4$.

$$E_1 \frac{\Delta t}{\Delta X^2} v_{i+1,j}^{l+1} + E_2 \frac{\Delta t}{\Delta Y^2} v_{i,j+1}^{l+1} - \left(1 + E_1 \frac{\Delta t}{\Delta X^2} + E_2 \frac{\Delta t}{\Delta Y^2} \right) v_{i,j}^{l+1}$$
$$= \left(V_1 \frac{\Delta t}{\Delta X} + V_2 \frac{\Delta t}{\Delta Y} + E_1 \frac{\Delta t}{\Delta X^2} + E_2 \frac{\Delta t}{\Delta Y^2} + k\Delta t - 1 \right) v_{i,j}^l$$
$$- \left(V_1 \frac{\Delta t}{\Delta X} + E_1 \frac{\Delta t}{\Delta X^2} \right) v_{i-1,j}^l - \left(V_2 \frac{\Delta t}{\Delta Y} + E_2 \frac{\Delta t}{\Delta Y^2} \right) v_{i,j-1}^l$$

(35b)

The v values are calculated by the following process:

$$0.5v_{2,1}^1 + 0.78v_{1,2}^1 - 1.022v_{1,1}^1$$
$$= 1.022v_{1,1}^0 - 0.75v_{0,1}^0 - 1.197v_{1,0}^0 = -292.05$$

$$0.5v_{2,2}^1 + 0.78v_{1,3}^1 - 1.022v_{1,2}^1$$
$$= 1.022v_{1,2}^0 - 0.75v_{0,2}^0 - 1.197v_{1,1}^0 = -112.5$$

$$0.5v_{2,3}^1 + 0.78v_{1,4}^1 - 1.022v_{1,3}^1$$
$$= 1.022v_{1,3}^0 - 0.75v_{0,3}^0 - 1.197v_{1,2}^0 = -112.5$$

$$0.5v_{2,4}^1 + 0.78v_{1,5}^1 - 1.022v_{1,4}^1$$
$$= 1.022v_{1,4}^0 - 0.75v_{0,4}^0 - 1.197v_{1,3}^0 = -112.5$$

$$0.5v_{3,1}^1 + 0.78v_{2,2}^1 - 1.022v_{2,1}^1$$
$$= 1.022v_{2,1}^0 - 0.75v_{1,1}^0 - 1.197v_{2,0}^0 = -179.55$$

$$0.5v_{3,2}^1 + 0.78v_{2,3}^1 - 1.022v_{2,2}^1$$
$$= 1.022v_{2,2}^0 - 0.75v_{1,2}^0 - 1.197v_{2,1}^0 = -112.5$$

$$0.5v_{3,3}^1 + 0.78v_{2,4}^1 - 1.022v_{2,3}^1$$
$$= 1.022v_{2,3}^0 - 0.75v_{1,3}^0 - 1.197v_{2,2}^0 = 0$$

$$0.5v_{3,4}^1 + 0.78v_{2,5}^1 - 1.022v_{2,4}^1$$
$$= 1.022v_{2,4}^0 - 0.75v_{1,4}^0 - 1.197v_{2,3}^0 = 0$$

When these values of u and v were calculated, the final solution of the Barakart-Clark method can be obtained by using average value of those two matrixes.

Results and discussion

Figure 1 shows the solutions of explicit, implicit and Barakat-Clark methods. It proves that the numerical solution obtained by the Barakat-Clark method approached the real world process even through the explicit and implicit results are stable in this case. Figure 2 shows the solutions obtained using the ADI and Barakat-Clark methods for the 2-D water quality modelling process. In Figure 2a, it is the results of ADI methods in three different times. Similarly, Figure 2b shows the results of the Barakat-Clark method. The ADI method clearly shows that the results are not as stable as the results of Barakat-Clark method. In fact, during the final time, the ADI result is beyond of boundary conditions. However,

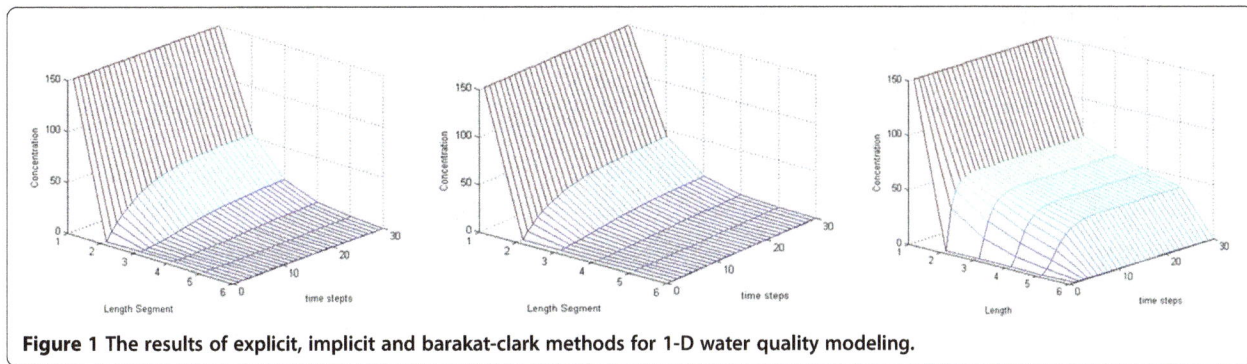

Figure 1 The results of explicit, implicit and barakat-clark methods for 1-D water quality modeling.

it may have been caused by a large space of x or y. However, under same steps and space of x and y, the Barakat-Clark method obtained the better accuracy results. The Barakat-Clark method uses less computing time to get those results. This method has the advantage of solving the three dimensions time-dependent equations which is not convenient for ADI method.

The data results (Tables 2, 3, 4, 5 and 6) of the above equations show the detailed numerical solutions of the water quality modelling process. The findings in Table 4 show smoother curves of water quality because of different method. Although the explicit method obtained stable results in this case, when X value is changed to a smaller value, the results will be extremely unstable. It means that the stability of the solution is depends on the size

of grid for explicit method. It is obvious to see that the solution of the Barakat-Clark method is more stable, even though the size of the grid was changed. This is because the Barakat-Clark numerical method has been considered average values or mean values of implicit and explicit data. Therefore, the Barakat-Clark methodology is more accurate than others in the case of the water quality simulation process.

Conclusions

In this study, the finite difference numerical methods are applied to the water quality modeling processes. They simulate the concentration process of pollutants in hydrological systems. As one of the existing numerical methods, the Barakat-Clark has obtained higher accuracy

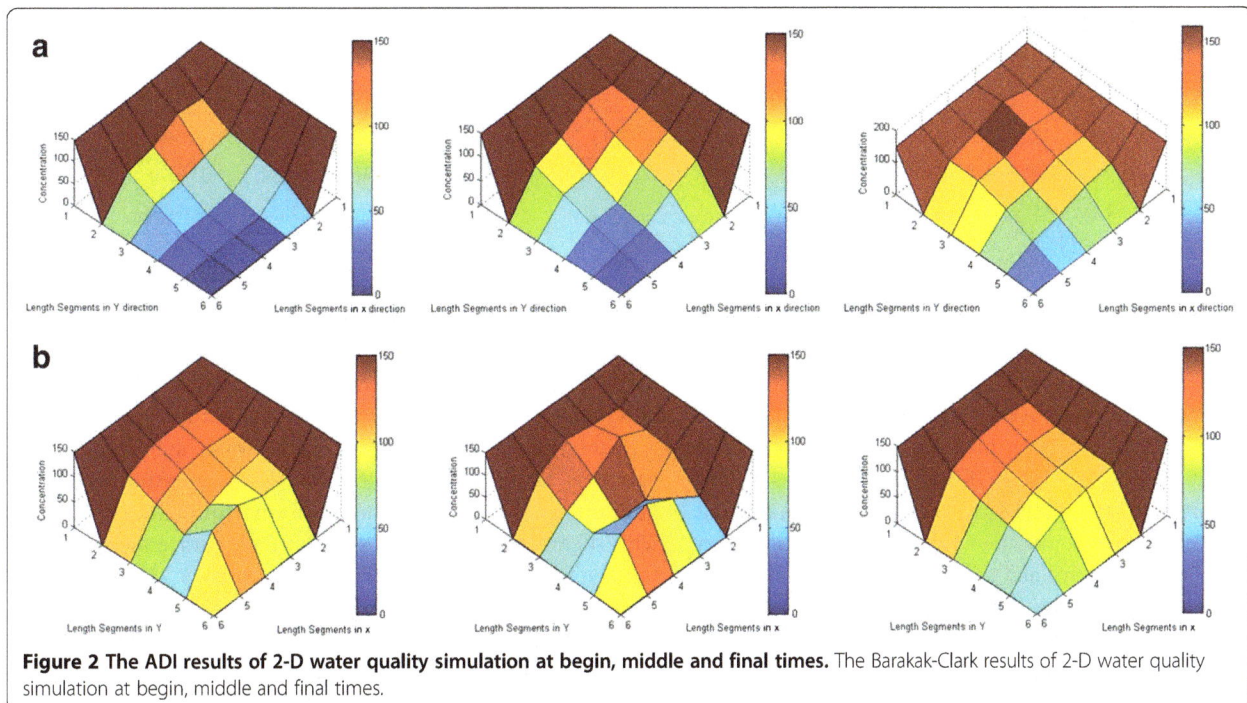

Figure 2 The ADI results of 2-D water quality simulation at begin, middle and final times. The Barakak-Clark results of 2-D water quality simulation at begin, middle and final times.

Table 2 Concentration results of pollutants using the explicit method for 1-D water quality modeling

Initial concentration	In 200 meters	In 400 meters	In 600 meters	In 800 meters	In 1000 meters
150	0	0	0	0	0
150	7.725	0	0	0	0
150	14.4303	0.397838	0	0	0
150	20.23717	1.088483	0.020489	0	0
150	25.2544	1.986332	0.073841	0.001055	0
150	29.57928	3.022264	0.166355	0.004719	0
150	33.29857	4.141085	0.299884	0.012663	0
150	36.48943	5.299292	0.473141	0.026436	0
150	39.2203	6.463141	0.682715	0.047313	0
150	41.5517	7.606981	0.923863	0.076227	0
150	43.53705	8.711823	1.191119	0.113744	0
150	45.22331	9.764117	1.47874	0.160073	0
150	46.65173	10.75472	1.781036	0.215098	0
150	47.85842	11.67799	2.092601	0.278429	0
150	48.8749	12.53111	2.408467	0.349445	0
150	49.72862	13.31337	2.724195	0.427354	0
150	50.44344	14.02577	3.035924	0.51124	0
150	51.04005	14.6705	3.340382	0.600106	0
150	51.5363	15.25066	3.634879	0.692922	0
150	51.94761	15.76992	3.917271	0.788652	0
150	52.28723	16.23236	4.185922	0.88629	0
150	52.56653	16.64226	4.439657	0.984874	0
150	52.79524	17.00393	4.677705	1.083513	0
150	52.98163	17.32166	4.899652	1.181391	0
150	53.13278	17.59962	5.105387	1.27778	0
150	53.25467	17.84177	5.295051	1.37204	0
150	53.35235	18.05189	5.468992	1.463626	0
150	53.4301	18.23348	5.627726	1.552081	0
150	53.49151	18.38978	5.771896	1.637034	0
150	53.53957	18.52378	5.902238	1.718198	0

Table 3 Concentration results of pollutants using the implicit method for 1-D water quality modeling

Initial concentration	In 200 meters	In 400 meters	In 600 meters	In 800 meters	In 1000 meters
150	0	0	0	0	0
150	6.605527	0.290887	0.01281	0.000565	0
150	12.24669	0.787725	0.045628	0.002495	0
150	17.05822	1.423646	0.101648	0.006618	0
150	21.15694	2.146537	0.181281	0.01366	0
150	24.64405	2.916286	0.283099	0.024177	0
150	27.60703	3.70248	0.404526	0.038536	0
150	30.12146	4.482505	0.542353	0.056907	0
150	32.2525	5.239974	0.693113	0.079283	0
150	34.05628	5.96343	0.853335	0.105505	0
150	35.58104	6.645285	1.019722	0.135292	0
150	36.86826	7.280959	1.189259	0.168269	0
150	37.95349	7.868175	1.359277	0.204	0
150	38.86719	8.406391	1.527482	0.242008	0
150	39.63541	8.896348	1.691955	0.281801	0
150	40.28042	9.339706	1.851141	0.322889	0
150	40.82121	9.738758	2.003824	0.3648	0
150	41.27396	10.09621	2.149094	0.407088	0
150	41.65245	10.415	2.286313	0.449343	0
150	41.96837	10.69817	2.41508	0.491198	0
150	42.23167	10.94878	2.535196	0.53233	0
150	42.45075	11.1698	2.64663	0.572458	0
150	42.63275	11.36411	2.749488	0.61135	0
150	42.78369	11.53442	2.843985	0.648814	0
150	42.90865	11.68326	2.930424	0.684701	0
150	43.01191	11.81298	3.009169	0.718899	0
150	43.09709	11.92575	3.08063	0.751328	0
150	43.16721	12.02354	3.145246	0.781942	0
150	43.22482	12.10814	3.203473	0.81072	0
150	43.27204	12.18115	3.255771	0.837664	0

results with stability than the other methods. By comparing the solution of Barakat-Clark method with the other numerical methods, its results better reflect the reality of the simulation without oversimplify solving the process. The water quality modeling process, which estimates the real world conditions, can be effectively solved by means of the Barakat-Clark method. It enables the managers of water authorities to know the concentration of pollution in surface water systems and to conveniently use those solutions as a reference in their hydrological systems. Consequently, the accurate of water quality simulation processes can be enhanced. The results of the case study indicates that the solution of Barakat-Clark method has proved that

this method is better by comparison than the others for water quality modeling. Solutions of the numerical method actually reflect a compromise between the finite difference method and requirement of accuracy.

In the real world, administrators of water management agencies are paying a high price for water quality to guarantee the safe conditions of environmental water. This means that authorities desire to accurate monitoring data to determine the stream or river situations. This may cause a higher economic cost. However, simulation values of water quality are reliable within the Barakat-Clark method. Its advantages have been clearly shown in the above case study. It shows that the results

Table 4 Concentration results of pollutants using the Barakat-Clark method for 1-D water quality modeling

Initial concentration	In 200 meters	In 400 meters	In 600 meters	In 800 meters	In 1000 meters
150	0	0	0	0	0
150	37.5	16.82825	7.971279	3.775869	0
150	52.48305	31.93705	18.66957	10.53269	0
150	60.00174	41.99309	28.2431	18.08965	0
150	63.00452	48.05255	35.37834	24.85119	0
150	64.50276	51.40358	40.16946	30.14466	0
150	65.10873	53.22341	43.16784	33.9338	0
150	65.4055	54.16237	44.95798	36.47639	0
150	65.52848	54.64781	45.98663	38.10145	0
150	65.587	54.88959	46.56245	39.101	0
150	65.61204	55.01092	46.87698	39.69728	0
150	65.62356	55.07013	47.04599	40.04403	0
150	65.62865	55.09924	47.13527	40.24148	0
150	65.63092	55.11326	47.18191	40.35188	0
150	65.63195	55.12005	47.20596	40.41268	0
150	65.6324	55.12329	47.21827	40.4457	0
150	65.63261	55.12484	47.22451	40.46343	0
150	65.6327	55.12558	47.22765	40.47285	0
150	65.63274	55.12592	47.22922	40.47781	0
150	65.63276	55.12609	47.23	40.4804	0
150	65.63277	55.12617	47.23038	40.48174	0
150	65.63277	55.1262	47.23057	40.48243	0
150	65.63278	55.12622	47.23067	40.48278	0
150	65.63278	55.12623	47.23071	40.48296	0
150	65.63278	55.12623	47.23073	40.48305	0
150	65.63278	55.12623	47.23075	40.48309	0
150	65.63278	55.12623	47.23075	40.48312	0
150	65.63278	55.12623	47.23075	40.48313	0
150	65.63278	55.12624	47.23075	40.48313	0
150	65.63278	55.12624	47.23075	40.48314	0

Table 6 Results of Barakat-Clark method for 2-D water quality modeling in final time (x/y direction)

Initial concentration	In 200 meters	In 400 meters	In 600 meters	In 800 meters	In 1000 meters
150	150	150	150	150	150
150	119.8325	127.1802	125.1696	101.3011	0
150	107.2249	110.1196	104.7594	77.7571	0
150	101.0918	99.1784	90.7855	64.4765	0
150	97.8035	92.4481	81.7728	56.4511	0
150	0	0	0	0	0

of the Barakat-Clark method is more reliable and more accurate with process stability. On the other hand, other numerical methods such as ADI method produced a different results under same circumstances. Therefore, the Barakat-Clark method can be considered as a better finite method in the 2-D water modeling system, and this paper is the first attempt to compare the Barakat-Clark 2-D method with other multiple numerical algorithms in the water quality modeling process. From the above applications, it is obvious that the Barakat-Clark method shares higher stability results with the same environmental conditions. However, in this study, the boundary conditions have been pre-defined. Therefore, further studies of this method have to correspond to the variation of boundary conditions in real world. Finally, this the 2-D Barakat-Clark method well reflects the accuracy of simulation process, and thus, it can provide an exact reference for water quality modeling in hydrological systems.

Competing interests

The authors declare that they have no competing interests.

Authors' contributions

Author LJ composed this paper, and others have revised it many times for publication. All authors read and approved the final manuscript.

Acknowledgements

This research was supported by the Program for Innovative Research Team (IRT1127), the Natural Science and Engineering Research Council of Canada, the MOE Key Project Program (311013), Natural Science Foundation of China (51109181) and the Scientific Research fund project (XKJJ201105) of Xiamen University of Technology. The writers are very grateful to the editor and the anonymous reviewers for their insightful comments and suggestions.

Author details

[1]College of Environmental Science and Engineering, Xiamen University of Technology, Xiamen, Fujian Province 361024, China. [2]College of Applied Mathematics, Xiamen University of Technology, Xiamen, Fujian Province 361024, China.

Table 5 Results of ADI method for 2-D water quality modeling in final time (x/y direction)

Initial concentration	In 200 meters	In 400 meters	In 600 meters	In 800 meters	In 1000 meters
150	150	150	150	150	150
150	124.9046	133.1181	96.9082	77.9531	0
150	117.1522	96.6831	60.7363	54.3516	0
150	102.0593	74.4782	34.3785	31.6676	0
150	80.1334	54.9374	21.4405	13.6352	0
150	0	0	0	0	0

References

Barakat HZ, Clark JA (1966) On the solution of the diffusion equations by numerical methods. J Heat Tran 11:421–427

Cash JR (2003) Efficient numerical methods for the solution of stiff initial-value problems and differential algebraic equations. Proc R Soc Lond 459:797–815

Christie MA, Bond DJ (1987) Detailed simulation of unstable processes in miscible flooding. Reserv Eng 2:514–522

Davis HT (1962) Induction to nonlinear differential and integral equations. Dover, New York

Evans DJ, Abdullah AR (1985) A new explicit method for diffusion-convection equation. Comp Maths Applis 11:145–154

Hoffman J (1992) Numerical methods for engineers and scientists. McGraw-Hill, New York

Kellogg RB (1964) An alternating direction method for operator equations. SIAM 12:848–854

Liu WC, Jantai K, Albert Y (2005) Modelling hydrodynamics and water quality in the separation waterway of the Yulin offshore industrial park, Tainwan. Environ Model Softw 20:309–328

Moiianty RK (2004) An unconditionally stable difference scheme for the one-space-dimensional linear hyperbolic equation. Appl Math Lett Letters 17:101–105

Munavalli GR, Mohan Kumar MS (2004) Modified Lagrangian method for modeling water quality in distribution systems. Water Res 38:973–2988

Na TY (1979) Computational methods in engineering boundary value problem. Academic Press, New York

Noye BJ, Hayman KJ (1992) Explicit two-level finite difference methods for the two-dimensional diffusion equation. J Comput Math 42:223–236

Richard JB, Robert S (1997) Expanded stability through higher accuracy for time-centered advection schemes. American Meteorological Society, Springer, Herdelberg, p 6

Rudi R, Matjaz C (1997) Hydrodynamic and water quality modeling: An experience. Ecol Model 101:195–207

Shampine LF, Gear CW (1979) A user's view of solving stiff ordinary differential equations. SIAM Rev 21:1

Shawgfeh N, Kaya D (2004) Comparing numerical methods for the solutions of systems of ordinary differential equations. Appl Math Lett 17:323–328

Stasa FL (1985) Applied finite element analysis for engineers. Holt Rinehart and Winston, New York

Taylor JR (1982) An introduction to error analysis. Mill Valley, CA

Timo T, Sampsa K, Ville K, Matthieu M, Peng CY (2013) Water quality analysis using an inexpensive device and a mobile phone. Env Syst Res 2:9

Walter MG (2006) Use of distribution system water quality models in support of water security. Security of Water Supply System, Springer, Heidelerg, pp 39–50

16

Robust interval quadratic programming and its application to waste management under uncertainty

Yongping Li[1*] and Guohe Huang[2]

Abstract

Background: No country has ever experienced as large or as fast an increase in municipal solid waste (MSW) quantities that China is now facing. The MSW generation rate in the City of Changchun continues to increase since it has been encountered swift urbanization, industrialization and economic development during the past decades.

Results: In this study, a robust interval quadratic programming method is developed for the planning of MSW management in the City of Changchun, China. The developed method can not only tackle uncertainties expressed as interval values, fuzzy sets, and their combinations, but also reflect economies-of-scale effects on waste disposal of cost.

Conclusions: The results are valuable for helping governmental officials more intuitive to know some basic situation, such as optimal waste-flow allocation, waste-flow routing, facility-capacity expansion, and system cost over the planning horizon. Results can also be used to generate decisions for supporting the city's long-term MSW management and planning, and thus help managers to identify desired MSW management policies in association with cost minimization under uncertainty.

Keywords: Environment, Management, Fuzzy sets, Policy analysis, Quadratic programming, Solid waste, Uncertainty

Background

For decades, massive urbanization and rapid development of global urban economy have increased municipal solid waste (MSW) generation rate. MSW management is crucial for environmental protection and public health and has become a major challenge confronted by the world, particularly for many urban regions of developing countries. For example, global waste generation rate has nearly doubled since 1960, from 2.7 to 4.4 pounds per capital per day, while more than 70% of MSW generated is disposed of at landfills (USEPA, 2007). Due to the waste management hierarchy, one of the greatest challenges that decision makers face is to figure out how to diversify the treatment options, increase the reliability of infrastructure systems, and leverage the redistribution of waste streams among landfilling, incineration, compost, recycling and other facilities (Chang and Davila, 2007). Consequently, many urban regions and countries have established various kinds of laws and regulations to enhance MSW management and planning. A large number of optimization techniques have been proposed for supporting decisions of MSW management and evaluating relevant operation and investment policies; they involve linear, dynamic, integer and multiobjective programming methods (Baetz, 1990; Lund et al. 1994; Masui et al. 2000; Kollikkathara et al. 2010; Cao and Huang, 2011).

The complexity of planning MSW management can be significantly compounded by the fact that many system components cannot be known with certainty beforehand. Hence, in many real-world applications, the quality of information produced by deterministic optimization techniques can be rendered highly questionable when the input data cannot be expressed with precision (Li and Huang, 2006; Li et al. 2011). The complexities could be further amplified not only by interactions among the uncertain parameters but also through additional economic implications. Such complexities have placed many MSW

* Correspondence: yongping.li@iseis.org
[1]MOE Key Laboratory of Regional Energy Systems Optimization, Resources and Environmental Research Academy, North China Electric Power University, Beijing 102206, China
Full list of author information is available at the end of the article

management problems beyond deterministic programming methods. Data imprecision can be addressed through optimization approaches such as fuzzy, stochastic and interval mathematical programming (Wilson and Baetz, 2001a, b; Huang et al. 2005a, b; El Hanandeh and El-Zein, 2010; Fan and Huang, 2012). Fuzzy mathematical programming considers uncertainties as fuzzy sets, and is effective in reflecting ambiguity and vagueness in resource availabilities. Robust programming based on the concept of fuzzy interval was able to deal with ambiguous coefficients in the optimization model and reflect the vague information of decision makers' implicit knowledge (Inuiguchi and Sakawa, 1998; Dubois et al. 2001; Ben-Tal and Nemirovski, 2002; Li et al. 2008). Moreover, this method delimits an uncertain decision space by specifying uncertainties through dimensional enlargement of the original fuzzy constraints, and thus enhances the robustness of the optimization process.

However, economies of scale (EOS) may affect the cost coefficients in a mathematical programming problem and make the relevant objective function nonlinear; difficulties arise due to system nonlinearities when these methods are applied to MSW management problems. Even if a nonlinear model was formulated, the modelers would prefer to convert it into equivalent linear forms and use linear methods for generating solutions (Ko and Chang, 2008; Li et al. 2009). Nonlinear mathematical models for real-world applications are hampered by a general lack of appropriate modeling solutions for effectively addressing uncertainties and nonlinearities simultaneously. Quadratic programming is useful for reflecting nonlinearity in cost objectives and has global optimum under a number of system conditions (Hillier and Lieberman 1986). Previously, a few studies incorporating uncertainty within MSW planning were reported, through introducing fuzzy and/or interval optimization methods into the quadratic programming framework to reflect uncertainty and nonlinearity (Chen and Huang, 2001). Fuzzy-based quadratic programming is incapable of dealing with ambiguous coefficients or decision makers' vague preferences; interval-based quadratic programming has difficulties in addressing uncertainties presented in terms of probabilistic or possibilistic distributions.

China has experienced a very rapid increase in its economy during the last two decades. However, in the absence of a comprehensively sustainable development scheme, this increase has brought severe environmental issues, such as water resources depletion and pollution, soil erosion, desertification, acid rain, sandstorms, forest depletion, and solid waste pollution. Among them, solid waste is becoming a critical issue, not only in terms of the impacts being created but also in terms of resources being consumed. No country has ever experienced as large or as fast an increase in solid waste quantities that

China is now facing. China produces around 29% of the world's solid waste each year, and with the economy continuing to grow rapidly, it is clear that China bears what may be the heaviest solid waste management burden in the world. It has been estimated that the amounts of industrial waste increased by 10% while at the same time municipal waste increased by 15% per year in China. In 2004, China surpassed the United States becoming as the world's largest waste generator, and by 2030 China's annual solid waste quantities will increase by another 150% - growing from about 190 million tons in 2004 to over 480 million tons in 2030 (Su et al. 2009). Therefore, development of systems analysis method for effective managing MSW and thus providing scientific bases for decision makers is desired.

The objective of this study is to develop a robust interval quadratic programming method for the planning of MSW management in the City of Changchun, China. Robust programming method will be incorporated within an interval quadratic programming framework for better accounting for uncertainties and nonlinearities. The developed method will then be applied to a case of long-term waste management planning. It can not only handle uncertainties expressed as fuzzy sets and interval values, but also deal with nonlinearities in the objective function to reflect the effect of EOS on waste management cost. The results can be used for generating a range of decision alternatives under various system conditions, and thus helping managers to identify desired waste-management policies.

Methods

Interval programming is a technique that readily processes interval data, thereby avoiding many problems encountered by other optimization methods (e.g. stochastic programming) when faced with parameter uncertainty. Since practitioners generally find it easier to specify estimates of fluctuation ranges than to determine appropriate distributional information, interval value proves especially meaningful in most practical settings. Consider an interval quadratic programming problem without $x_j x_k$ $(j \neq k)$ terms as follows (Chen and Huang, 2001):

$$Min \; f^{\pm} = \sum_{j=1}^{n} \left[c_j^{\pm} x_j^{\pm} + d_j^{\pm} \left(x_j^{\pm} \right)^2 \right] \tag{1a}$$

subject to:

$$\sum_{j=1}^{n} a_{ij}^{\pm} x_j^{\pm} \leq b_i^{\pm}, \quad i = 1, 2, ..., m \tag{1b}$$

$$x_j^{\pm} \geq 0, j = 1, 2, ..., n \tag{1c}$$

where $a_{ij}^{\pm}$, $b_i^{\pm}$, $c_j^{\pm}$, $d_j^{\pm}$ and $x_j^{\pm}$ are interval parameters/variables; the '-' and '+' superscripts represent lower and upper bounds of an interval parameter/variable, respectively. If the quadratic programming problem satisfies the Kuhn-Tucker conditions (Kuhn and Tucker, 1951) or has a concave objective function, it will then have a global optimum. Such a problem can then be transformed into two deterministic submodels that correspond to lower and upper bounds of the objective-function value, based on an interactive algorithm (Chen and Huang, 2001). However, model (1) has difficulties in reflecting uncertainties presented as fuzzy sets; moreover, it may lead to infeasibility when the model's right-hand-side parameters have large intervals.

Robust programming involves the optimization of a precise objective function subject to a fuzzy decision space delimited by constraints with fuzzy coefficients and fuzzy capacities (Inuiguchi and Sakawa, 1998). By delimiting the uncertain decision space through dimensional enlargement of the original fuzzy constraints, this approach can enhance the robustness of the optimization effort. A general robust programming problem can be defined as follows:

$$Min \ f = CX \tag{2a}$$

subject to:

$$AX \underset{\sim}{\leq} B \tag{2b}$$

$$X \geq 0 \tag{2c}$$

where $A \in \{\Re\}^{m \times n}$, $B \in \{\Re\}^{m \times 1}$, $C \in \{R\}^{1 \times n}$, $X \in \{R\}^{n \times 1}$, $\Re$ denotes a set of fuzzy parameters and variables, R denotes a set of deterministic numbers, and $\underset{\sim}{\leq}$ means fuzzy inequality. Let constraints in (2b) take the following specific form:

$$A_1 x_1 \oplus A_2 x_2 \oplus \cdots \oplus A_n x_n \underset{\sim}{\leq} B \tag{3}$$

where A_j ($j = 1, 2, ..., n$) and B are fuzzy subsets, and symbol $\oplus$ denotes addition of fuzzy subsets. Fuzziness of the decision space is due to uncertainties in coefficients A_j and B. Letting U_j and V be base variables imposed by fuzzy subsets A_j and B, we have:

$$\mu_{Aj}: \quad U_j - \ [0, 1] \tag{4a}$$

$$\mu_B: \quad V - \ [0, 1] \tag{4b}$$

where μ_{Aj} denotes the possibility of consuming a specific amount of resource by activity j, and μ_B indicates the possible availability of resource B. Fuzzy subset N can be expressed as the following L-R fuzzy numbers (Dubois and Prade, 1978):

$$\mu_N(x) = \begin{cases} F_L\left(\dfrac{u - x}{\beta}\right), & \text{if } -\infty < x < u, \ \beta > 0 \\ 1, & \text{if } x = u \\ F_R\left(\dfrac{x - u}{\delta}\right), & \text{if } u < x < +\infty, \ \delta > 0 \end{cases} \tag{5a}$$

where u is the mean value of N; β and δ are the left and right spreads, respectively; F_L and F_R are the shape

functions. For a linear case, fuzzy subset N can be defined as follows:

$$\mu_N(x) = \begin{cases} 0, & \text{if } x < \underline{a} \text{ or } x > \bar{a} \\ 1, & \text{if } x = u \\ 1 - 2|u - x|/(\bar{a} - \underline{a}), & \text{if } \underline{a} \leq x \leq \bar{a} \end{cases} \tag{5b}$$

where $[\underline{a}, \ \bar{a}]$ is an interval imposed by fuzzy subset N. According to the concept of level set (fuzzy α-cut) and the representation theorem, constraints in (3) can be represented as follows:

$$(\underset{\sim}{A}_1)_\alpha x_1 \oplus (\underset{\sim}{A}_2)_\alpha x_2 \oplus \cdots \oplus (\underset{\sim}{A}_n)_\alpha x_n \subseteq \underset{\sim}{B}_\alpha, \ \alpha \in [0, 1] \tag{6a}$$

where

$$(\underset{\sim}{A}_j)_\alpha = \{a_j \in \underset{\sim}{U}_j \mid \mu_{A_j}(a_j) \geq \alpha\} \tag{6b}$$

$$\underset{\sim}{B}_\alpha = \{b \in \underset{\sim}{V} \mid \mu_B(b) \geq \alpha\} \tag{6c}$$

Assume that the fuzzy subsets in Equation (3) are finite and have the following characteristic:

$$\{\mu_{A_j}(a_j) \mid a_j \in \underset{\sim}{U}_j\} = \{\alpha_1, \alpha_2, \cdots, \alpha_k\} \tag{7}$$

where $0 \leq \alpha_1 \leq \alpha_2 \leq ... \leq \alpha_k \leq 1$. Then, for each α_s ($s = 1, 2, ..., k$), constraints in (6a) become:

$$(\underset{\sim}{A}_1)_{\alpha_s} x_1 \oplus (\underset{\sim}{A}_2)_{\alpha_s} x_2 \oplus \cdots \oplus (\underset{\sim}{A}_n)_{\alpha_s} x_n \subseteq \underset{\sim}{B}_{\alpha_s}, \ \alpha_s \in [0, 1] \tag{8}$$

where $(\underset{\sim}{A}_j)_{\alpha_s}$ ($j = 1, 2, ..., n$; $s = 1, 2, ..., k$) and $\underset{\sim}{B}_{\alpha_s}$ constitute convex and non-empty fuzzy sets. Then, fuzzy constraints in (8) can be replaced by the following $2k$ precise inequalities, where k denotes the number of α-cut levels (Luhandjula and Gupta, 1996):

$$\bar{a}_1^s x_1 + \bar{a}_2^s x_2 + \cdots + \bar{a}_n^s x_n \leq \bar{b}^s, s = 1, 2, ..., k \tag{9a}$$

$$\underline{a}_1^s x_1 + \underline{a}_2^s x_2 + \cdots, s = 1, 2, ..., k + \underline{a}_n^s x_n \geq \underline{b}^s \tag{9b}$$

with

$$\bar{a}_j^s = \sup\left(a_j^s\right), \ a_j^s \in (\underset{\sim}{A}_j)_{\alpha_s} \tag{9c}$$

$$\underline{a}_j^s = \inf\left(a_j^s\right), \ a_j^s \in (\underset{\sim}{A}_j)_{\alpha_s} \tag{9d}$$

$$\bar{b}^s = \sup(b^s), \ b^s \in \underset{\sim}{B}_{\alpha_s} \tag{9e}$$

$$\underline{b}^s = \inf(b^s) b^s \in \underset{\sim}{B}_{\alpha_s} \tag{9f}$$

where $\sup(t)$ and $\inf(t)$ denote the superior and inferior limits among set t, respectively. Therefore, for a fuzzy robust linear program with m fuzzy constraints, the decision space for problem (2b) can be delimited by the following deterministic constraints based on the relations as defined in (6a) to (9f):

$$\sum_{j=1}^{n} \left(\bar{a}_{ij}^{s} x_j \right) \leq \bar{b}_i^{s}, i = 1, 2, ..., m; s = 1, 2, ..., k \quad (10a)$$

$$\sum_{j=1}^{n} \left(\underset{\sim}{a}_{ij}^{s} x_j \right) \geq \underset{\sim}{b}_i^{s}, i = 1, 2, ..., m; s = 1, 2, ..., k \quad (10b)$$

$$x_j \geq 0, j = 1, 2, ..., n \quad (10c)$$

Obviously, the robust programming model can be converted into a deterministic version through transforming the m imprecise constraints into $2\,km$ precise inclusive ones that correspond to k α-cut levels, such that robustness of the optimization process could be enhanced (Li et al. 2008). However, the robust programming method may become inapplicable when uncertainties and nonlinearities exist in the objective function; besides, it has difficulties in dealing with uncertainties that cannot be presented as membership functions. Therefore, robust programming will be incorporated within an interval quadratic programming framework in response to the above challenges. This leads to a robust interval quadratic programming model as follows:

$$\text{Min } f^{\pm} = \sum_{j=1}^{n} \left[c_j^{\pm} x_j^{\pm} + d_j^{\pm} \left(x_j^{\pm} \right)^2 \right] \quad (11a)$$

subject to:

$$\sum_{j=1}^{n} a_{rj}^{\pm} x_j^{\pm} \leq b_r^{\pm}, \ r = 1, 2, ..., m_1 \quad (11b)$$

$$\sum_{j=1}^{n} \underset{\sim}{a}_{tj}^{\pm} x_j^{\pm} \underset{\sim}{\leq} \tilde{b}_t^{\pm}, \ t = m_1 + 1, m_1 + 2, ..., m \quad (11c)$$

$$x_j^{\pm} \geq 0, \ j = 1, 2, ..., n \quad (11d)$$

where $\underset{\sim}{a}_{tj}^{\pm}$ and $\underset{\sim}{b}_t^{\pm}$ denote a set of intervals with vague lower and upper bounds. Assume that no intersection exists between the fuzzy sets at the two bounds. Letting $\underset{\sim}{a}_{tj}^{-}$ and $\underset{\sim}{a}_{tj}^{+}$ be lower and upper bounds of $\underset{\sim}{a}_{tj}^{\pm}$, we have $\underset{\sim}{a}_{tj}^{\pm} = [\underset{\sim}{a}_{tj}^{-}, \underset{\sim}{a}_{tj}^{+}]$.

Since nonlinearity exists in the objective function, it is difficult in defining the bounds for cost coefficients and decision variables corresponding to f^- and f^+. Consequently, two situations need to be considered. When cost coefficients $c_j^{\pm}$ and $d_j^{\pm}$ have different signs (i.e. when $c_j^{\pm} \geq 0$, then $d_j^{\pm}$ would be ≤ 0, and vice versa), the optimal bound distribution for $x_j^{\pm}$ can be identified based on a derivative algorithm proposed by Chen and Huang

(2001). Firstly, let all left- and/or right-hand-side coefficients be equal to their mid-values. Then, model (11) can be converted into a robust deterministic quadratic programming problem as follows:

$$\text{Min } f_{\text{mv}} = \sum_{j=1}^{n} \left[\left(c_j \right)_{\text{mv}} \left(x_j \right)_{\text{mv}} + \left(d_j \right)_{\text{mv}} \left(x_j \right)_{\text{mv}}^2 \right] \quad (12a)$$

subject to:

$$\sum_{j=1}^{n} \left(a_{rj} \right)_{\text{mv}} \left(x_j \right)_{\text{mv}} \leq \left(b_r \right)_{\text{mv}}, \forall r \quad (12b)$$

$$\sum_{j=1}^{n} \left(\bar{a}_{tj}^{s} \right)_{\text{mv}} \left(x_j \right)_{\text{mv}} \leq \left(\bar{b}_t^{s} \right)_{\text{mv}}, \forall t; s = 1, 2, \cdots, k \quad (12c)$$

$$\sum_{j=1}^{n} \left(\underset{\sim}{a}_{tj}^{s} \right)_{\text{mv}} \left(x_j \right)_{\text{mv}} \geq \left(\underset{\sim}{b}_t^{s} \right)_{\text{mv}}, \forall t; s = 1, 2, \cdots, k \quad (12d)$$

$$\left(x_j \right)_{\text{mv}} \geq 0, \ j = 1, 2, ..., n \quad (12e)$$

where $\left(c_j \right)_{\text{mv}}$, $\left(d_j \right)_{\text{mv}}$, $\left(a_{rj} \right)_{\text{mv}}$ and $\left(b_r \right)_{\text{mv}}$ are mid-values of $c_j^{\pm}$, $d_j^{\pm}$, $a_{rj}^{\pm}$ and $b_r^{\pm}$ [e.g. $\left(c_j \right)_{\text{mv}} = \left(c_j^{-} + c_j^{+} \right)/2$]; $\left(\underset{\sim}{a}_{tj}^{s} \right)_{\text{mv}}$, $\left(\bar{a}_{tj}^{s} \right)_{\text{mv}}$, $\left(\underset{\sim}{b}_t^{s} \right)_{\text{mv}}$ and $\left(\bar{b}_t^{s} \right)_{\text{mv}}$ are mid-values of $\underset{\sim}{a}_{tj}^{\pm}$ and $\underset{\sim}{b}_t^{\pm}$ [e.g. $\left(\underset{\sim}{a}_{tj}^{s} \right)_{\text{mv}} = \left(\underset{\sim}{a}_{tj}^{-s} + \underset{\sim}{a}_{tj}^{+s} \right)/2$]. The solutions for model (12) are $X_{\text{mv opt}} = \left\{ \left(x_j \right)_{\text{mv opt}} \big| \forall j \right\}$, where $\left(x_j \right)_{\text{mv opt}} \in \left[x_{j \text{ opt}}^{-}, x_{j \text{ opt}}^{+} \right], \forall j$.

Secondly, the optimal bound distribution for $x_j^{\pm}$ can be identified according to the following criteria:

$$f_j^{+} \left(x_{j \text{ opt}}^{+} \right) \geq f_j^{+} \left(x_{j \text{ opt}}^{-} \right), \text{ when } 2d_j^{+} \left(x_j \right)_{\text{mv opt}} + c_j^{+} > 0 \quad (13a)$$

$$f_j^{+} \left(x_{j \text{ opt}}^{+} \right) \leq f_j^{+} \left(x_{j \text{ opt}}^{-} \right), \text{ when } 2d_j^{+} \left(x_j \right)_{\text{mv opt}} + c_j^{+} < 0 \quad (13b)$$

When criterion (13a) is satisfied, then x_j^{+} corresponds to f^+; when criterion (13b) holds, then x_j^{-} corresponds to f^+. Assume criterion (13a) is satisfied, we can convert model (11) into two submodels (corresponding to f^- and f^+) as follows:

Submodel (1)

$$\text{Min } f^- = \sum_{j=1}^{k_1} \left[c_j^- x_j^- + d_j^- \left(x_j^- \right)^2 \right]$$
$$+ \sum_{j=k_1+1}^{n} \left[c_j^- x_j^+ + d_j^- \left(x_j^+ \right)^2 \right] \quad (14a)$$

subject to:

$$\sum_{j=1}^{k_1} |a_{rj}|^+ \text{Sign}\left(a_{rj}^+ \right) x_j^- + \sum_{j=k_1+1}^{n} |a_{rj}|^- \text{Sign}\left(a_{rj}^- \right) x_j^+ \leq b_r^-, \forall r$$

$$(14b)$$

$$\sum_{j=1}^{k_1} \overline{\{|a_{tj}|^+ \text{Sign}(a_{tj}^+)\}}^s x_j^- + \sum_{j=k_1+1}^{n} \overline{\{|a_{tj}|^- \text{Sign}(a_{tj}^-)\}}^s x_j^+ \leq \overline{b_t}^{-s},$$
$$\forall t; s = 1, 2, \cdots, k$$

$$(14c)$$

$$\sum_{j=1}^{k_1} \{|a_{tj}|^+ \text{Sign}(a_{tj}^+)\}^s x_j^- + \sum_{j=k_1+1}^{n} \{|a_{tj}|^- \text{Sign}(a_{tj}^-)\}^s x_j^+ \geq \underline{b_t}^{-s},$$
$$\forall t; s = 1, 2, \cdots, k$$

$$(14d)$$

$$0 \leq x_j^- \leq (x_j)_{\text{mv opt}}, \quad j = 1, 2, \cdots, k_1 \quad (14e)$$

$$x_j^+ \geq (x_j)_{\text{mv opt}}, \quad j = k_1 + 1, k_1 + 2, \cdots, n \quad (14f)$$

Submodel (2)

$$\text{Min } f^+ = \sum_{j=1}^{k_1} [c_j^+ x_j^+ + d_j^+ \left(x_j^+ \right)^2]$$
$$+ \sum_{j=k_1+1}^{n} [c_j^+ x_j^- + d_j^+ \left(x_j^- \right)^2] \quad (15a)$$

subject to:

$$\sum_{j=1}^{k_1} |a_{rj}|^- \text{Sign}\left(a_{rj}^- \right) x_j^+ + \sum_{j=k_1+1}^{n} |a_{rj}|^+ \text{Sign}\left(a_{rj}^+ \right) x_j^- \leq b_r^+, \forall r$$

$$(15b)$$

$$\sum_{j=1}^{k_1} \overline{\{|a_{tj}|^- \text{Sign}(a_{tj}^-)\}}^s x_j^+ + \sum_{j=k_1+1}^{n} \overline{\{|a_{tj}|^+ \text{Sign}(a_{tj}^+)\}}^s x_j^- \leq \overline{b_t}^s,$$
$$\forall t; s = 1, 2, \cdots, k$$

$$(15c)$$

$$\sum_{j=1}^{k_1} \{|a_{tj}|^- \text{Sign}(a_{tj}^-)\}^s x_j^+ + \sum_{j=k_1+1}^{n} \{|a_{tj}|^+ \text{Sign}(a_{tj}^+)\}^s x_j^- \geq \underline{b_t}^{+s},$$
$$\forall t; s = 1, 2, \cdots, k$$

$$(15d)$$

$$x_j^+ \geq (x_j)_{\text{mv opt}}, \quad j = 1, 2, \cdots, k_1 \quad (15e)$$

$$0 \leq x_j^- \leq (x_j)_{\text{mv opt}}, \quad j = k_1 + 1, k_1 + 2, \cdots, n \quad (15f)$$

Solving the above two submodels, we can obtain the solutions for model (11): $x_{j\text{opt}}^{\pm} = \left[x_{j\text{opt}}^-, x_{j\text{opt}}^+ \right], \forall j$ and $f_{\text{opt}}^{\pm} = \left[f_{\text{opt}}^-, f_{\text{opt}}^+ \right]$. When cost coefficients $c_j^{\pm}$ and $d_j^{\pm}$ have the same sign (i.e., when $c_j^{\pm} \geq 0$, then $d_j^{\pm} \geq 0$, and vice versa), model (11) can be directly transformed into two submodels, which correspond to the lower and upper bounds of the objective function value, respectively. Then, each submodel can be converted into a conventional linear program, provided that condition (7) is satisfied for each fuzzy constraint.

Case study

The case study will focus on the planning of waste-flow allocation and waste-management facility development/expansion for the City of Changchun, which is the capital of Jilin Province and located in the northeast of China. The region covers an area of approximately 379.9 square kilometers, including five main districts (abbreviated as D1 to D5, as shown in Figure 1). The city has a population of 2.78 million, where the households generate residential wastes of approximately 3060 tonnes per day. The city's MSW collection and disposal system consists of six transfer stations, one waste-to-energy facility (i.e. Xinghua incinerator), and one landfill. Each district is responsible for its own curbside garbage collection, using either its own force or a contracted service. The city is responsible for the disposal of the collected wastes through the use of the transfer stations and waste management facilities. The majority of wastes collected through curbside pick-up from districts are delivered to the transfer stations, where some recyclable wastes received are recycled; then the others are compacted and hauled to the landfill or incinerator. The current operating capacities of transfer stations are approximately 2200 tonnes per day, less than the amount of waste generated. This leads to some wastes

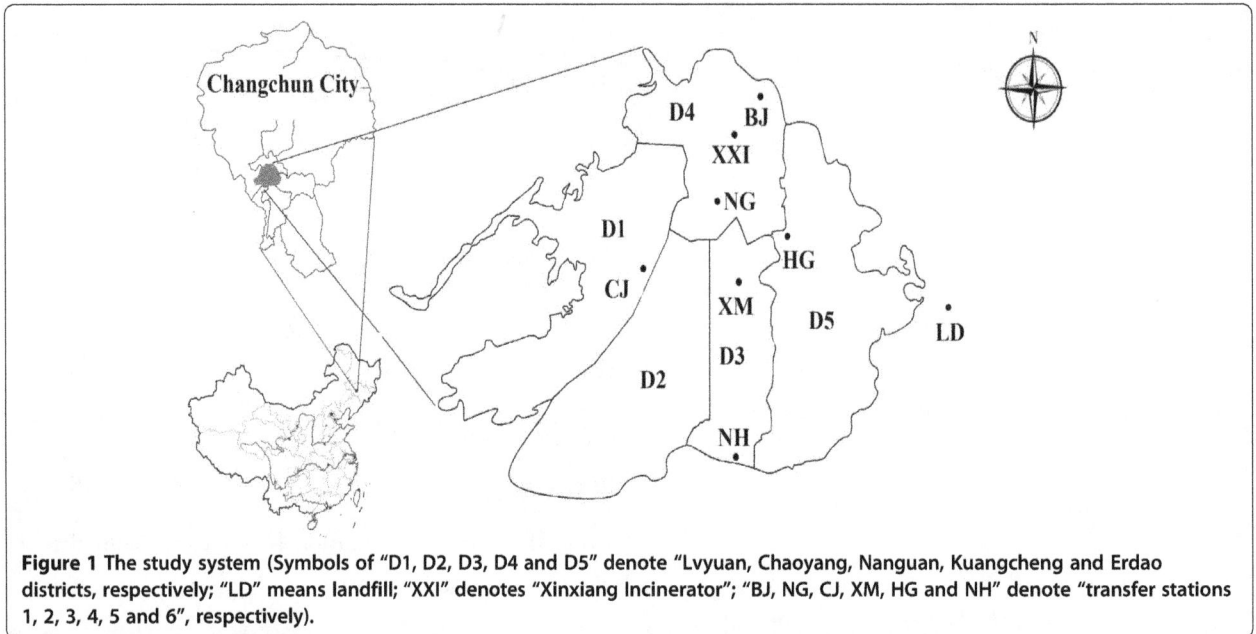

Figure 1 The study system (Symbols of "D1, D2, D3, D4 and D5" denote "Lvyuan, Chaoyang, Nanguan, Kuangcheng and Erdao districts, respectively; "LD" means landfill; "XXI" denotes "Xinxiang Incinerator"; "BJ, NG, CJ, XM, HG and NH" denote "transfer stations 1, 2, 3, 4, 5 and 6", respectively).

directly being hauled to the landfill or the incinerator. Figure 2 shows the routine of waste allocation and disposal of. Source reduction and recycling are the two most desirable operations to achieve waste minimization (Tai et al. 2011); therefore, collecting recyclables at the source by residents. This minimization of total waste could benefit the stages of transportation and treatment. However, the quantity of recyclables collected by residents and junkmen greatly exceeds the quantity of recyclables in the municipal recycling system.

The MSW generated typically include paper, yard waste, food waste, plastics, metals, glass, wood and other items. Similar to many other cities in China, the study city mainly relies on the use of landfill for handling its solid waste. The landfill is used directly to satisfy waste

disposal demand or alternatively to provide capacity for the other facilities' residue disposals. Before 2008, approximately 2800 to 3000 tonnes per day of the waste (i.e. over 90% of total residential wastes) were buried at the landfill with simple pretreatment, especially in the areas of rural–urban fringe and countryside. In addition, the amount of residential waste diverted from landfill is still low (i.e. less than 17% of total wastes generated by households); increasing the waste diversion, separation and recovery rate and thus reducing the wastes to landfill are becoming an important goal. In 2009, the city built a waste-to-energy facility with a daily capacity of about 500 tonnes (i.e. the Xinxiang incinerator) to help reduce the amount of wastes that ends up at the landfill. The incinerator can generate approximately 51 million kwh per year and gain profits through the sale of electricity and environmentally-friendly building materials. Over the last ten years, the city has achieved significant improvements in waste management. Regional policies and guidelines for collection, transport, treatment and disposal of municipal and industrial wastes in environmentally safe ways have been established. However, the city has been unable to keep up with the growing demand for waste service coverage, environmental requirements for safe disposal systems, and rationalization of cost-effectiveness in service delivery. The capacities of waste management at regional and central levels, the abilities of the local scientists and technicians, and the environmental protection consciousness of the local communities are required to improve immediately to meet the requirements of environmental protection and regional sustainability.

Precise data is hard to be obtained due to temporal and spatial variations in MSW system conditions. The

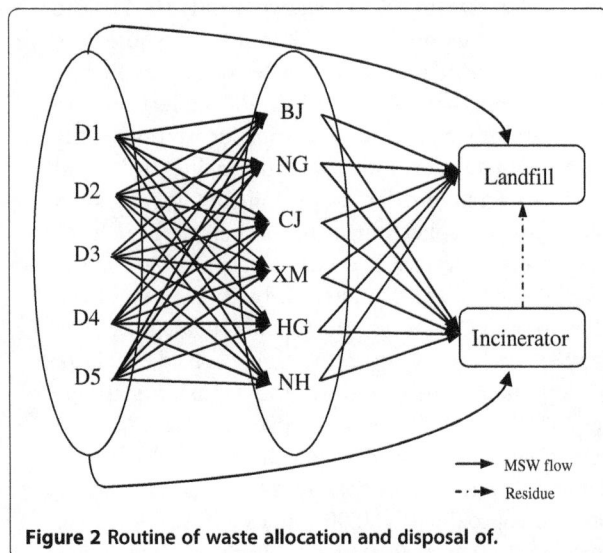

Figure 2 Routine of waste allocation and disposal of.

city's future MSW generation rate is uncertain since the city has been encountered rapid increase in the amount of MSW as a result of swift urbanization, industrialization and economic development during the past decades. The average waste-generation rate estimated may keep increasing to 1.30 kg/capita/day in 2020, while the city's population growth rate is 0.8% per year (CUPC, 2004); correspondingly, in 2020, the residential sector could generate around 4000 tonnes of wastes per day. This tendency could result in insufficient capacities of waste-management facilities to meet the overall waste disposal demand. Also, intrinsic fluctuations of many impact factors, such as waste compositions and operation conditions, may result in uncertainties associated with the capacities of waste management facilities. Waste collection cost can also be affected by vehicle types, collection efficiencies, oil prices, and collection routes; the operation cost may be related to labor fees, equipment prices, energy prices, and management expenses; these can result in uncertainties in estimating the costs for waste collection, transportation and disposal of. For example, collection cost within a collection area depends on the type of vehicle used (and costs associated with it) and the efficiency of collection. The efficiency of collection in turn depends on factors such as type of vehicle, crew size, collection routes, collection frequency, and local conditions. The waste transportation cost is related to vehicle movements outside the collection areas during a working day. Consequently, interactions exist among the waste flows and their transportation costs; particularly, when waste flows are high or hauling distances are long, the effects of EOS may become significant. In general, the EOS effects in terms of waste transportation/operation can be expressed as a sizing model with a power law (Thuesen et al. 1977):

$$C_t = C_{re}(X_t/X_{re})^m \qquad (16)$$

where X_t is decision variable for waste flow (tonne/day), X_{re} is a reference waste flow (tonne/day), C_t is the transportation/treatment cost for waste flow X_t ($/tonne), C_{re} is a known cost for reference waste flow X_{re} ($/tonne), and m is an EOS exponent ($0 < m < 1$).

Table 1 presents waste-generation rates for the five districts. The study time span is 15 years, which is further divided into three 5-year periods. Tables 2 and 3 show transportation costs for waste flows from districts to transfer stations, districts to landfill/incinerator, and transfer stations to landfill/incinerator. The nonlinear relationships in transportation cost were approximated by inexact linear functions. In this study, the EOS exponent for transportation cost is 0.8 ~ 0.9. Figure 3 shows the curve of waste flow vs transportation cost from district 1 to landfill in period 1. Table 4 shows the other modeling inputs,

Table 1 Waste-generation rate

District	k = 1	k = 2	k = 3
D1	[590, 670]	[610, 700]	[638, 722]
D2	[765, 870]	[795, 907]	[825, 941]
D3	[630, 719]	[656, 747]	[683, 776]
D4	[486, 552]	[511, 574]	[526, 597]
D5	[403, 457]	[419, 476]	[436, 494]

implying that a variety of uncertainties presented as different formats (e.g. deterministic and fuzzy intervals) exist in the modeling parameters.

In general, uncertainties and nonlinearities exist in the city's MSW management activities, which can affect the optimization processes and the decision schemes generated. Therefore, based on the robust interval quadratic programming method, the study problem can be formulated as follows:

$$\text{Min } f^{\pm} = \sum_{i=1}^{2}\sum_{j=1}^{5}\sum_{k=1}^{3} L_k X_{ijk}^{\pm}\left[\left(\alpha_{ijk}^{\pm} X_{ijk}^{\pm} + \beta_{ijk}^{\pm}\right) + OP_{ik}^{\pm}\right]$$
$$+ \sum_{r=1}^{6}\sum_{j=1}^{5}\sum_{k=1}^{3} L_k X_{rjk}^{\pm}\left[\left(\alpha_{rjk}^{\pm} X_{rjk}^{\pm} + \beta_{rjk}^{\pm}\right) + OP_{rk}^{\pm}\right]$$
$$+ \sum_{i=1}^{2}\sum_{r=1}^{6}\sum_{k=1}^{3} L_k X_{irk}^{\pm}\left[\left(\alpha_{irk}^{\pm} X_{irk}^{\pm} + \beta_{irk}^{\pm}\right) + OP_{ik}^{\pm}\right]$$
$$+ \sum_{k=1}^{3} L_k \left\{ \left(\sum_{j=1}^{5} FE_1^{\pm} X_{2jk}^{\pm} + \sum_{r=1}^{6} FE_2^{\pm} X_{2rk}^{\pm}\right)\right.$$
$$\left[\alpha_{2k}^{\pm}\left(\sum_{j=1}^{5} FE_1^{\pm} X_{2jk}^{\pm} + \sum_{r=1}^{6} FE_2^{\pm} X_{2rk}^{\pm}\right) + \beta_{2k}^{\pm}\right]$$
$$\left. + OP_{1k}^{\pm} \right\}$$
$$- \sum_{k=1}^{3} L_k RE_{2k}^{\pm}\left(\sum_{j=1}^{5} \eta_1^{\pm} X_{2jk}^{\pm} + \sum_{r=1}^{6} \eta_2^{\pm} X_{2rk}^{\pm}\right)$$
$$- \sum_{j=1}^{5}\sum_{r=1}^{6} L_k RE_{rk}^{\pm}\eta_3^{\pm} X_{rjk}^{\pm} + \sum_{r=1}^{6}\sum_{k=1}^{3} VTC_{rk}^{\pm}\cdot ETC_{rk}^{\pm}$$
$$+ \sum_{k=1}^{3} VLC^{\pm}\cdot ELC^{\pm} + \sum_{k=1}^{3} VIC_k^{\pm}\cdot EIC_k^{\pm}$$

$$(17a)$$

subject to:

$$\sum_{k=1}^{k'} L_k DF^{\pm}\left(\sum_{j=1}^{5} X_{1jk}^{\pm} + \sum_{r=1}^{6} X_{1rk}^{\pm}\right.$$
$$\left. + FE_1^{\pm}\sum_{j=1}^{5} X_{2jk}^{\pm} + FE_2^{\pm}\sum_{r=1}^{6} X_{2rk}^{\pm}\right) \le TLC^{\pm}$$
$$+ \sum_{k=1}^{k'} ELC_k^{\pm}, \quad k'$$
$$= 1, 2, 3 \qquad (17b)$$

Table 2 Shipping cost for waste to facilities

	k = 1	k = 2	k = 3
Waste transport cost (to landfill) ($/t)			
D1	[8.44-0.0073x, 10.34-0.0090x]	[8.87-0.0076x, 10.86-0.0093x]	[9.57-0.0082x, 11.73-0.0100x]
D2	[6.74-0.0059x, 9.41-0.0082x]	[7.07-0.0061x, 9.88-0.0085x]	[7.64-0.0065x, 10.67-0.0091x]
D3	[5.60-0.0049x, 8.40-0.0073x]	[5.88-0.0051x, 8.82-0.0076x]	[6.35-0.0054x, 9.52-0.0081x]
D4	[7.30-0.0063x, 9.59-0.0083x]	[7.67-0.0066x, 10.07-0.0087x]	[8.28-0.0071x, 10.88-0.0093x]
D5	[0.98-0.0009x, 6.74-0.0058x]	[1.03-0.0009x, 7.07-0.0061x]	[1.11-0.0009x, 7.64-0.0065x]
BJ	[3.26-0.0028x, 3.75-0.0032x]	[3.42-0.0029x, 3.93-0.0034x]	[3.69-0.0031x, 4.25-0.0036x]
NG	[2.88-0.0025x, 3.31-0.0029x]	[3.02-0.0026x, 3.47-0.0029x]	[3.26-0.0028x, 3.75-0.0032x]
CJ	[3.45-0.0030x, 3.98-0.0034x]	[3.62-0.0031x, 4.18-0.0036x]	[3.91-0.0033x, 4.51-0.0038x]
XM	[2.51-0.0022x, 2.88-0.0025x]	[2.64-0.0022x, 3.03-0.0026x]	[2.85-0.0024x, 3.27-0.0028x]
HG	[2.49-0.0022x, 2.87-0.0025x]	[2.62-0.0030x, 3.01-0.0026x]	[2.82-0.0024x, 3.25-0.0028x]
NH	[3.37-0.0029x, 3.88-0.0034x]	[3.54-0.0030x, 4.08-0.0035x]	[3.83-0.0033x, 4.40-0.0037x]
Waste transport cost (to incinerator) ($/t)			
D1	[3.27-0.0028x, 5.77-0.0050x]	[3.43-0.0029x, 6.06-0.0052x]	[3.71-0.0032x, 6.54-0.0056x]
D2	[4.17-0.0036x, 6.26-0.0054x]	[4.38-0.0038x, 6.57-0.0056x]	[4.73-0.0040x, 7.10-0.0060x]
D3	[3.45-0.0030x, 6.66-0.0058x]	[3.62-0.0031x, 6.99-0.0061x]	[3.91-0.0033x, 7.55-0.0064x]
D4	[1.04-0.0009x, 2.30-0.0020x]	[1.09-0.0009x, 2.42-0.0021x]	[1.17-0.0010x, 2.61-0.0022x]
D5	[2.83-0.0025x, 7.15-0.0062x]	[2.97-0.0025x, 7.51-0.0065x]	[3.21-0.0027x, 8.11-0.0069x]
BJ	[0.48-0.0004x, 0.55-0.0005x]	[0.50-0.0004x, 0.57-0.0005x]	[0.54-0.0005x, 0.62-0.0005x]
NG	[0.78-0.0007x, 0.89-0.0008x]	[0.82-0.0007x, 0.94-0.0008x]	[0.88-0.0007x, 1.02-0.0009x]
CJ	[1.92-0.0017x, 2.21-0.0019x]	[2.01-0.0017x, 2.31-0.0020x]	[2.17-0.0018x, 2.50-0.0021x]
XM	[1.74-0.0015x, 2.01-0.0017x]	[1.83-0.0016x, 2.10-0.0018x]	[1.98-0.0017x, 2.27-0.0019x]
HG	[1.35-0.0012x, 1.55-0.0013x]	[1.42-0.0012x, 1.63-0.0014x]	[1.53-0.0013x, 1.76-0.0015x]
NH	[3.37-0.0029x, 3.88-0.0034x]	[3.54-0.0030x, 4.08-0.0035x]	[3.82-0.0033x, 4.40-0.0037x]

$$\sum_{j=1}^{5} X_{1jk}^{\pm} + \sum_{r=1}^{6} X_{1rk}^{\pm} + \sum_{j=1}^{5} FE_1^{\pm} X_{2jk}^{\pm}$$

$$+ \sum_{r=1}^{6} FE_2^{\pm} X_{2rk}^{\pm} \le DLC^{\pm} \qquad (17c)$$

$$(1 + \widetilde{\theta^{\pm}}) \left(\sum_{j=1}^{5} X_{2jk}^{\pm} + \sum_{r=1}^{6} X_{2rk}^{\pm} \right) \le \widetilde{IC}^{\pm}$$

$$+ \sum_{k=1}^{k'} EIC_k^{\pm}, k'$$

$$= 1, 2, 3 \qquad (17d)$$

$$\sum_{j=1}^{5} X_{rjk}^{\pm} \le TC_r^{\pm} + \sum_{k=1}^{k'} ETC_{rk}^{\pm}, \forall r; k' = 1, 2, 3 \qquad (17e)$$

$$\sum_{j=1}^{5} X_{rjk}^{\pm} (1 - \eta_4^{\pm}) = \sum_{i=1}^{2} X_{irk}^{\pm}, \forall r, k \qquad (17f)$$

$$\sum_{i=1}^{2} X_{ijk}^{\pm} + \sum_{r=1}^{6} X_{rjk}^{\pm} \ge WG_{jk}^{\pm}, \forall j, k \qquad (17g)$$

$$0 \le \sum_{k=1}^{3} ELC_k^{\pm} \le MELC \qquad (17h)$$

$$0 \le EIC_k^{\pm} \le MEIC_k, \forall k \qquad (17i)$$

$$0 \le ETC_{rk}^{\pm} \le METC_k, \forall r, k \qquad (17j)$$

$$\sum_{r=1}^{6} ETC_{rk}^{\pm} \le TETC_k, \forall k \qquad (17k)$$

$$X_{ijk}^{\pm}, X_{rjk}^{\pm}, X_{irk}^{\pm} \ge 0, \forall i, j, r, k \qquad (17l)$$

The detailed nomenclatures for the variables and parameters are provided in the Appendix. The objective is to minimize the sum of the expenses for collecting, shipping and disposing wastes as well as costs for expanding transfer stations, landfill and incinerator. The constraints define the interrelationships

Table 3 Transportation cost for waste from district to transfer station ($/t)

	k = 1	k = 2	k = 3
D1-BJ	[3.49-0.0030x, 5.38-0.0047x]	[3.66-0.0032x, 5.65-0.0049x]	[3.96-0.0034x, 6.10-0.0052x]
D2-BJ	[4.73-0.0041x, 6.44-0.0056x]	[4.97-0.0043x, 6.76-0.0058x]	[5.37-0.0046x, 7.30-0.0062x]
D3-BJ	[3.39-0.0029x, 5.67-0.0049x]	[3.56-0.0031x, 5.96-0.0051x]	[3.84-0.0033x, 6.43-0.0055x]
D4-BJ	[1.25-0.0019x, 5.36-0.0047x]	[1.31-0.0011x, 3.06-0.0026x]	[1.41-0.0012x, 3.30-0.0028x]
D5-BJ	[2.24-0.0019x, 5.36-0.0047x]	[2.35-0.0020x, 5.62-0.0048x]	[2.54-0.0022x, 6.08-0.0052x]
D1-NG	[1.74-0.0015x, 4.42-0.0038x]	[1.83-0.0016x, 4.64-0.0040x]	[2.00-0.0017x, 5.01-0.0043x]
D2-NG	[1.50-0.0013x, 4.29-0.0037x]	[1.57-0.0013x, 4.51-0.0039x]	[1.70-0.0014x, 4.87-0.0041x]
D3-NG	[1.25-0.0011x, 4.24-0.0037x]	[1.31-0.0011x, 4.45-0.0038x]	[1.41-0.0012x, 4.80-0.0041x]
D4-NG	[0.50-0.0004x, 2.05-0.0018x]	[0.52-0.0004x, 2.15-0.0018x]	[0.57-0.0005x, 2.32-0.0020x]
D5-NG	[1.49-0.0013x, 5.35-0.0046x]	[1.57-0.0013x, 5.61-0.0048x]	[1.70-0.0014x, 6.06-0.0052x]
D1-CJ	[0.25-0.0002x, 2.91-0.0025x]	[0.26-0.0002x, 3.06-0.0026x]	[0.28-0.0002x, 3.30-0.0028x]
D2-CJ	[0.50-0.0004x, 3.22-0.0028x]	[0.52-0.0004x, 3.38-0.0029x]	[0.57-0.0005x, 3.65-0.0031x]
D3-CJ	[2.24-0.0019x, 3.37-0.0031x]	[2.35-0.0020x, 3.54-0.0030x]	[2.54-0.0022x, 3.83-0.0032x]
D4-CJ	[1.74-0.0015x, 3.52-0.0031x]	[1.83-0.0016x, 3.69-0.0032x]	[1.98-0.0017x, 4.00-0.0034x]
D5-CJ	[2.99-0.0026x, 6.14-0.0053x]	[3.14-0.0027x, 6.45-0.0055x]	[3.39-0.0029x, 6.97-0.0059x]
D1-XM	[2.99-0.0026x, 3.22-0.0028x]	[3.14-0.0027x, 3.38-0.0029x]	[3.39-0.0029x, 3.65-0.0031x]
D2-XM	[1.74-0.0015x, 2.91-0.0025x]	[1.83-0.0016x, 3.06-0.0026x]	[1.98-0.0017x, 3.30-0.0028x]
D3-XM	[0.54-0.0004x, 2.45-0.0021x]	[0.56-0.0005x, 2.58-0.0022x]	[0.61-0.0005x, 2.78-0.0024x]
D4-XM	[1.50-0.0015x, 5.65-0.0049x]	[1.57-0.0013x, 2.74-0.0024x]	[1.70-0.0014x, 2.96-0.0025x]
D5-XM	[1.74-0.0015x, 5.65-0.0049x]	[1.83-0.0016x, 5.94-0.0051x]	[1.98-0.0017x, 6.41-0.0055x]
D1-HG	[4.99-0.0043x, 5.98-0.0052x]	[5.24-0.0045x, 6.28-0.0054x]	[5.66-0.0048x, 6.78-0.0058x]
D2-HG	[2.51-0.0022x, 3.36-0.0029x]	[2.64-0.0023x, 3.53-0.0030x]	[2.85-0.0024x, 3.81-0.0032x]
D3-HG	[1.01-0.0009x, 3.02-0.0026x]	[1.06-0.0009x, 3.17-0.0027x]	[1.14-0.0010x, 3.42-0.0029x]
D4-HG	[1.26-0.0011x, 2.91-0.0025x]	[1.33-0.0011x, 3.06-0.0026x]	[1.44-0.0012x, 3.30-0.0028x]
D5-HG	[0.58-0.0005x, 2.15-0.0019x]	[0.61-0.0005x, 2.25-0.0019x]	[0.66-0.0006x, 2.43-0.0021x]
D1-NH	[4.73-0.0041x, 5.37-0.0047x]	[4.97-0.0043x, 5.64-0.0049x]	[5.37-0.0048x, 6.09-0.0052x]
D2-NH	[1.75-0.0015x, 3.35-0.0029x]	[1.84-0.0016x, 3.52-0.0030x]	[1.99-0.0017x, 3.80-0.0032x]
D3-NH	[0.53-0.0004x, 3.22-0.0029x]	[0.55-0.0005x, 3.38-0.0029x]	[0.60-0.0005x, 3.65-0.0031x]
D4-NH	[5.31-0.0046x, 6.46-0.0056x]	[5.57-0.0048x, 6.78-0.0058x]	[6.02-0.0051x, 7.33-0.0062x]
D5-NH	[1.74-0.0015x, 5.79-0.0050x]	[1.83-0.0016x, 6.08-0.0052x]	[1.98-0.0017x, 6.56-0.0056x]

among the decision variables and the waste generation/management conditions. In detail, constraints (17 b) to (17 e) denote that the wastes disposed by each treatment facility (i.e. landfill, incinerator and transfer stations) must not exceed their existing and expanded capacities; constraint (17 f) means that the mass balance of waste flows at transfer stations, where the volume of wastes can be reduced and various useful wastes can be recycled; constraint (17 g) denotes that the waste flows disposed by the waste-management facilities must be over the total waste generation amounts; constraints (17 h) to (17 k) regulate the expansion scales for waste-management facilities; constraint (17 l) stipulates that the decision variables are non-negative.

Result analysis

Figure 4 presents the results for waste allocated from districts to different transfer stations to be pre-treated over the planning horizon. In detail, in period 1, district 1 would use three transfer stations (i.e. namely NG, CJ and XM) to pre-treat and shift its wastes, while the amounts of wastes to NG, CJ and XM would be 44, [330.0, 343.2] and [216.0, 227.2] t/d (i.e. tonne/day), respectively; district 2 would ship [765, 813] t/d of waste to BJ; wastes from district 3 to XM, HG and NH would be 64, [256.0, 277.2], and [310.0, 322.4] t/d, respectively; district 4 would employ BJ and NG to pre-treat its waste (i.e. 55 and [431 450] t/d of wastes, respectively); district 5 would only use HG to shift its wastes (i.e. 274 t/d). The solutions of waste shipped to transfer stations in

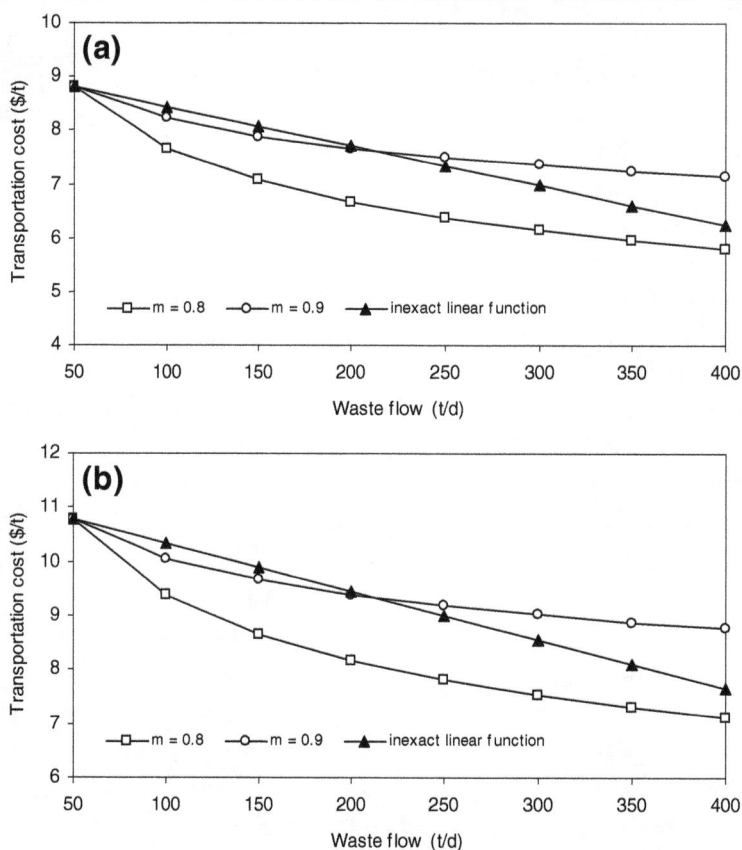

Figure 3 (a) Lower- and (b) upper-bound cost line.

periods 2 and 3 can be similarly interpreted based on the results as shown in Figure 4. In general, in period 1, the majority of wastes would be first transported to transfer stations to be pre-treated (i.e. [2874, 3268] t/d of wastes, occupying [87.8, 95.5]% of total wastes generated in the five districts), while a small fraction of wastes would be directly hauled to landfill and incinerator; in periods 2 and 3, all wastes generated would be first shipped to transfer stations and then to landfill and/or incinerator. The transfer stations have many advantages in waste transportation and treatment such as (i) decreasing vehicle traffic going to and from landfill and

Table 4 Technical modeling inputs

Residue generation rate when waste is shipped from district to incinerator	[0.25, 0.30]
Residue generation rate when waste is shipped from TR to incinerator	[0.12, 0.15]
Density factor for waste disposed of at landfill (m^3/t)	[1.47, 1.82]
Volume of the existing landfill (10^3 m^3)	[76, 80]
Daily landfill capacity (t/day)	[1950, 2020]
Daily incinerator capacity (t/day)	[[440, 470], [490, 520]]
Safety coefficient for incinerator	[[0.09, 0.13], [0.17, 0.21]]
Electricity-generation rate when waste is shipped from district to incinerator (KWh/t)	[280, 300]
Electricity-generation rate when waste is shipped from TR to incinerator (KWh/t)	[370, 400]
Material recycling rate at TR	[0.16, 0.20]
Mass loss rate at TR	[0.22, 0.25]
Maximum expansion capacity for landfill (10^6 m^3)	18
Maximum expansion capacity once time for incinerator (t/day)	250
Maximum expansion capacity once time for transfer station (t/day)	600

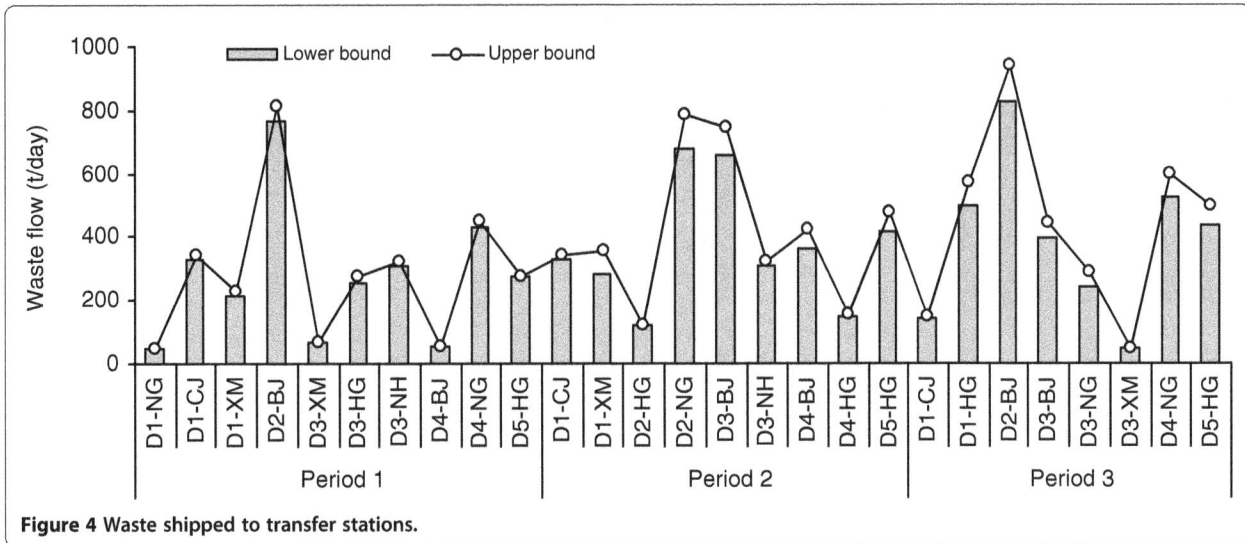

Figure 4 Waste shipped to transfer stations.

incinerator and thus reducing transportation cost, (ii) recycling various useful wastes, (iii) providing an inspection area where wastes can be viewed and unacceptable materials be removed, (iv) providing an effective control on dumping site at the landfill, (v) reducing the volume of wastes buried at the landfill, and (vi) raising the efficiency of incinerating wastes at the incinerator and reducing air-pollutant emissions. Besides, transfer stations are also more convenient for both MSW collectors and individual users since they are closer and easier to access than the landfill and incinerator sites.

Figure 5 shows the results of waste from transfer stations to landfill and incinerator over the planning span time. For example, in period 1, all wastes at the BJ, XM, HG and NH would be hauled to the landfill (i.e. [639.6, 651.0], 218.4, 413.4 and 241.8 t/d, respectively); all

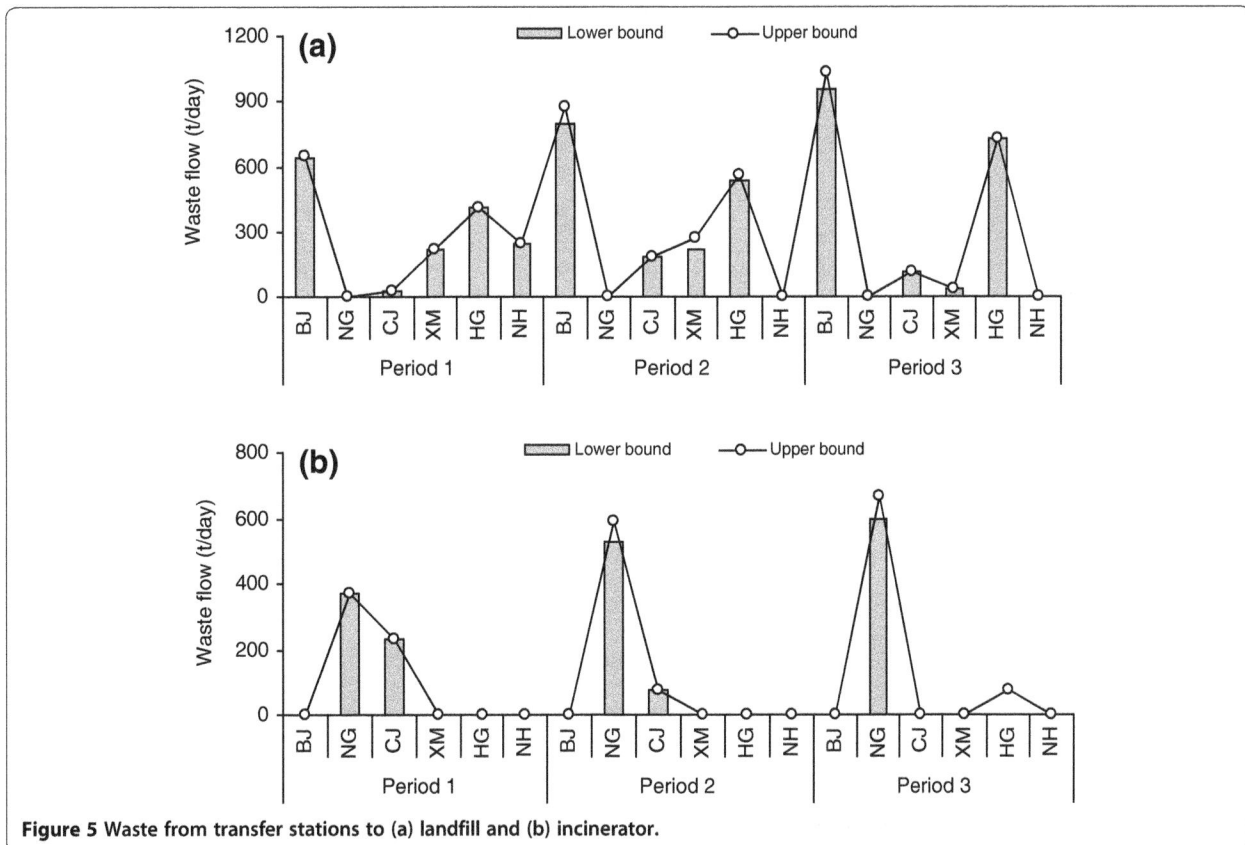

Figure 5 Waste from transfer stations to (a) landfill and (b) incinerator.

wastes at NG would be shipped to the incinerator; wastes of CJ allocated to the landfill and incinerator would be 27.9 and 229.5 t/d, respectively. In summary, landfill would be the main choice for disposal of the city's wastes; incinerator would help treat very small fraction of wastes. Figure 6 presents the amounts of wastes finally disposed at the landfill and incinerator. The amounts of wastes disposed of by the landfill would be [1670.1, 1903.5] t/d in period 1, [1733.0, 1889.3] t/d in period 2, and [1824.2, 1909.3] t/d in period 3; the wastes treated by incinerator would be [600.0, 647.0], [600.0, 664.0] and [600.0, 738.2] t/d in periods 1, 2 and 3, respectively. Approximately [73.6, 74.4]% of total wastes would be finally disposed of at the landfill over the planning horizon.

Since the city's MSW generation rates are increasing due to population expansion and economy development. Moreover, the available capacities of waste-management facilities may reduce with time periods (e.g. the cumulative capacity of landfill). This tendency could often result in insufficient capacities of waste-management facilities to meet the overall waste disposal of demand in future. The results indicate that capacity expansion for waste management facilities (including landfill, incinerator and transfer station) is required. The landfill would be expanded once in period 1, with an incremental volume of 18 million cubic meters. For transfer station and incinerator, there would be two expansion options corresponding to lower- and upper-bound objective function values (i.e. f^- and f^+), respectively. Figure 7 shows the optimal expansion scheme for the incinerator over the planning horizon. When the decision scheme tends toward f^- under advantageous system conditions, the incinerator would be expanded once in period 1 with an increment of 209.0 t/d; conversely, when the scheme tends toward f^+ under more demanding conditions, the incinerator would be expanded three times over the planning horizon, and the total expansion capacity is 364.4 t/d (i.e. with capacities of 263.2 t/d in period

1, 18.6 t/d in period 2, and 82.7 t/d in period 3). Transfer stations would be expanded to satisfy increasing waste-generation amount and waste pre-treatment requirement, as shown in Figure 8. For example, BJ would be expanded with capacities of [200.0, 288.0] t/d in period 1, [200.0, 214.2] t/d in period 2, and [200.0, 214.0] t/d in period 3. The capacities of BJ, NG, CJ, XM, HG and NH would reach to [1180.0, 1336.2], [739.2, 915.3], [305.0, 368.2], [260.0, 376.8], [900.0, 1096.2] and [280.0, 352.4] t/d, respectively. Consequently, the total transfer capacities would achieve [3664.2, 4445.1] t/d at the end of planning horizon (after capacity expansion), and lower and upper bounds respectively correspond to advantageous and demanding conditions.

Discussion and conclusion

A robust interval quadratic programming method has been developed through incorporating techniques of robust programming and interval quadratic programming within a general optimization framework. The developed method can not only tackle uncertainties expressed as interval values, fuzzy sets, and their combinations, but also deal with nonlinearities in the objective function such that economies-of-scale effects can be reflected. Furthermore, the uncertain decision space has been delimited through dimensional enlargement of the original fuzzy constraints, leading to enhanced robustness for the optimization process. Compared with the conventional quadratic programming methods, the developed method can address more uncertainties without unrealistic simplifications or information losses, such that the robustnesses of the optimization processes and solutions can be enhanced. The developed method has been applied to planning long-term municipal solid waste (MSW) management in the City of Changchun, China. The results have been generated, which are valuable for helping governmental officials more intuitive to know some basic situation under

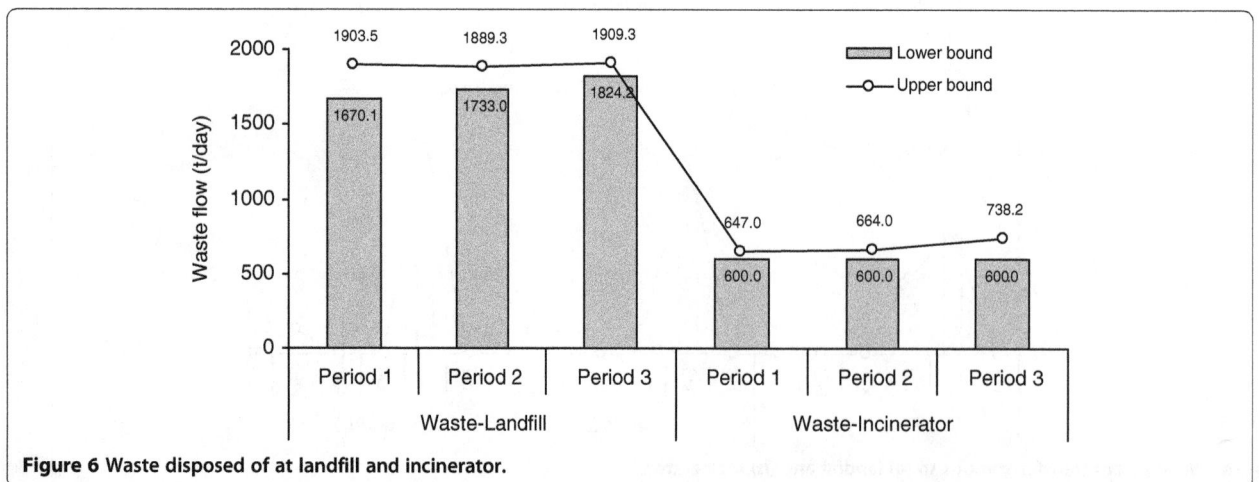

Figure 6 Waste disposed of at landfill and incinerator.

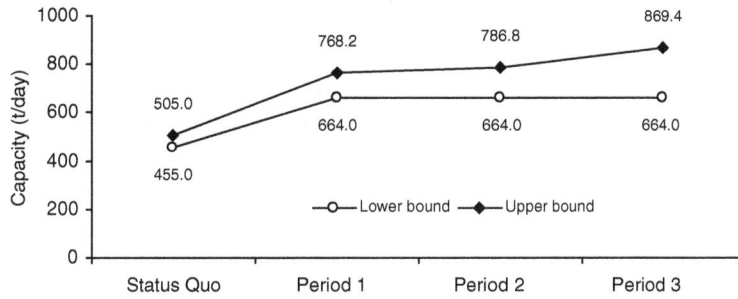

Figure 7 Capacity expansion scheme for incinerator.

complex uncertainties, such as optimal waste-flow allocation, waste-flow routing, facility-capacity expansion, and system cost. They can be used to further generate decisions for supporting long-term MSW management and planning activities in the city, and thus help managers to identify desired MSW policies in association with cost minimization under uncertainty.

The results indicate that, in the future 15 years, the city's majority of wastes would be disposed of at the landfill due to its relatively low operation cost and low capital for facility development/expansion. However, pollutant emissions from landfill site can take a number of forms: gaseous emissions of volatile organic compounds (VOCs), airborne particulate matter and leachate. For

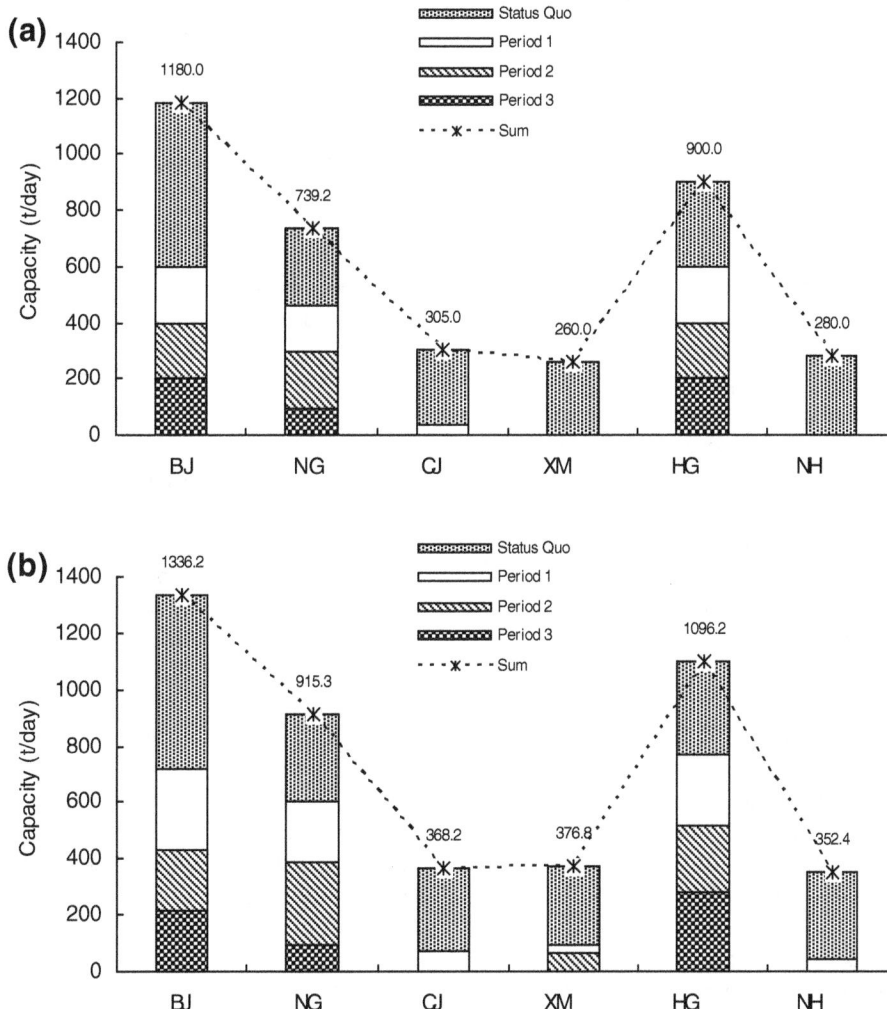

Figure 8 (a) Lower- and (b) upper-bound capacity-expansion schemes for transfer stations.

example, surface water could be polluted by rainwater flowing through solid waste piles. Groundwater could be contaminated by leachate from landfill sites where solid wastes are disposed of. The polluted surface water and groundwater can further affect the drinking water safety; leachate containing hazardous materials can enter soil and further reside in the agricultural products that make our foods poisonous. Secondly, landfills are the first and/or second largest contribution of methane (CH_4) source (e.g. in 2006, the amount of CH_4 released from landfills was 5985 Gg, occupying 23% of total US anthropogenic methane emissions) (USEPA, 2007). Recently, there is an increasing concern for CH_4, as a major greenhouse gas, while its global warming potential is about 23 on a 100-year time horizon (Mor et al. 2006; Chen et al. 2010). Thirdly, conflicts exist in the urban land resources due to the rapid population growth and swift economy development. Particularly, for the City of Changchun, the serious scarcity of land near urban centers leads to waste disposed of at landfill more and more noneconomic. Therefore, issues of land resource consumption, surface water/groundwater contamination, and greenhouse gas effect may imply higher environmental penalties than the savings obtained from waste buried.

The results also indicate that the city has to expand the incinerator to treat its more and more MSW over the planning horizon. Waste incineration can also generate considerable pollutant emissions (e.g. acid gases, metals and various organic compounds) that can present potential human health hazards. Incinerator emissions are complex and depend on the type of waste, the design of the incinerator, combustion conditions, and pollution control equipment. A number of pollutants (e.g. acid gases, metals and various organic compounds, including dioxins) are associated with health hazards and have thus raised serious concerns in the city. Therefore, research efforts focused on vulnerability analysis and risk assessment of human health and ambient environment for the city's incinerator are desired. Evaluation of the risk effects of air pollution on human health requires a series of assessment activities such as emission and dispersion of pollutants in the atmosphere, exposure of humans to pollutants, and adverse effects of the pollutants on human health; this will be of challenge for many waste managers. In addition, utilization of source-separated collection is one of the key steps in the city's future MSW management. Source-separated collection begins at the sources of MSW and involves the whole process of collection, transportation, disposal and recycling, which enables waste minimization, resource utilization, and hazardous waste disposal of. This also requires the local government to establish standards and regulations for the source separated collection.

Appendix
Nomenclatures

I   type of waste disposal facility, where i = 1 for landfill, and i = 2 for INCINERATOR

j name of district, and j = 1, 2, 3, 4, 5

r name of district, and r = 1, 2, 3, 4, 5, 6

k planning period, and k = 1, 2, 3

L_k length of period k (day)

$\alpha_{ijk}^{\pm}$ slope of transportation cost curve for waste from district j to facility i during period k

$\alpha_{rjk}^{\pm}$ slope of transportation cost curve for waste from district j to transfer station r in period k

$\beta_{rjk}^{\pm}$ Y-intersect of transportation cost curve for waste from district j to transfer station r in period k

$\alpha_{rjk}^{\pm}$ slope of transportation cost curve for waste from transfer station r to facility i in period k

$\beta_{rjk}^{\pm}$ Y-intersect of transportation cost curve for waste from transfer station r to facility i in period k

$\alpha_{2k}^{\pm}$ slope of transportation cost curve for residue from INCINERATOR to landfill in period k

$\beta_{2k}^{\pm}$ Y-intersect of transportation cost curve for residue from INCINERATOR to landfill in period k

$\widetilde{\theta}^{\pm}$ Safety coefficient of INCINERATOR

$DF^{\pm}$ Density factor for waste from district to landfill (m^3/t)

$FE_1^{\pm}$ Residue rate when waste is shipped from district to INCINERATOR (% of incoming waste)

$FE_2^{\pm}$ Residue rate when waste is shipped from transfer station to INCINERATOR (% of incoming waste)

$DLC^{\pm}$ Daily capacity of disposing waste at the landfills (t/day)

$\widetilde{IC}^{\pm}$ Capacity of the INCINERATOR (tonne/day)

$OP_{ik}^{\pm}$ Operating cost of waste disposal facility i during period k ($/t)

$OP_{rk}^{\pm}$ Operating cost of transfer station r during period k ($/t)

$RE_{2k}^{\pm}$ Revenue from sale electricity generated at INCINERATOR in period k ($/$10^3$ KWh)

$RE_{rk}^{\pm}$ Revenue from recycling waste at transfer station r in period k ($/$10^3$ KWh)

$TLC^{\pm}$ Volume of existing landfill (m^3)

$TR_{irk}^{\pm}$ Transportation cost for waste flow from transfer station r to facility i in period k ($/t)

$\eta_1^{\pm}$ Electricity-generation rate when waste is shipped from district to INCINERATOR (KWh/t)

$\eta_2^{\pm}$ Electricity-generation rate when waste is shipped from TR to INCINERATOR (KWh/t)

$\eta_3^{\pm}$ Material recycling rate at the transfer station (%)

$\eta_4^{\pm}$ Mass loss ratio at the transfer station (%)

$VLC_k^{\pm}$ Variable cost for expanding landfill in period k ($/$m^3$)

$VTC_{rk}^{\pm}$ Variable cost for expanding transfer station r in period k ($/t)

$VIC_k^{\pm}$ Variable cost for expanding INCINERATOR in period k ($/t)

$WG_{jk}^{\pm}$ Amount of waste generated in district j in period k (t/day)

$X_{ijk}^{\pm}$ Waste flow from district j to landfill or INCINERATOR in period k (t/day)

$X_{jrk}^{\pm}$ Waste flow from district j to transfer station r in period k (t/day)

$X_{irk}^{\pm}$ Waste flow from transfer station r to landfill or INCINERATOR in period k (t/day)

$ELC^{\pm}$ Volume of landfill to be expanded (m^3)

$EIC_k^{\pm}$ Capacity expanded for INCINERATOR in period k (t/day)

$ETC_{rk}^{\pm}$ Capacity expanded for transfer station r in period k (t/day)

$MELC$ Maximum volume of landfill is allowed to be expanded (m^3)

$MEIC_k$ Maximum allowance of capacity expansion for INCINERATOR in period k (t/day)

$METC_k$ Maximum allowance of capacity expansion for each transfer station in period k (t/day)

$TETC_k$ Maximum allowance of capacity expansion for total transfer stations in period k (t/day)

Competing interests
The author declares that they have no competing interests.

Authors' contributions
YL developed a robust interval quadratic programming method and applied it to planning municipal solid waste management in the City of Changchun. GH participated in the method development and performed the result analysis. Both authors read and approved the final manuscript.

Acknowledgements
This Research was supported by the Natural Sciences Foundation of Beijing (8122038), the Program for Changjiang Scholars and Innovative Research Team in University (IRT1127), and the Program for New Century Excellent Talents in University (NCET-10-0376). The authors are grateful to the editors and the anonymous reviewers for their insightful comments and suggestions.

Author details
[1]MOE Key Laboratory of Regional Energy Systems Optimization, Resources and Environmental Research Academy, North China Electric Power University, Beijing 102206, China. [2]MOE Key Laboratory of Regional Energy Systems Optimization, Resources and Environmental Research Academy, North China Electric Power University, Beijing 102206, China.

References
Baetz BW (1990) Optimization/simulation modeling for waste management capacity planning. Journal of Urban Planning and Development (ASCE) 116(2):59–79

Ben-Tal A, Nemirovski A (2002) Robust optimization – methodology and applications. Mathematic Program Search B 92:453–480

Cao MF, Huang GH (2011) Scenario-based methods for interval linear programming problems. Journal of Environmental Informatics 17(2):65–74

Chang NB, Davila E (2007) Minimax regret optimization analysis for a regional solid waste management system. Waste Manag 27:820–832

Changchun Urban Planning Council (2004) Master plan of Changchun City (2005–2020). Changchun, China (in Chinese)

Chen MJ, Huang GH (2001) A derivative algorithm for inexact quadratic program-application to environmental decision-making under uncertainty. European Journal of Operational Research 128:570–586

Chen WT, Li YP, Huang GH, Chen X, Li YF (2010) A two-stage inexact-stochastic programming model for planning carbon dioxide emission trading under uncertainty. Applied Energy 87:1033–1047

Dubois D, Prade H (1978) Operations on fuzzy number. International Journal of Systems Science 9:613–626

Dubois D, Prade H, Sabbadin R (2001) Decision-theoretic foundations of qualitative possibility theory. European Journal of Operational Research 128:459–478

El Hanandeh A, El-Zein A (2010) Life-cycle assessment of municipal solid waste management alternatives with consideration of uncertainty: SIWMS development and application. Waste Manag 30:902–911

Fan YR, Huang GH (2012) A robust two-step method for solving interval linear programming problems within an environmental management context. Journal of Environmental Informatics 19(1):1–12

Hillier FS, Lieberman GJ (1986) Introduction to operations research, 4th edn. Holden-Day, Oakland, CA

Huang GH, Chi GF, Li YP (2005a) Long-term planning of an integrated solid waste management system under uncertainty – I. Model development. Environmental Engineering Science 22(6):823–834

Huang GH, Chi GF, Li YP (2005b) Long-term planning of an integrated solid waste management system under uncertainty – II. A north American case study. Environmental Engineering Science 22(6):835–853

Inuiguchi M, Sakawa M (1998) Robust optimization under softness in a fuzzy linear programming problem. International Journal of Approximate Reasoning 18:21–34

Kollikkathara N, Feng H, Yu DL (2010) A system dynamic modeling approach for evaluating municipal solid waste generation, landfill capacity and related cost management issues. Waste Manag 30:2194–2203

Ko AS, Chang NB (2008) Optimal planning of co-firing alternative fuels with coal in a power plant by grey nonlinear mixed integer programming model. J Environ Manage 88:11–27

Kuhn HW, Tucker AW (1951) Nonlinear programming. In: Neyman J (ed) Proceedings of the Second Berkeley Symposium on Mathematical Statistics and Probability, University of California Press. Berkeley, CA, pp 481–492

Li YP, Huang GH (2006) Minimax regret analysis for municipal solid waste management: An interval-stochastic programming approach. J Air Waste Manag Assoc 56(7):931–944

Li YP, Huang GH, Nie XH, Nie SL (2008) A two-stage fuzzy robust integer programming approach for capacity planning of environmental management systems. European Journal of Operational Research 189:399–420

Li YP, Huang GH, Yang ZF, Nie SL (2009) 0–1 Piecewise linearization approach for inexact nonlinear programming – application to environmental management under uncertainty. Canadian Journal of Civil Engineering 36:1071–1084

Li YP, Huang GH, Sun W (2011) Management of uncertain information for environmental systems using a multistage fuzzy-stochastic programming model with soft constraints. Journal of Environmental Informatics 18(1):28–37

Luhandjula MK, Gupta MM (1996) On fuzzy stochastic optimization. Fuzzy Set Syst 81:47–55

Lund JR, Tchobanoglous G, Anex R, Lawver R (1994) Linear programming for analysis of material recovery facilities. Journal of Environmental Engineering (ASCE) 120(5):1082–1094

Masui T, Morita T, Kyogoku J (2000) Analysis of recycling activities using a multi-sectoral economic model with material flow. European Journal of Operational Research 122(2):405–415

Mor S, Ravindra K, De Visscher A, Dahiya RP, Chandra A (2006) Municipal solid waste characterization and its assessment for potential methane generation: A case study. Sci Total Environ 371:1–10

Su J, Huang GH, Xi BD, Li YP, Qin XS, Huo SL, Jiang YH (2009) A hybrid inexact optimization approach for solid waste management in the city of Foshan, China. J Environ Manage 91(2):389–402

Tai J, Zhang WQ, Che Y, Feng D (2011) Municipal solid waste source-separated collection in China: A comparative analysis. Waste Manag 31:1673–1682

Thuesen HG, Fabrycky WJ, Thuesen GJ (1977) Engineering Economy. Prentice-Hall, Englewood Cliffs, New Jersey

USEPA (2007) Municipal Solid Waste in the United States: 2007 Facts and Figures.,

Wilson BG, Baetz BW (2001a) Modeling municipal solid waste collection systems
    using derived probability distributions I: Model development. Journal of
    Environmental Engineering (ASCE) 127(11):1031–1038
Wilson BG, Baetz BW (2001b) Modeling municipal solid waste collection systems
    using derived probability distributions II: Extensions and applications. Journal
    of Environmental Engineering (ASCE) 127(11):1039–1047

Permissions

List of Contributors

Wencong Yue
State Key Laboratory of Water Environment Simulation, School of Environment, Beijing Normal University, No. 19, XinJieKouWai St., HaiDian District, Beijing 100875, China

Yanpeng Cai
State Key Laboratory of Water Environment Simulation, School of Environment, Beijing Normal University, No. 19, XinJieKouWai St., HaiDian District, Beijing 100875, China
Institute for Energy, Environment and Sustainable Communities, University of Regina, 120, 2 Research Drive, Regina, Saskatchewan S4S 7H9, Canada

Qiangqiang Rong
State Key Laboratory of Water Environment Simulation, School of Environment, Beijing Normal University, No. 19, XinJieKouWai St., HaiDian District, Beijing 100875, China

Lei Cao
Environmental Development Centre of Ministry of Environmental Protection, No. 1 YuhuinanluChaoyang District, Beijing 100029, China

Xumei Wang
Environmental Development Centre of Ministry of Environmental Protection, No. 1 YuhuinanluChaoyang District, Beijing 100029, China

Timo Toivanen
VTT Technical Research Centre of Finland, P.O. Box 1000, FI-02044 VTT Espoo, Finland

Sampsa Koponen
Finnish Environment Institute, P.O. Box 140, Helsinki, Finland

Ville Kotovirta
VTT Technical Research Centre of Finland, P.O. Box 1000, FI-02044 VTT Espoo, Finland

Matthieu Molinier
VTT Technical Research Centre of Finland, P.O. Box 1000, FI-02044 VTT Espoo, Finland

Peng Chengyuan
VTT Technical Research Centre of Finland, P.O. Box 1000, FI-02044 VTT Espoo, Finland

Asmaa Mourhir
School of Sciences and Engineering, Al Akhawayn University in Ifrane, PO.Box 2083, Ifrane 53000, Morocco

Tajjeeddine Rachidi
School of Sciences and Engineering, Al Akhawayn University in Ifrane, PO.Box 1881, Ifrane 53000, Morocco

Mohammed Karim
Faculté des Sciences, Université Sidi Mohammed Ben Abdellah, Faculté des Sciences Dhar el Mehraz, Fès, Morocco

Khoshrooz Kazemi
Faculty of Engineering and Applied Science, Memorial University of Newfoundland, St. John's, Newfoundland and Labrador A1B 3X5, Canada

Baiyu Zhang
Faculty of Engineering and Applied Science, Memorial University of Newfoundland, St. John's, Newfoundland and Labrador A1B 3X5, Canada

Leonard M Lye
Faculty of Engineering and Applied Science, Memorial University of Newfoundland, St. John's, Newfoundland and Labrador A1B 3X5, Canada

Weiyun Lin
Faculty of Engineering and Applied Science, Memorial University of Newfoundland, St. John's, Newfoundland and Labrador A1B 3X5, Canada

Ronald Taylor Locke
Department of Electrical & Computer Engineering, Boston University, 8 St Mary's St., 02215 Boston MA, USA

Ioannis Ch Paschalidis
Department of Electrical & Computer Engineering, Boston University, 8 St Mary's St., 02215 Boston MA, USA

Alireza Soffianian
Department of Natural Resources, Isfahan University of Technology, Isfahan 84156-93111, Iran

Elham Sadat Madani
Environmental Sciences, Department of Natural Resources, Isfahan University of Technology, Isfahan 84156-93111, Iran

Mahnaz Arabi
Environmental Sciences, Department of Natural Resources, Isfahan University of Technology, Isfahan 84156-93111, Iran
Department of Agriculture, Payam-e-Noor University, Ardestan 83818-98951, Iran

Jinxin Zhu
Institute for Energy, Environment and Sustainable Communities, University of Regina 3737 Wascana Parkway, Regina, SK S4S 0A2, Canada

Gail Krantzberg
Engineering and Public Policy in the School of Engineering Practice, McMaster University, 1280 Main St., W Hamilton, ON L8S 4 L7, Canada

Wei Li
MOE Key Laboratory of Regional Energy Systems Optimization, S&C Resources and Environmental Research Academy, North China Electric Power University, Beijing 102206, China

Xiaoyu Liu
Research Assistant, MOE Key Laboratory of Regional Energy and Environmental Systems Optimization, Resources and Environmental Research Academy, North China Electric Power University, Beijing 102206, China

Guanzhong Sun
Research Assistant, MOE Key Laboratory of Regional Energy and Environmental Systems Optimization, Resources and Environmental Research Academy, North China Electric Power University, Beijing 102206, China

Ling Ji
Research Assistant, Research Institute of Technology Economics Forecasting and Assessment, School of Economics and Management, North China Electric Power University, Beijing 102206, China

Guohe Huang
MOE Key Laboratory of Regional Energy Systems Optimization, S&C Resources and Environmental Research Academy, North China Electric Power University, Beijing 102206, China

Maki Ito
Environmental Systems Engineering, Faculty of Engineering and Applied Science, University of Regina, 3737 Wascana Parkway, Regina SK, S4S 0A2, Canada

Shahid Azam
Environmental Systems Engineering, Faculty of Engineering and Applied Science, University of Regina, 3737 Wascana Parkway, Regina SK, S4S 0A2, Canada

Eiman Tamah Al-Shammari
Kuwait University, Kuwait, Kuwait

Motahareh Armin
Department of Systems Design Engineering, University of Waterloo, Waterloo, Ontario N2L3G1, Canada

Keith W Hipel
Department of Systems Design Engineering, University of Waterloo, Waterloo, Ontario N2L3G1, Canada
Centre for International Governance Innovation, Waterloo, Ontario N2L 6C2, Canada

Mitali De
School of Business and Economics, Wilfrid Laurier University, Waterloo, Ontario N2L3C5, Canada

Jiapei Chen
Faculty of Engineering and Applied Science, University of Regina, S4S 0A2 Regina, SK, Canada

Larry M Deschaine
HydroGeoLogic, Inc., Reston, VA 20190, USA
Department of Energy & Environment, Chalmers University of Technology, Göteborg, SE 412 96, Sweden

Theodore P Lillys
HydroGeoLogic, Inc., Reston, VA 20190, USA

János D Pintér
PCS Inc, Halifax, NS, Canada

Zheng Wang
College of Environmental & Energy Engineering, Beijing University of Technology, Beijing 100022 China
China Institute of Water Resources and Hydropower Research, Beijing 100038 China

Shuiyuan Cheng
College of Environmental & Energy Engineering, Beijing University ofTechnology, Beijing 100022 China

Lei Liu
Department of Civil and Resource Engineering, Dalhousie University, Halifax, NS B3H 4R2 Canada

Xiurui Guo
College of Environmental & Energy Engineering, Beijing University ofTechnology, Beijing 100022 China

Yuan Chen
College of Environmental & Energy Engineering, Beijing University ofTechnology, Beijing 100022 China

Cuihong Qin
College of Environmental & Energy Engineering, Beijing University ofTechnology, Beijing 100022 China

Ruixia Hao
College ofArchitecture & Civil Engineering, Beijing University of Technology, Beijing100022 China

Jin Lu
Department of Water Environment, China Institute of WaterResources and Hydropower Research, Beijing 100038 China

Jijun Gao
Department of Water Environment, China Institute of WaterResources and Hydropower Research, Beijing 100038 China

Lei Jin
College of Environmental Science and Engineering, Xiamen University of Technology, Xiamen, Fujian Province 361024, China

Aili Yang
College of Environmental Science and Engineering, Xiamen University of Technology, Xiamen, Fujian Province 361024, China

Li Wang
College of Applied Mathematics, Xiamen University of Technology, Xiamen, Fujian Province 361024, China

Haiyan Fu
College of Environmental Science and Engineering, Xiamen University of Technology, Xiamen, Fujian Province 361024, China

Yongping Li
MOE Key Laboratory of Regional Energy Systems Optimization, Resources and Environmental Research Academy, North China Electric Power University, Beijing 102206, China

Guohe Huang
MOE Key Laboratory of Regional Energy Systems Optimization, Resources and Environmental Research Academy, North China Electric Power University, Beijing 102206, China